Astrophysics and Space Science Proceedings

For further volumes:
http://www.springer.com/series/7395

Highlights of Spanish Astrophysics V

J. M. Diego
Editor

Instituto de Física de Cantabria, Spain

L. J. Goicoechea
Editor

Universidad de Cantabria, Spain

J. I. González-Serrano
Editor

Instituto de Física de Cantabria, Spain

J. Gorgas
Editor

Universidad Complutense de Madrid, Spain

Springer

Editors

Jose M. Diego
Instituto de Física de Cantabria
Avda. de Los Castros, s/n
39005 Santander Cantabria
Spain
jdiego@ifca.unican.es

J. Ignacio González-Serrano
Instituto de Física de Cantabria
Avda. de Los Castros, s/n
39005 Santander Cantabria
Spain
gserrano@ifca.unican.es

Luis J. Goicoechea
Universidad de Cantabria
Depto. Física Moderna
Avda. de Los Castros, s/n
39005 Santander Cantabria
Spain
goicol@unican.es

Javier Gorgas
Universidad Complutense de Madrid
Depto. Astrofísica
Ciudad Universitaria, s/n
28040 Madrid
Spain
jgorgas@fis.ucm.es

Additional material to this book can be downloaded from http://extra.springer.com

ISSN 1570-6591 e-ISSN 1570-6605
ISBN 978-3-642-11249-2 e-ISBN 978-3-642-11250-8
DOI 10.1007/978-3-642-11250-8
Springer Heidelberg Dordrecht London New York

Library of Congress Control Number: 2010921801

Cover design: eStudio Calamar S.L.

Printed on acid-free paper

Springer is part of Springer Science+Business Media (www.springer.com)

Preface

Astronomy is a scientific discipline that has developed a rapid and impressive growth in Spain. Thirty years ago, Spain occupied a purely anecdotal presence in the international context, but today it occupies the eighth position in the world in publication of astronomical articles, and, among other successes, owns and operates ninety per cent of the world's largest optical telescope GTC (*Gran Telescopio Canarias*).

The Eighth Scientific Meeting of the Spanish Astronomical Society (*Sociedad Española de Astronomía*, SEA), held in Santander in July 7–11 2008, whose proceedings are in your hands, clearly shows the enthusiasm, motivation and quality of the present Spanish astronomical community. The event brought together 322 participants, who represent almost 50% of Spanish professional astronomers. This percentage, together with the continuously increasing, with respect to previous SEA meetings, number of oral presentations and poster contributions (179 and 127 respectively), confirms that the SEA conferences have become a point of reference to assess the interests and achievements of astrophysical research in Spain.

The most important and current topics of modern Astrophysics were taken into account at the preliminary meeting, as well as the number and quality of participants and their contributions, to select the invited speakers and oral contributors. We took a week to enjoy the high quality contributions submitted by Spanish astronomers to the Scientific Organizing Committee. The selection was difficult. We wish to acknowledge the gentle advice and commitment of the SOC members.

The contents of these Proceedings reflect the broad interests of the Spanish astronomical community, but we want to emphasize two important aspects. In the first place, although only 15 years ago our community played a passive role using the observing facilities provided by other countries, in the last few years we have witnessed a spectacular increase in the active participation of the community in handling instruments and in the creation of groups in our research centers, mainly propelled by the GTC enterprise and other international projects. For example, in the first volume of *Highlights of Spanish Astrophysics* of only ten years ago, 15% of the contributions were about instruments. In this volume, the contributions in the Observatories (real and virtual) and Instrumentation category constitute 30% of the total number. Secondly, in this meeting, for the first time, parallel sessions devoted to *Teaching and Outreach of Astronomy* were organized. The success of these sessions, with a high number of participants, reflects the involvement of our community in the celebration of the International Year of Astronomy 2009 and our commitment to return to the society the knowledge that we are gathering.

Welcoming friends at home is always a pleasure. But welcoming about half of the Spanish astronomical community in Santander was, in addition, a challenge. Preparations for this big event, the largest ever organized in Astronomy by the community at the Instituto de Física de Cantabria (CSIC-UC) and at the University of Cantabria, took about a year and a half. For the Local Organizing Committee, the event was a success, and we believe that it was also perceived as such by the Spanish astronomers who visited Santander. A number of activities were organized inconnection with the meeting, devoted mostly to enhance the link of Astronomy with society. That included public talks, school contests, industry displays and articles and interviews in newspapers, among other activities. We also believe this extra component of the meeting, not recorded in this book, was worth the effort that we invested in it. Publishing the proceedings of the meeting is the last part of the activities of the LOC, so this formally closes their contribution to this series. We are now eagerly waiting to help our colleagues in and around Madrid to organize the next one in 2010.

This meeting was possible thanks to the financial support of governmental institutions, universities, research centers and private Spanish companies. The Society is indebted to the host institutions (the Instituto de Física de Cantabria and the University of Cantabria), and to the wonderful and friendly city of Santander, and its city council, for making us feel at home. It was a wonderful experience.

José Miguel Rodríguez Espinosa - SEA President
Emilio J. Alfaro - Chairman of the SOC
Xavier Barcons - Chairman of the LOC

José M. Diego
Luis J. Goicoechea
José Ignacio González Serrano
Javier Gorgas
Editors

Contents

Scientific Organizing Committee

Emilio J. Alfaro (Chairman, Instituto de Astrofísica de Andalucía)
Ruth Carballo (University of Cantabria)
Manuel Collados (Instituto de Astrofísica de Canarias)
Francisco Garzón (Instituto de Astrofísica de Canarias)
José Ignacio González Serrano (Instituto de Física de Cantabria, IFCA, CSIC-UC)
Margarita Hernanz (Institut de Ciencies de l'Espai, ICE-CSIC)
Enrique Martínez González (Instituto de Física de Cantabria, IFCA, CSIC-UC)
José Luis Ortiz (Instituto de Astrofísica de Andalucía)

Local Organizing Committee

Xavier Barcons (Chairman, IFCA, CSIC-UC)
Javier Bussons (IFCA, CSIC-UC)
Ruth Carballo (University of Cantabria)
Francisco Carrera (IFCA, CSIC-UC)
Maite Ceballos (IFCA, CSIC-UC)
Amalia Corral (IFCA, CSIC-UC)
José María Diego (IFCA, CSIC-UC)
Jacobo Ebrero (IFCA, CSIC-UC)
Rodrigo Gil-Merino (IFCA, CSIC-UC)
Luis J. Goicoechea (University of Cantabria)
José Ignacio González Serrano (IFCA, CSIC-UC)
Marcos López Caniego (University of Cambridge)
Enrique Martínez González (IFCA, CSIC-UC)
Carmen Morales (LAEFF-INTA)
Ángel Ruiz (IFCA, CSIC-UC)
María Concepción Ruiz (IFCA, CSIC-UC)

Sponsored by

Ministerio de Educación y Ciencia (MEC through Programa Nacional de Astronomía y Astrofísica)
Consejo Superior de Investigaciones Científicas (CSIC)
Consejería de Educación del Gobierno de Cantabria
Universidad de Cantabria (UC)
Sociedad Regional Cantabria de I+D+i (IDICAN)
Instituto Nacional de Técnicas Aeroespaciales (INTA)
Instituto de Astrofísica de Canarias (IAC)

Banco Santander S.A.
Computadoras, Redes e Ingeniería S.A. (EADS)
Enabling Grids for e-Science Project (EGEE)
Fractal S.L.N.E.
Grupo GMV
Iberlaser S.A.
IDOM Ingeniería y Consultoría
Ingeniería y Servicios Aeroespaciales S.A. (INSA)
NTE, S.A.
SENER Ingeniería y Sistemas S.A.
SERCO Group PLC
Tribuna de Astronomía
Tecnologías de Telecomunicaciones y de la Información (TTI)

List of participants

Aberasturi Vega, Miriam,	Universidad Complutense de Madrid
Alacid Polo, José Manuel,	INSA-LAEFF / Spanish VO
Alberdi Odriozola, Antxon,	Instituto de Astrofísica de Andalucía (CSIC)
Alfaro Navarro, Emilio J.,	Instituto de Astrofísica de Andalucía (CSIC)
Alfonso Garzón, Julia,	Instituto de Estructura de la Materia (CSIC)
Alonso Fernández, José,	TTI-NORTE
Alonso Sobrino, Roi,	Laboratoire d'Astrophysique de Marseille - OAMP
Alves, Joao,	Calar Alto Observatory
Amo Baladrón, M. Aránzazu,	DAMIR(IEM, CSIC)
Andrade Baliño, Manuel,	Observatorio Astronómico R.M. Aller
Angles Alcazar, Daniel,	University of Puerto Rico
Antoja Castelltort, Teresa,	Universitat de Barcelona
Aparicio Villegas, Teresa,	Instituto de Astrofísica de Andalucía (CSIC)
Arnalte Mur, Pablo,	Observatori Astronmic de la Universitat de València
Arregi Bengoa, Jesús,	Escuela Universitaria de Ingeniería de Vitoria-Gasteiz
Arregui Uribe-Echevarria, Iñigo,	Departament de Fisica, Universitat de les Illes Balears
Asensio Ramos, Andrés,	Instituto de Astrofísica de Canarias (IAC)
Badenes Montoliu, Carles,	Princeton University
Bakos, Judit,	Instituto de Astrofísica de Canarias (IAC)
Balaguer Núñez, Lola,	Dept. d'Astronomia i Meteorologia, Universitat de Barcelona
Balcells Comas, Marc,	Instituto de Astrofísica de Canarias (IAC)
Bañó Esplugues, Gisela,	LAEFF - INTA
Barcons Jáuregui, Xavier,	Instituto de Física de Cantabria (CSIC-UC)
Barrado Izagirre, Naiara,	EHU-UPV
Barreiro, Belén,	Instituto de Física de Cantabria (CSIC-UC)
Barret, Didier,	S.F.2A
Barro Calvo, Guillermo,	Dept. de Astrofísica, Universidad Complutense de Madrid
Battaner López, Eduardo,	Universidad de Granada
Bayo Aran, Amelia,	LAEFF - INTA
Bellot Rubio, Luis Ramón,	Instituto de Astrofísica de Andalucía (CSIC)
Benavidez, Paula G.,	Universitat d'Alacant
Benítez Lozano, Narciso,	Instituto de Matemáticas y Física Fundamental (CSIC)
Bernabeu Pastor, Guillermo,	Dept. Física, Ing. de Sistemas y Teoría de la Señal, U. de Alicante
Breitfellner, Michel,	SERCO
Buenrostro Leiter, Valeria,	Instituto de Astrofísica de Canarias (IAC)

Buitrago Alonso, Fernando, University of Nottingham
Caballero García, María Dolores, LAEFF - INTA
Caballero Hernández, José Antonio, Dept. de Astrofísica, Universidad Complutense de Madrid
Cabañas, Ana, IBERLASER
Cabrera Lavers, Antonio, Instituto de Astrofísica de Canarias (IAC)
Calogero, Simone Carmelo, Universidad de Granada
Campa Romero, Julia, CIEMAT
Campo Bagatín, Adriano, Universitat d'Alacant
Campos, Isabel, Instituto de Física de Cantabria (CSIC-UC)
Cantó Doménech, José, Escuela Politécnica Superior de Alcoi, U. Politécnica de Valencia
Carballo Bello, Julio Alberto, Instituto de Astrofísica de Canarias (IAC)
Carballo Fidalgo, Ruth, Dept. Matemática Aplicada y CC de la Computación, Univ. de Cantabria
Carbonell Huguet, Marc, Dept. de Matemtiques i Informtica, Univ. de les Illes Balears
Cárdenas Vázquez, M. Concepción, Instituto de Astrofísica de Andalucía (CSIC)
Cardiel López, Nicolás, Dept. de Astrofísica, Universidad Complutense de Madrid
Carrasco Martínez, José Manuel, Dept. d'Astronomia i Meteorologia, Universitat de Barcelona
Carrera Troyano, Francisco J., Instituto de Física de Cantabria (CSIC-UC)
Casalta, Joan Manel, NTE
Casares Velázquez, Jorge, Instituto de Astrofísica de Canarias (IAC)
Castander Serentill, Francisco Javier, Institut de Ciencies de l'Espai (IEEE-CSIC)
Castañeda Fernández, Héctor, Instituto de Astrofísica de Canarias (IAC)
Castillo Morales, Ãfrica, Dept. de Astrofísica, Universidad Complutense de Madrid
Castro Rodríguez, Nieves, Instituto de Astrofísica de Canarias (IAC)
Castro Rodríguez, Norberto, Instituto de Astrofísica de Canarias (IAC)
Catalá Poch, María Asunción, Dept. d'Astronomia i Meteorologia, Universitat de Barcelona
Catalán Ruiz, Silvia, Institut de Ciencies de l'Espai (IEEE-CSIC)
Ceballos Merino, Maite, Instituto de Física de Cantabria (CSIC-UC)
Cenarro Lagunas, Javier, Instituto de Astrofísica de Canarias (IAC)
Cepa Nogué, Jordi, Instituto de Astrofísica de Canarias (IAC)
Collados Vera, Manuel, Instituto de Astrofísica de Canarias (IAC)
Colomé Ferrer, Josep, Institut de Ciencies de l'Espai (IEEE-CSIC)
Colomer Sanmartín, Francisco, Observatorio Astronómico Nacional
Comerón Limbourg, Sébastien, Instituto de Astrofísica de Canarias (IAC)
Contreras González, José Luis, Dpt. de Física Atómica, Universidad Complutense de Madrid
Corral Ramos, Amalia, Instituto de Física de Cantabria (CSIC-UC)
Corral Santana, Jesús M., Instituto de Astrofísica de Canarias (IAC)
Costado Dios, Teresa, Instituto de Astrofísica de Canarias (IAC)
Crespo Chacón, Inés, Dept. de Astrofísica, Universidad Complutense de Madrid
Cristóbal Hornillos, David, Instituto de Astrofísica de Andalucía (CSIC)
Cruz Rodríguez, Marcos, Instituto de Física de Cantabria (CSIC-UC)
Cuesta Crespo, Luis, Centro de Astrobiología (CSIC-INTA)
Curto Martin, Andrés, Instituto de Física de Cantabria (CSIC-UC)
Davó Guillén, María José, Universitat d'Alacant
de Avillez, Miguel A., SPA
de Cea del Pozo, Elsa, Institut de Ciencies de l'Espai (IEEE-CSIC)

de Gregorio Monsalvo, Itziar,	ESO - European Southern Observatory
de León Cruz, Julia María,	Instituto de Astrofísica de Canarias (IAC)
de Lorenzo-Cáceres Rodríguez, Adriana,	Instituto de Astrofísica de Canarias (IAC)
de Rojas, Cristina,	SERCO
de Ugarte Postigo, Antonio,	ESO - European Southern Observatory
de Zeeuw, Tim,	ESO - European Southern Observatory
del Puerto Varela, Carmen,	Instituto de Astrofísica de Canarias/Museo de la Ciencia y el Cosmos
Delgado Sánchez, Antonio Jesús,	Instituto de Astrofísica de Andalucía (CSIC)
Diago Nebot, Pascual David,	Observatori Astronmic de la Universitat de València
Díaz López, Cristina,	Dept. de Astrofísica, Universidad Complutense de Madrid
Díaz Medina, Antonio J.,	Universitat de les Illes Balears
Diego Rodríguez, José M.,	Instituto de Física de Cantabria (CSIC-UC)
Djupvik, Anlaug Amanda,	Nordic Optical Telescope
Domingo Garau, Albert,	LAEFF - INTA
Domínguez Palmero, Lilian,	Instituto de Astrofísica de Canarias (IAC)
Domínguez Quintero, Rosa María,	Instituto de Astrofísica de Canarias (IAC)
Ebrero Carrero, Jacobo,	Instituto de Física de Cantabria (CSIC-UC)
Eiroa, Carlos,	Universidad Autónoma de Madrid
Eliche Moral, María del Carmen,	Dept. de Astrofísica, Universidad Complutense de Madrid
Escartín Vigo, José Antonio,	Dept. Física i Enginyeria Nuclear (UPC)
Fabricius, Claus,	IEEC/Dept. d'Astronomia i Meteorologia, Universitat de Barcelona
Falco, Emilio,	Whipple Observatory, Smithsonian Institution
Fernández Barba, David,	Institut de Ciències del Cosmos (ICCUB)
Fernández Lorenzo, Mirian,	Instituto de Astrofísica de Canarias (IAC)
Fernández Soto, Alberto,	Universitat de Valencia
Figueras Siñol, Francesca,	Universidad de Barcelona
Florido Navío, Estrella,	Universidad de Granada
Fonseca González, María Victoria,	Dept. de Física Atómica, Universidad Complutense de Madrid
Font Ribera, Andreu,	Institut d'Estudis Espacials de Catalunya (IEEC-CSIC)
Forteza Ferrer, Pep,	Universitat de les Illes Balears
Funke, Bernd,	Instituto de Astrofísica de Andalucía (CSIC)
Gallego Maestro, Jesús,	Dept. de Astrofísica, Universidad Complutense de Madrid
Gálvez Ortiz, María Cruz,	Dept. de Astrofísica, Universidad Complutense de Madrid
García Ãlvarez, David,	Gran Telescopio CANARIAS
García García, Miriam,	Instituto de Astrofísica de Canarias (IAC)
García Senz, Domingo,	Dept. Física i Enginyeria Nuclear (UPC)
García Torres, Miguel,	INTA-LAEFF / Spanish VO
García Vargas, María Luisa,	FRACTAL SLNE
Garzón López, Francisco,	Instituto de Astrofísica de Canarias (IAC)
Gaztañaga , Enrique,	Institut de Ciencies de l'Espai (IEEE-CSIC)
Gil de Paz, Armando,	Dept. de Astrofísica, Universidad Complutense de Madrid
Gil-Merino Rubio, Rodrigo,	Instituto de Física de Cantabria (CSIC-UC)
Giménez Cañete, Ãlvaro,	Centro de Astrobiología (CSIC-INTA)
Godano Moreno, Ricardo,	Metamorfosis Polipoética
Goicoechea Santamaría, Luis Julián,	Universidad de Cantabria

Gómez de Castro, Ana Inés,	Facultad de CC Matemáticas, Universidad Complutense de Madrid
Gómez Flechoso, María Ãngeles,	Universidad Autónoma de Madrid
Gómez Velarde, Gabriel,	Gran Telescopio CANARIAS
Gómez, Ismael,	CRISA
González Fernández, Carlos,	Instituto de Astrofísica de Canarias (IAC)
González Martín, Omaira,	Instituto de Astrofísica de Andalucía (CSIC)
González Pérez, Violeta,	Institut de Ciencies de l'Espai (IEEE-CSIC)
González Serrano, José Ignacio,	Instituto de Física de Cantabria (CSIC-UC)
González Solares, Eduardo,	Instituto de Astronomia (IoA)
González, Paloma,	IBERLASER
Gorgas García, Javier,	Dept. de Astrofísica, Universidad Complutense de Madrid
Guirado Puerta, José Carlos,	Universidad de Valencia
Gutiérrez Sánchez, Raúl,	INTA-LAEFF / Spanish VO
Guzmán Llorente, Rafael,	Universidad de Florida
Hebrero Domínguez, Gema,	Space Plasmas & AStroparticle Group, Dpto. Física, Univ. Alcalá
Hempel, Ãngela,	Instituto de Astrofísica de Canarias (IAC)
Hernán Obispo, María Magdalena,	Dept. de Astrofísica, Universidad Complutense de Madrid
Hernández Fernández, Jonathan,	Instituto de Astrofísica de Andalucía (CSIC)
Hernanz Carbó, Margarita,	Institut de Ciencies de l'Espai (IEEE-CSIC)
Herranz Muñoz, Diego,	Instituto de Física de Cantabria (CSIC-UC)
Herrero Davó, Artemio,	Instituto de Astrofísica de Canarias (IAC)
Huertas Company, Marc,	LESIA - Observatoire de Paris-Meudon
Hueso Alonso, Ricardo,	Escuela T. Superior de Ingeniería, Universidad del Pais Vasco
Husillos Rodríguez, César,	Instituto de Astrofísica de Andalucía (CSIC)
Isasi Parache, Yago,	Universidad de Barcelona
Izquierdo Gómez, Jaime,	Dept. de Astrofísica, Universidad Complutense de Madrid
Jiménez Luján, Florencia,	Instituto de Física de Cantabria (CSIC-UC)
Jiménez Monferrer, Sergio,	Universidad de Valencia
Jiménez Teja, Yolanda,	Instituto de Astrofísica de Andalucía (CSIC)
Jordi Nebot, Carme,	Dept. d'Astronomia i Meteorologia, Universitat de Barcelona, ICC-IEEC
Julbe López, Francesc,	ESO - European Southern Observatory
Khomenko, Elena,	Instituto de Astrofísica de Canarias (IAC)
Labiano Ortega, Ãlvaro,	Instituto de Estructura de la Materia (CSIC)
Lanz Oca, Luis Fernando,	Instituto de Física de Cantabria (CSIC-UC)
Lara López, Luisa María,	Instituto de Astrofísica de Andalucía (CSIC)
Lara López, Maritza Arlene,	Instituto de Astrofísica de Canarias (IAC)
Legarreta Etxagibel, Jon,	Escuela Ingeniería Técnica Industrial de Bilbao
Licandro Goldaracena, Javier,	Instituto de Astrofísica de Canarias (IAC)
Llorente, Jesús Salvador,	SENER
López del Fresno, Mauro,	INTA-LAEFF / Spanish VO
López Hermoso, Rosario,	Universidad de Barcelona
López Martín, Luis,	Instituto de Astrofísica de Canarias (IAC)
López Moreno, José Juan,	Instituto de Astrofísica de Andalucía (CSIC)
López San Juan, Carlos,	Instituto de Astrofísica de Canarias (IAC)
López Santiago, Javier,	Dept. de Astrofísica, Universidad Complutense de Madrid

López-Caniego Alcarria, Marcos, Cavendish Laboratory, University of Cambridge
Lorente Balanza, Rosario, European Space Astronomy Centre (ESAC)
Lorenzo Espinosa, Javier, Universitat d'Alacant
Luna Bennasar, Manuel, Universidad de las Islas Baleares
Luri Carrascoso, Xavier, Universidad de Barcelona
Maldonado Prado, Jesús, Universidad Autónoma de Madrid
Maldonado, Manuel, FRACTAL SLNE
Mampaso, Antonio, Instituto de Astrofísica de Canarias (IAC)
Manrique Oliva, Alberto, Universidad de Barcelona
Marco Tobarra, Amparo, Universitat d'Alacant
Marcos Arenal, Pablo, Universidad Complutense de Madrid
Mármol Queraltó, Esther, Dept. de Astrofísica, Universidad Complutense de Madrid
Márquez Pérez, Isabel, Instituto de Astrofísica de Andalucía (CSIC)
Martí Ribas, Josep, Universidad de Jaén
Martí Vidal, Iván, Departamento de Astronomía y Astrofisíca, Universidad de Valencia
Martín Guerrero de Escalante, Eduardo, Instituto de Astrofísica de Canarias (IAC)
Martín Hernández, José Manuel, Dept. de Astrofísica, Universidad Complutense de Madrid
Martín Manjón, Mariluz, Universidad Autónoma de Madrid
Martínez Arnáiz, Raquel M., Dept. de Astrofísica, Universidad Complutense de Madrid
Martínez Badenes, Vicent, Instituto de Astrofísica de Andalucía (CSIC)
Martínez Delgado, David, Instituto de Astrofísica de Canarias (IAC)
Martínez Delgado, Ismael, Instituto de Astrofísica de Canarias (IAC)
Martínez García, Vicent J., Observatori Astronmic de la Universitat de València
Martínez González, Enrique, Instituto de Física de Cantabria (CSIC-UC)
Martínez Núñez, Silvia, Dept. Física, Ing. de Sistemas y Teoría de la Señal, Univ. de Alicante
Martínez Roger, Carlos, Instituto de Astrofísica de Canarias (IAC)
Martínez Serrano, Francisco Jesús, Universidad Miguel Hernández
Martínez Valpuesta, Inma, Instituto de Astrofísica de Canarias (IAC)
Martínez Vaquero, Luis Alberto, Universidad Autónoma de Madrid
Mas Hesse, José Miguel, Centro de Astrobiología (CSIC-INTA)
Masana Fresno, Eduard, Dept. d'Astronomia i Meteorologia, Universitat de Barcelona
Mateos, Silvia, University of Leicester
Mayor, Michel, Univ. Genève
Mendigutía Gómez, Ignacio, LAEFF - INTA
Miralles Caballero, Daniel, Instituto de Estructura de la Materia (CSIC)
Miralles Torres, Juan Antonio, Universitat d'Alacant
Miravent, Carlos Enrique, SENER
Molina Cuevas, Antonio, Instituto de Astrofísica de Andalucía (CSIC)
Molino Benito, Alberto, Instituto de Astrofísica de Andalucía (CSIC)
Mollá Lorente, Mercedes, CIEMAT
Montes Gutiérrez, David, Dept. de Astrofísica, Universidad Complutense de Madrid
Montesinos Comino, Benjamín, Instituto de Astrofísica de Andalucía (CSIC), LAEFF/INTA
Mora Fernández, Alcione, Universidad Autónoma de Madrid
Morales Durán, Carmen, LAEFF - INTA
Morales Peralta, Juan Carlos, Institut d'Estudis Espacials de Catalunya (IEEC-CSIC)

Muñoz Marín, Víctor Manuel, Instituto de Astrofísica de Andalucía (CSIC)
Muñoz Mateos, Juan Carlos, Dept. de Astrofísica, Universidad Complutense de Madrid
Murga Llano, Gazika, IDOM
Najarro de la Parra, Francisco, DAMIR(IEM, CSIC)
Negueruela Díez, Ignacio, Universitat d'Alacant
Ojero Pascual, Eduardo, European Space Astronomy Centre (ESAC)
Oliver Herrero, Ramón, Universitat de les Illes Balears
Ortiz Gil, Amelia, Observatori Astronmic de la Universitat de València
Ortiz Moreno, José Luis, Instituto de Astrofísica de Andalucía (CSIC)
Oscoz Abad, Alejandro, Instituto de Astrofísica de Canarias (IAC)
Otí Floranes, Héctor, LAEFF - INTA
Oya Vallejo, Igor, Departamento de Física Atómica, Universidad Complutense de Madrid
Padilla Torres, Carmen Pilar, Instituto de Astrofísica de Canarias (IAC)
Pan Fernández, Jorge, IDOM
Pascual Ramírez, Sergio, Dept. de Astrofísica, Universidad Complutense de Madrid
Pascual Sánchez, J.-Fernando, Universidad de Valladolid
Pedraz Marcos, Santos, Calar Alto Observatory
Peña Bielva, M. Ãngel, TTI-NORTE
Peralta Calvillo, Javier, Grupo de Ciencias Planetarias (UPV/EHU)
Perea Duarte, Jaime, Instituto de Astrofísica de Andalucía (CSIC)
Pereira Santaella, Miguel, Instituto de Estructura de la Materia (CSIC)
Pérez García, Ana María, Instituto de Astrofísica de Canarias (IAC)
Pérez González, Pablo G., Dept. de Astrofísica, Universidad Complutense de Madrid
Pérez Hoyos, Santiago, Universidad del País Vasco
Pérez Martín, Isabel, Departamento de Física Teórica y del Cosmos, Universidad de Granada
Pérez Martínez, Ricardo, INSA / XMM Science Operation Centre
Planelles Mira, Susana, Departamento de Astronomía y Astrofisíca, Universidad de Valencia
Ponente, Pier Paolo, Instituto de Física de Cantabria (CSIC-UC)
Pons Botella, José Antonio, Departament de Fisica, Universitat d'Alacant
Portell de Mora, Jordi, Dept. d'Astronomia i Meteorologia, Universitat de Barcelona
Povic, Mirjana, Instituto de Astrofísica de Canarias (IAC)
Prieto Muñoz, Mercedes, Instituto de Astrofísica de Canarias (IAC)
Prieto Prieto, Almudena, Instituto de Astrofísica de Canarias (IAC)
Puschmann, Klaus Gerhard, Instituto de Astrofísica de Canarias (IAC)
Rabaza Castillo, Ovidio, Instituto de Astrofísica de Andalucía (CSIC)
Ramos Almeida, Cristina, Instituto de Astrofísica de Canarias (IAC)
Rial Lesaga, Samuel, Departament de Fisica, Universitat de les Illes Balears
Ribas Rubio, Salvador J., Dept. d'Astronomia i Meteorologia, Universitat de Barcelona
Ribó Gomis, Marc, Universitat de Barcelona
Rísquez Oneca, Daniel, LAEFF - INTA
Rizzo Caminos, Ricardo, LAEFF - INTA
Rodes Roca, José Joaquín, Universitat d'Alacant
Rodrigo Blanco, Carlos, INTA-LAEFF / Spanish VO
Rodríguez Espinosa, José Miguel, Instituto de Astrofísica de Canarias (IAC)
Rodríguez Frías, María Dolores, Universidad de Alcalá de Henares

Rodríguez Gómez, Diego,	SENER
Rodríguez Hidalgo, Inés,	IAC / Dpto. de Astrofísica, Univ. de La Laguna
Rojas Palenzuela, José Felix,	Universidad del Pais Vasco
Romero Gómez, Merce,	Laboratoire d'Astrophysique de Marseille - OAMP
Ros Ibarra, Eduardo,	Max-Planck-Institut fr Radioastronomie
Rosenberg González, Alfred,	Instituto de Astrofísica de Canarias (IAC)
Rubiño Martin, José Alberto,	Instituto de Astrofísica de Canarias (IAC)
Rubio da Costa, Fátima,	University of Glasgow. Departament of Physics and Astronomy
Ruiz Camuñas, Ángel,	Instituto de Física de Cantabria (CSIC-UC)
Ruiz Cobo, Basilio,	Instituto de Astrofísica de Canarias (IAC)
Ruiz del Mazo, José Enrique,	Instituto de Astrofísica de Andalucía (CSIC)
Ruiz Zelmanovitch, Natalia,	Universidad Complutense de Madrid
Ruiz, Patricia,	Instituto Nacional Tecnica Aeroespacial
Sabater Montes, José,	Instituto de Astrofísica de Andalucía (CSIC)
Sabau Graziati, Lola,	Instituto Nacional Tecnica Aeroespacial
Sáez Milán, Diego Pascual,	Departamento de Astronomía y Astrofisíca, Universidad de Valencia
Salvador Solé, Eduard,	Dept. d'Astronomia i Meteorologia, Universitat de Barcelona
Sánchez Colín, Ãngel,	Instituto de Física de Cantabria (CSIC-UC)
Sánchez Contreras, Carmen,	Instituto de Estructura de la Materia (CSIC)
Sánchez de Miguel, Alejandro,	Dept. de Astrofísica, Universidad Complutense de Madrid
Sánchez Doreste, Néstor,	Instituto de Astrofísica de Andalucía (CSIC)
Sánchez Janssen, Rubén,	Instituto de Astrofísica de Canarias (IAC)
Sánchez Lavega, Agustín,	Escuela T. Superior de Ingeniería, Universidad del Pais Vasco
Sánchez Portal, Miguel,	Herschel Science Centre, INSA/ESAC, ESA
Sánchez Romero, Oscar,	Universidad de Granada
Sánchez Sánchez, Sebastián Francisco,	Calar Alto Observatory
Sánchez-Blázquez, Patricia,	University of Central Lancashire
Sancho de la J. Llabrés, Llucia,	Universitat de les Illes Balears
Santander García, Miguel,	Isaac Newton Group
Sarro Baro, Luis Manuel,	UNED
Schoch, Pedro J.,	GMV
Serra, Sinue,	Universitat de Barcelona
Sintes Olives, Alicia M.,	Universitat de les Illes Balears
Solano Márquez, Enrique,	INSA-LAEFF / Spanish VO
Soler Juan, Roberto,	Departament de Fisica, Universitat de les Illes Balears
Soria Ruiz, Rebeca,	Observatorio Astronómico Nacional
Suades Sol, Moisès,	Institut de Ciencies de l'Espai (IEEE-CSIC)
Suso López, Julia,	Observatori Astronmic de la Universitat de València
Talavera Iniesta, Antonio,	European Space Astronomy Centre (ESAC)
Tejerina, Eduardo,	Instituto de Física de Cantabria (CSIC-UC)
Toloba Jurado, Elisa,	Dept. de Astrofísica, Universidad Complutense de Madrid
Torra Roca, Jordi,	Dept. d'Astronomia i Meteorologia, Universitat de Barcelona
Torrelles Arnedo, José María,	Institut de Ciencies de l'Espai (IEEE-CSIC)
Trias Cornellana, Miquel,	Universitat de les Illes Balears
Trujillo Cabrera, Ignacio,	Instituto de Astrofísica de Canarias (IAC)

Ulla Miguel, Ana,	Universidade de Vigo
Ullán Nieto, Aurora,	Centro de Astrobiología (CSIC-INTA)
Valdivielso Casas, Luisa,	Instituto de Astrofísica de Canarias (IAC)
Vallejo, Juan Carlos,	GMV
Vaquerizo Gallego, Juan Ãngel,	LAEFF - INTA
Vazdekis Vazdekis, Alexandre,	Instituto de Astrofísica de Canarias (IAC)
Velasco Trasmonte, Almudena,	INSA-LAEFF / Spanish VO
Verdes-Montenegro Atalaya, Lourdes,	Instituto de Astrofísica de Andalucía (CSIC)
Vicente Martínez, Belén,	Instituto de Astrofísica de Canarias (IAC)
Vielva, Patricio,	Instituto de Física de Cantabria (CSIC-UC)
Viironen, Kerttu,	Instituto de Astrofísica de Canarias (IAC)
Vilardell Sallés, Francesc,	Universitat de Barcelona
Villar Martín, Montserrat,	Instituto de Astrofísica de Andalucía (CSIC)
Villaver Sobrino, Eva,	Space Telescope Science Institute/ ESA
Viñas Bastart, Jordi,	Facultad de Física, Universitat de Barcelona
Yun, Joao,	Centro de Astronomia e Astrofisica da Universidade de Lisboa
Zurita Muñoz, Almudena,	Dept. de Física Teórica y del Cosmos, Universidad de Granada

Contributors

M. Abelleira Departamento de Astrofísica y C.C. de la Atmósfera, Facultad de C.C. Físicas, Universidad Complutense de Madrid, Avda. Complutense s/n, 28040 Madrid, Spain, mabelleira@astrax.fis.ucm.es

Miriam Aberasturi Laboratorio de Astrofísica Espacial y Física Fundamental, INSA, Apdo. 78, 28691 Villanueva de la Cãnada, Madrid, Spain, miriam@laeff.inta.es

Carlos Abia Departamento de Física Teórica y del Cosmos, Universidad de Granada, Spain, cabia@ugr.es

F.J. Aceituno Instituto de Astrofísica de Andalucía (IAA-CSIC), C. Bajo de Huétor 50, 18080 Granada, Spain

J.A. Acosta-Pulido Instituto de Astrofísica de Canarias, 38205 La Laguna, Spain

P.A.R. Ade Department of Physics and Astronomy, Cardiff University, 5 The Parade, Cardiff CF24 3AA, UK

J. Aguirre Centro de Astrobiología, Instituto Nacional de Técnica Aerospacial, Ctra. de Torrejón a Ajalvir, Km 4, 28850-Torrejón de Ardoz, Madrid, Spain, aguirreaj@inta.es

B. Aja Departamento de Ingenieria de COMunicaciones (DICOM), Laboratorios de I+D de Telecomunicaciones, Plaza de la Ciencia s/n, 39005 Santander, Spain

J. Manuel Alacid SVO/LAEFF-CAB/INTA-CSIC, Apdo 78, 28691 Villanueva de la Cãnada, Madrid, Spain, jmalacid@laeff.inta.es

A. Alberdi IAA (CSIC), Camino bajo de Huétor 50, 18008 Granada, Spain, Antxon@iaa.es

Emilio J. Alfaro Instituto de Astrofísica de Andalucía, CSIC, Granada, Spain, emilio@iaa.es

Julia Alfonso-Garzón CAB/LAEFF (CSIC-INTA), POB 78 28691 Villanueva de la Cãnada, Madrid, Spain, julia@laeff.inta.es

C. Almeida Institute for Computational Cosmology, Department of Physics, University of Durham, South Road, Durham DH1 3LE, UK

A. Alonso-Herrero Instituto de Estructura de la Materia, CSIC, 28006 Madrid, Spain
and
Steward Observatory, University of Arizona, Tucson, AZ 85721, USA

R. Álvarez ESAC (ESA), Madrid, Spain, Ruben.Alvarez@sciops.esa.int

Johannes Andersen Nordic Optical Telescope, Apdo 474, 38700 Santa Cruz de La Palma, Spain, ja@not.iac.es

M. Andrade Astronomical Observatory R.M. Aller (USC), Santiago de Compostela, Spain, manuel.andrade@usc.es

M. Andreev Terskol Branch of Institute of Astronomy RAS, Russia

G. Anglada-Escudé Departament d 0Astronomia i Meteorologia, Universitat de Barcelona, Martí i Franqués 1, Barcelona 08028, Spain
and
Department of Terrestrial Magnetism, Carnegie Institution of Washington, 5241 Broad Branch Road, NW, Washington, DC 20015-1305, USA

D. Anglés Department of Physics, University of Puerto Rico, Rio Piedras Campus, Box 23343, Puerto Rico, USA, daniel.angles@gmail.com

T. Antoja Departament d'Astronomia i Meteorologia and IEEC-UB, Institut de Ciències del Cosmos de la Universitat de Barcelona, Martí i Franquès, 1, 08028 Barcelona, Spain, tantoja@am.ub.es

María Arévalo SVO/LAEFF-CAB/INTA-CSIC, PO Box 78, 28691 Villanueva de la Cãnada, Madrid, Spain, maria.arevalo@laeff.inta.es

J. Ariño IDOM, Avda. Lehendakari Aguirre, 3, 48014 Bilbao, Spain

Pablo Arnalte-Mur Observatori Astronòmic, Universitat de València, Paterna, Spain
and
Departament d'Astronomia i Astrofísica, Universitat de València, Apartat de Correus 22085, 46071 València, Spain, pablo.arnalte@uv.es

J. Arregi Grupo de Ciencias Planetarias, Departament de Fisica Aplicada I, Escuela Universitaria Ingenieria, Universidad del Pais Vasco/Euskal Herriko Unibertsitatea, Vitoria-Gasteiz, Spain, jesus.arregi@ehu.es

I. Arregui Departament de Física, Universitat de les Illes Balears, 07122 Palma de Mallorca, Spain, inigo.arregui@uib.es

Santiago Arribas DAMIR-IEM-CSIC, Serrano 121, 28006 Madrid, Spain, arribas@damir.iem.csic.es

E. Artal Departamento de Ingenieria de COMunicaciones (DICOM), Laboratorios de I+D de Telecomunicaciones, Plaza de la Ciencia s/n, 39005 Santander, Spain

Yago Ascasibar Astrophysikalisches Institut Potsdam, Germany
and
Universidad Autonoma de Madrid, Spain, yago.ascasibar@uam.es

R. Azzollini Instituto de Astrofísica de Canarias, C/Vía Láctea s/n, 38205 La Laguna, S/C de Tenerife, Spain, ruyman@iac.es

Judit Bakos Instituto de Astrofísica de Canarias, Calle Vía Lactea, 38200 La Laguna, Tenerife, Spain, jbakos@iac.es

L. Balaguer-Núñez Dpt. d'Astronomia i Meteorologia, Universitat de Barcelona, ICC, Avda. Diagonal 647, 08028 Barcelona, Spain, lola.balaguer@am.ub.es

Marc Balcells Instituto de Astrofísica de Canarias, C. Vía Åactea s/n, Tenerife, Spain
and
Universidad de La Laguna, Tenerife, Spain, balcells@iac.es

Y.Y. Balega Special Astrophysical Observatory (Russia), Karachaevo-Cherkesia, Russia, balega@sao.ru

J.L. Ballester Departament de Física, Universitat de les Illes Balears, 07122 Palma de Mallorca, Spain

A. Ballu Ecole Nationale Supérieure de Physique, Université Louis Pasteur, Strasbourg, France

Isabelle Baraffe école Normale Supérieure de Lyon, CRAL (UMR CNRS 5574), Université de Lyon, France

Xavier Barcons Ministerio de Ciencia e Innovación, Spain
and
Instituto de Física de Cantabria (CSIC-UC), Avenida de los Castros, 39005 Santander, Spain, barcons@ifca.unican.es

N. Barrado-Izagirre Universidad del Pais Vasco, Alda. Urkijo s/n, 48013 Bilbao, Spain, naiara.barrado@ehu.es

David Barrado y Navascués Laboratorio de Astrofísica Espacial y Física Fundamental, INTA, Madrid, Spain, barrado@laeff.inta.es

R.B. Barreiro Instituto de Física de Cantabria, CSIC-Universidad de Cantabria, Avda. de los Castros s/n, 39005 Santander, Spain, barreiro@ifca.unican.es

J.A. Barrio Departamento de Física Atómica, UCM, Madrid, Spain

G. Barro Departamento de Astrofísica y Ciencias de la Atmósfera, Facultad de C.C. Físicas, Universidad Complutense de Madrid, 28040 Madrid, Spain, gbc@astrax.fis.ucm.es

E. Battaner Departamento de Física Teórica y del Cosmos, Edif. Mecenas, planta baja, Campus Fuentenueva, Universidad de Granada, 18071 Granada, Spain.
and
Instituto de Física Teórica y Computacional Carlos I, Granada, Spain, battaner@ugr.es

R. Battye Jodrell Bank Centre for Astrophysics, School of Physics and Astronomy, University of Manchester, Oxford Road, Manchester M13 9PL, UK

C.M. Baugh Institute for Computational Cosmology, Department of Physics, University of Durham, South Road, Durham DH1 3LE, UK

A. Bayo LAEFF-CAB, INTA-CSIC, P.O. Box 78, 28691 Villanueva de la Cãnada, Madrid, Spain, abayo@laeff.inta.es

J.P. Beaulieu Institut d'Astrophysique de Paris (IAP), CNRS (UMR 7095), Paris, France, beaulieu@iap.fr

J.E. Beckman Instituto de Astrofísica de Canarias, 38200 La Laguna, Spain, jeb@iac.es
and
Consejo Superior de Investigaciones Científicas, Spain

L.R. Bellot Rubio Instituto de Astrofísica de Andalucía (CSIC), Granada, Spain, lbellot@iaa.es

G. Bendo Astrophysics Group, Imperial College, Blackett Laboratory, London

C.R. Benn Isaac Newton Group, Apartado 321, 38700 Santa Cruz de La Palma, Spain, crb@ing.iac.es

V.J.S. Béjar Instituto de Astrofísica de Canarias, Calle Vía Láctea s/n, 38200 La Laguna, Tenerife, Spain

G. Bernabéu Universitat d'Alacant, Spain, bernabeu@ua.es

G. Bergond Instituto de Astrofísica de Andalucía, CSIC, Apdo. 3004, 18080 Granada, Spain, gilles@iaa.es

A. Berihuete Universidad de Cádiz, Spain

D. Bermejo-Pantaleón Instituto de Astrofísica de Andalucía, CSIC, Granada, Spain, diego@iaa.es

J.J. Bock Jet Propulsion Laboratory, Pasadena, CA 91109-8099, USA

S. Boissier Observatoire Astronomique de Marseille-Provence, Laboratoire d'Astrophysique de Marseille, Cedex, France
and
Centre National de la Recherche Scientifique, France

A. Bongiovanni Instituto de Astrofísica de Canarias, 38205 La Laguna, Spain, bongio@iac.es

D. Bose (for the MAGIC Collaboration) Departamento de Física Atómica, UCM, Madrid, Spain

A. Boselli Observatoire Astronomique de Marseille-Provence, Laboratoire d'Astrophysique de Marseille, Cedex, France
and
Centre National de la Recherche Scientifique, France

H. Bouy Instituto de Astrofísica de Canarias, Calle Vía Láctea s/n, 38200 La Laguna, Tenerife, Spain

E. Bravo Departament de Fisica i Enginyeria Nuclear (UPC), Barcelona, Spain, eduardo.bravo@upc.edu

M. Bremer Inst. de Radioast. Milim. (IRAM), 300 rue de la Piscine, 38406 Saint Martin d'Héres, France

F. Bresolin Institute of Astronomy, 2680 Woodlawn Drive, Honolulu, HI 96822, USA

C. Brunt School of Physics, University of Exeter, Stocker Road, Exeter EX4 4QL, UK

V. Buat Observatoire Astronomique de Marseille-Provence, Laboratoire d'Astrophysique de Marseille, Cedex, France
and
Centre National de la Recherche Scientifique, France

Fernando Buitrago School of Physics and Astronomy, University of Nottingham, NG7 2RD, UK, ppxfb@nottingham.ac.uk

N. Bursov Special Astroph. Obs. (SAO-RAS), Nizhnij Arkhyz, Karachai-Cirkassian Rep. 369167, Russia

I. Caballero Departamento de Astrofísica y Ciencias de la Atmósfera, Facultad de Física, Universidad Complutense de Madrid, 28040 Madrid, Spain, caballero@astrax.fis.ucm.es

José A. Caballero Departamento de Astrofísica y Ciencias de la Atmósfera, Facultad de Física, Universidad Complutense de Madrid, 28040 Madrid, Spain, caballero@astrax.fis.ucm.es

M.D. Caballero-García (on behalf of a large collaboration team) Centro de Astrobiología – LAEFF (CSIC–INTA), Apartado 78, 28691 Villanueva de la Cãnada, Madrid, Spain

Anna Cabre Institut de Ciencies de l'Espai (IEEC/CSIC), Barcelona, www.ice.cat, cabre@ice.cat

A. Cabrera IAC, Spain, acabrera@iac.es

J. Cairol Observatori Astronòmic del Garraf, Barcelona, Catalunya, Spain

M.D. Caballero-García Institute of Astronomy, Madingley road, Cambridge CB3 0HA, UK, mcaballe@ast.cam.ac.uk

J. Calvo Departamento de Matemática Aplicada, Univ. de Granada, Granada, Spain, juancalvo@ugr.es

D. Calzetti Department of Astronomy, University of Massachusetts, Amherst, MA, USA

I. Campos IFCA (CSIC-UC), Avda Los Castros, Santander, Spain, iscampos@ifca.unican.es

J.L. Cano Departamento de Ingenieria de COMunicaciones (DICOM), Laboratorios de I+D de Telecomunicaciones, Plaza de la Ciencia s/n, 39005 Santander, Spain

M. Caramazza INAF – Osservatorio Astronomico di Palermo, Piazza del Parlamento 1, 90134 Palermo, Italy, mcarama@astropa.unipa.it
and
Dip. di Scienze Fisiche e Astronomiche, Sez. di Astronomia, Università di Palermo, Piazza del Parlamento 1, 90134 Palermo, Italy

R. Carballo Departamento de Matemática Aplicada y Ciencias de la Computación, Univ. de Cantabria. ETSI Caminos, Canales y Puertos, Avda de los Castros s/n, 39005 Santander, Spain, carballor@unican.es

J.A. Carballo-Bello Instituto de Astrofísica de Canarias (IAC), Spain, jacb@iac.es

M.C. Cárdenas Instituto de Astrofísica de Andalucía (IAA-CSIC), P.O. Box 3004, 18080 Granada, Spain, conchi@iaa.es

Nicolás Cardiel Departamento de Astrofísica, Facultad de Ciencias Físicas, Universidad Complutense de Madrid, Spain, ncl@astrax.fis.ucm.es

E. Carmona Max-Planck-Institut für Physik, München, Germany

J.M. Carrasco University of Barcelona, ICC-IEEC, Martí i Franquès 1, 08028 Barcelona, Spain, carrasco@am.ub.es

Francisco J. Carrera Instituto de Física de Cantabria (CSIC-UC), Avenida de los Castros, 39005 Santander, Spain, carreraf@ifca.unican.es

J. Casares Instituto de Astrofísica de Canarias, 38200 La Laguna, Tenerife, Spain, jcv@iac.es

F.J. Casas Instituto de Fisica de Cantabria (IFCA), CSIC-Univ. de Cantabria, Avda. los Castros s/n, 39005 Santander, Spain

Angelo Cassatella INAF-IFSI, Via del Fosso del Cavaliere 100, 00133 Roma, Italy

F.J. Castander Instituto de Estudios Espaciales de Cataluña, Spain

H. Castãneda Instituto de Astrofísica de Canarias, 38205 La Laguna, Spain, hcastane@iac.es

A. Castillo-Morales Departamento de Astrofísica y C.C. de la Atmósfera, Facultad de C.C. Físicas, Universidad Complutense de Madrid, Avda. Complutense s/n, 28040 Madrid, Spain, acm@astrax.fis.ucm.es

F.J. Castander (for the DUNE and EIC collaborations) Institut de Ciències de l'Espai, Campus UAB, Facultat de Ciències, 08192 Bellaterra, Barcelona, Spain, fjc@ieec.fcr.es

N. Castro Instituto de Astrofísica de Canarias, C/Vía Láctea s/n, 38200 Tenerife, Spain, norberto@iac.es

A.J. Castro-Tirado Instituto de Astrofísica de Andalucía (IAA-CSIC), C. Bajo de Huétor 50, 18080 Granada, Spain

S. Catalán Institut d'Estudis Espacials de Catalunya, c/Gran Capità 2–4, 08034 Barcelona, Spain, catalan@ieec.uab.es
and
Institut de Ciències de l'Espai, CSIC, Facultat de Ciències, UAB, 08193 Bellaterra, Spain

M. Teresa Ceballos Instituto de Física de Cantabria (CSIC-UC), Santander, Spain, ceballos@ifca.unican.es

A.J. Cenarro Instituto de Astrofísica de Canarias, C/Vía Láctea s/n, 38205 La Laguna, Tenerife, Spain, cenarro@iac.es

Jordi Cepa Nogué Instituto de Astrofísica de Canarias, 38200 La Laguna, Tenerife, Spain
and
Departamento de Astrofísica, Facultad de Física, Universidad de La Laguna, La Laguna, Spain, jcn@iac.es

Gilles Chabrier école Normale Supérieure de Lyon, CRAL (UMR CNRS 5574), Université de Lyon, France

E.L. Chapin Department of Physics and Astronomy, University of British Columbia, 6224 Agricultural Road, Vancouver, BC, Canada V6T 1Z1

M.V. Charcos-Llorens University of Florida, 211 Bryant Space Science Center, Gainesville, FL 32611, USA, mcharcos@astro.ufl.edu

B. Ruiz Cobo Instituto de Astrofísica de Canarias, Tenerife, Spain, brc@iac.es

Angel Colin Instituto de Fsica de Cantabria (CSIC-UC), Av. Los Castros s/n, 39005 Santander, Spain, acolin@ifca.unican.es

Luis Colina DAMIR-IEM-CSIC, Serrano 121, 28006 Madrid, Spain, colina@damir.iem.csic.es
and
Steward Observatory, University of Arizona, Tucson, AZ 85721, USA

J. Colomé Institut d'Estudis Espacials de Catalunya (IEEC), Gran Capità 2–4 (Edifici Nexus), 08034 Barcelona, Spain
and
Institut de Ciències de l'Espai (CSIC-IEEC), Facultat Ciències, Universitat Autònoma de Barcelona, Torre C-5 Parell, 2ona planta, 08193 Bellaterra, Spain

S. Comerón Instituto de Astrofísica de Canarias, 38200 La Laguna, Spain, sebastien@iac.es

Christopher J. Conselice School of Physics and Astronomy, University of Nottingham, NG7 2RD, UK, conselice@nottingham.ac.uk

J.L. Contreras Departamento de Física Atómica, UCM, Madrid, Spain

M. Cornide Departamento de Astrofísica y CC. de la Atmósfera, Facultad de Física, Universidad Complutense de Madrid, Avda. Complutense s/n, 28040 Madrid, Spain

A. Corral Instituto de Física de Cantabria (CSIC-UC), 39005 Santander, Spain
and
INAF-OAB, via Brera 28, 20121 Milan, Italy, amalia.corral@brera.inaf.it

J.M. Corral-Santana Instituto de Astrofísica de Canarias (IAC), Spain, jcorral@iac.es

M.T. Costado (on behalf of the MAGIC Collaboration) Instituto de Astrofísica de Canarias, 38205 La Laguna, Tenerife, Spain
and
Departamento de Astrofísica, Universidad de La Laguna, 38206 La Laguna, Tenerife, Spain, mcostado@iac.es

Luis Cuesta Crespo (and the Group of Robotic Telescopes at CAB) Centro de Astrobiología, INTA, Carretera de Ajalvir, km 4, 28850 Torrejón de Ardoz, Madrid, Spain, cuestacl@inta.es

I. Crespo-Chacón Departamento de Astrofísica, Facultad de Ciencias Físicas, Universidad Complutense de Madrid, 28040 Madrid, Spain, icc@astrax.fis.ucm.es

D. Cristobal-Hornillos Instituto de Astrofísica de Andalucía (IAA), Granada, Spain

M. Cruz IFCA, CSIC-Univ. de Cantabria, Avda. los Castros s/n, 39005 Santander, Spain, cruz@ifca.unican.es

R. Cunniffe Instituto de Astrofísica de Andalucía (IAA-CSIC), C. Bajo de Huétor 50, 18080 Granada, Spain

A. Curto Instituto de Física de Cantabria, CSIC-Universidad de Cantabria, Avda. de los Castros s/n, 39005 Santander, Spain, curto@ifca.unican.es
and
Departamento de Física Moderna, Universidad de Cantabria, Santander, Spain

D.A. Dale Department of Physics and Astronomy, University of Wyoming, Laramie, WY, USA

R. Davies Jodrell Bank Centre for Astrophysics, School of Physics and Astronomy, University of Manchester, Oxford Road, Manchester M13 9PL, UK

R. Davis Jodrell Bank Centre for Astrophysics, School of Physics and Astronomy, University of Manchester, Oxford Road, Manchester M13 9PL, UK

E. de Castro Departamento de Astrofísica y CC. de la Atmósfera, Facultad de Física, Universidad Complutense de Madrid, Avda. Complutense s/n, 28040 Madrid, Spain

E. de Cea del Pozo Institut de Ciències de l'Espai (IEEC-CSIC), Campus UAB, Torre C5, 2a planta, 08193 Barcelona, Spain, decea@ieec.uab.es

Juan de Dios Santander-Vela Instituto de Astrofísica de Andalucia, CSIC, Camino Bajo de Huétor 50, Apdo 3004, 18080 Granada, Spain
and
Red temática Observatorio Virtual Español, Spain, jdsant@iaa.es

Richard de Grijs Department of Physics and Astronomy, The University of Sheffield, UK, R.DeGrijs@sheffield.ac.uk

L. Sancho de la Jordana Universitat de les Illes Balears, Cra. Valldemossa km 7.5, 07122 Palma de Mallorca, Spain, llucia.sancho@uib.es

L. de la Fuente Departamento de Ingenieria de COMunicaciones (DICOM), Laboratorios de I+D de Telecomunicaciones, Plaza de la Ciencia s/n, E-39005 Santander, Spain

Adriana de Lorenzo-Cáceres Instituto de Astrofísica de Canarias, C/Vía Láctea S/N, 38205 La Laguna, Spain, adlcr@iac.es

R. de los Reyes Dpto. Física Atómica, UCM, Madrid, Spain

Alejandro Sánchez de Miguel Departamento de Astrofísica y C.C. de la Atmósfera, Facultad de C.C. Físicas, Universidad Complutense de Madrid, Avda. Complutense s/n, 28040 Madrid, Spain, alex@astrax.fis.ucm.es

Nicolas de Séréville Institut de Physique Nuclaire (IPN), Orsay, France, deserevi@ipno.in2p3.fr

A. de Ugarte Postigo European Southern Observatory (ESO), Casilla 19001, Santiago 19, Chile, adeugart@eso.org

Jonas Debosscher Instituut voor Sterrenkundem Katholieke Universiteit Leuven, Celestijnenlaan 200D BUS 2401, 3001 Leuven, Belgium, jonas.debosscher@ster.kuleuven.ac.be

Carmen del Puerto Instituto de Astrofísica de Canarias, 38200 La Laguna, Spain
and
Museo de la Ciencia y el Cosmos, C/Vía Láctea s/n, 38200 La Laguna, Spain, cpv@museosdetenerife.org

C. Delgado (on behalf of the MAGIC Collaboration) Centro de Investigaciones Energéticas, Medioambientales y Tecnológicas, Madrid, Spain, carlos.delgado@ciemat.es

R. Demarco Department of Physics and Astronomy, The Johns Hopkins University, Baltimore, USA

R. Deshpande University of Central Florida, Physics Department, 4000 Central Florida Blvd, Orlando, USA

M.J. Devlin Department of Physics and Astronomy, University of Pennsylvania, 209 South 33rd Street, Philadelphia, PA 19104, USA

P.D. Diago Observatori Astronòmic de la Universitat de València, Ed. Instituts d'Investigació, Polígon La Coma, 46980 Paterna, València, Spain, Pascual.Diago@uv.es

A.I. Díaz Universidad Autónoma de Madrid, Madrid, Spain, angeles.diaz@uam.es

C. Díaz Departamento de Astrofísica y C.C. de la Atmósfera, Facultad de C.C. Físicas, Universidad Complutense de Madrid, Avda. Complutense s/n, 28040 Madrid, Spain

S. Dicker Department of Physics and Astronomy, University of Pennsylvania, 209 South 33rd Street, Philadelphia, PA 19104, USA

Jose M. Diego IFCA, Universidad de Cantabria-CSIC, 39005 Santander, Spain, jdiego@ifca.unican.es

Anlaug Amanda Djupvik Nordic Optical Telescope, Apdo 474, 38700 Santa Cruz de La Palma, Spain, amanda@not.iac.es

J.A. Docobo Astronomical Observatory R.M. Aller (USC), Santiago de Compostela, Spain, joseangel.docobo@usc.es

A. Domingo (The INTEGRAL–OMC Scientific Archive) Centro de Astrobiología – LAEFF (CSIC–INTA), POB 78, Apartado 78, 28691 Villanueva de la Cañada, Madrid, Spain, albert@laeff.inta.es

R.M. Domínguez Instituto de Física de Cantabria, Spain, dominguez@ifca.unican.es

R. Domínguez Tenreiro Univ. Autónoma de Madrid, 28049 Cantoblanco, Madrid, Spain, rosa.dominguez@uam.es

Lilian Domínguez-Palmero Instituto de Astrofísica de Canarias, 38200 La Laguna, Tenerife, Spain, ldp@iac.es

J. Donley University of Arizona, USA

S. Dye School of Physics and Astronomy, Cardiff University, Queens Buildings, The Parade, Cardiff CF24, 3AA, UK

Jacobo Ebrero Instituto de Física de Cantabria (CSIC-UC), Avenida de Los Castros, 39005 Santander, Spain, ecarrero@ifca.unican.es

M.L. Edwards University of Florida, 211 Bryant Space Science Center, Gainesville, FL 32611, USA, edwards@astro.ufl.edu

S.S. Eikenberry University of Florida, 211 Bryant Space Science Center, Gainesville, FL 32611, USA, eiken@astro.ufl.edu

Carlos Eiroa Departamento de Física Teórica, C-XI, Facultad de Ciencias, Universidad Autónoma de Madrid, Cantoblanco, 28049 Madrid, Spain, carlos.eiroa@uam.es
and
Spanish Virtual Observatory Thematic network, Spain

M.C. Eliche-Moral Departamento de Astrofísica y C.C. de la Atmósfera, Facultad de C.C. Físicas, Universidad Complutense de Madrid, Avda. Complutense s/n, 28040 Madrid, Spain, cem@astrax.fis.ucm.es

José Antonio Escartín Vigo UPC, Jordi Girona 1-3 B5, 08034 Barcelona, Spain, jose.antonio.escartin@upc.edu

Daniel Espada Academia Sinica, Institute for Astrononomy and Astrophysics, Taiwan, despada@asiaa.sinica.edu.tw
and
Harvard-Smithsonian Center for Astrophysics, 160 Concord Ave. M-223, Cambridge, MA 02138, USA, despada@cfa.harvard.edu

Pedro Esparza Departamento de Química Inorgánica, Facultad de Química, Universidad de la Laguna, Calle Francisco Sánchez s/n, 38204 La Laguna, Tenerife, Spain

Victor Espigares IAA-CSIC, Camino Bajo de Huétor 50, 18008 Granada, Spain, espigar@iaa.es

N. Espino Departamento de Astrofísica y C.C. de la Atmósfera, Facultad de C.C. Físicas, Universidad Complutense de Madrid, Avda. Complutense s/n, 28040 Madrid, Spain

Gisela Bañó Esplugues LAEFF-INTA, ESAC, 28691 Villanueva de la Cañada, Madrid, Spain

B. Etxeita IDOM, Avda. Lehendakari Aguirre, 3, 48014 Bilbao, Spain

A.C. Fabian (on behalf of a large collaboration team) Institute of Astronomy, Madingley road, Cambridge CB3 0HA, UK

J. Fabregat Observatori Astronòmic de la Universitat de València, Ed. Instituts d'Investigació, Polígon La Coma, 46980 Paterna, València, Spain
and
GEPI, Observatoire de Paris, CNRS, Université Paris Diderot, Place Jules Janssen, 92195 Meudon Cedex, France

C. Fabricius University of Barcelona, ICC-IEEC, Martí i Franquès 1, 08028 Barcelona, Spain

Emilio E. Falco F. L. Whipple Observatory, Smithsonian Institution, P.O. Box 6369, Amado, AZ 85645, USA, falco@cfa.harvard.edu

Jesús Falcón-Barroso European Space Agency/ESTEC, Keplerlaan 1, Postbus 299, 2200 Noordwijk, The Netherlands, jfalcon@rssd.esa.int

T.A. Fatkhullin Special Astroph. Obs. (SAO-RAS), Nizhnij Arkhyz, Karachai-Cirkassian Rep. 369167, Russia

D. Fernández Institut d'Estudis Espacials de Catalunya (IEEC), Gran Capità 2–4 (Edifici Nexus), 08034 Barcelona, Spain
and
Departament d'Astronomia i Meteorologia, Institut de Ciències del Cosmos (UB-IEEC), Universitat de Barcelona, Av.Diagonal 647, 08028 Barcelona, Spain, tantoja@am.ub.es

M. Fernández Instituto de Astrofísica de Canarias, 38205 La Laguna, Spain

M.J. Fernández-Figueroa Departamento de Astrofísica, Facultad de Ciencias Físicas, Universidad Complutense de Madrid, 28040 Madrid, Spain, mjf@astrax.fis.ucm.es

Mirian Fernández Lorenzo IAC, c/Vía Láctea S/N, La Laguna, Spain, mfernan@iac.es

A. Fernández-Soto (on behalf of the ALHAMBRA Core Team) Instituto de Fsica de Cantabria (CSIC-UC), Av. de los Castros s/n, 39005 Santander, Spain, fsoto@ifca.unican.es

The ALHAMBRA Core Team M. Moles (PI), J.A.L. Aguerri, E. Alfaro, N. Benítez, T. Broadhurst, J. Cabrera-Caño, F.J. Castander, J. Cepa, M. Cerviño, D. Cristóbal-Hornillos, R.M. González Delgado, L. Infante, I. Márquez, V.J. Martínez, J. Masegosa, A. del Olmo, J. Perea, F. Prada, J.M. Quintana and S.F. Sánchez

P. Ferrero Thüringer Landessternwarte Tautenburg, Sternwarte 5, 07778 Tautenburg, Germany

O. Le Fèvre LAM, CNRS-Université de Provence, 38 rue Frédéric Joliot-Curie, 13388 Marseille Cedex 13, France

Francesca Figueras Departament d'Astronomia i Meteorologia and IEEC-UB, Institut de Ciencies del Cosmos de la Universitat de Barcelona, Marti i Franques 1, 08028 Barcelona, Spain, cesca@am.ub.es

E.L. Fitzpatrick Department of Astronomy and Astrophysics, Villanova University, 800 Lancaster Avenue, Villanova, PA 19085, USA

Estrella Florido Departamento de Física Teórica y del Cosmos, Universidad de Granada, Granada, Spain
and
Instituto Carlos I, Spain, estrella@ugr.es

M.V. Fonseca Dpto. Física Atómica, UCM, Madrid, Spain

A. Font-Ribera Institut de Ciències de l'Espai (IEEC-CSIC), Campus UAB, Bellaterra 08193, Spain, font@ieec.uab.es

X. Francisco Institut d'Estudis Espacials de Catalunya (IEEC), Gran Capità 2–4 (Edifici Nexus), 08034 Barcelona, Spain

J. French University College, Belfield, Dublin 4, Ireland

Carmen Blasco Fuertes European Space Research and Technology Centre (ESTEC), Keplerlaan 1, SRE-SA, 2201 AZ Noordwijk, The Netherlands, cblasco@rssd.esa.int

B. Funke Instituto de Astrofísica de Andalucía, CSIC, Granada, Spain, bernd@iaa.es

C. Gabriel ESAC (ESA), Madrid, Spain, Carlos.Gabriel@sciops.esa.int

David Galadí Centro Astronómico Hispano Alemán, Jesús Durbán Remón 2, 04004 Almería, Spain
and
Centro Astronómico Hispano Alemán (CAHA), 18008 Granada, Spain, dgaladi@caha.es

José Gallardo Departamento de Astronomía, Universidad de Chile, Casilla 36-D, Santiago, Chile

J. Gallego Ministerio de Ciencia e Innovación, Spain
and
Departamento de Astrofísica y Ciencias de la Atmósfera, Facultad de C.C. Físicas, Universidad Complutense de Madrid (UCM), Avda. Complutense s/n, 28040 Madrid, Spain, jgm@astrax.fis.ucm.es

M.C. Gálvez Dpto. de Astrofísica y CC. de la Atmósfera, Facultad de Física, Universidad Complutense de Madrid, Avda. Complutense s/n, 28040 Madrid, Spain
and
Centre for Astrophysics Research, Science and Technology Research Institute, University of Hertfordshire, College Lane, Hatfield, Hertfordshire AL10 9AB, UK, M.Galvez-Ortiz@herts.ac.uk

Emilio García Instituto de Astrofísica de Andalucía, CSIC, Camino Bajo de Huétor 50, Apdo. 3004, 18080 Granada, Spain, garcia@iaa.es

Miriam García Instituto de Astrofísica de Canarias, C/Vía Láctea S/N, 38200 La Laguna, Tenerife, Spain, mgg@iac.es

D. García-Alvarez Imperial College London, Blackett Laboratory, Prince Consort Road, London SW7 2AZ, UK
and
Instituto de Astrofísica de Canarias and GTC Project Office, 38205 La Laguna, Tenerife, Spain, d.garcia-alvarez@imperial.ac.uk

E. García-Berro Institut d'Estudis Espacials de Catalunya, c/Gran Capità 2–4, 08034 Barcelona, Spain
and
Departament de Física Aplicada, Escola Politècnica Superior de Castelldefels, Universitat Politècnica de Catalunya, Avda. del Canal Olímpic s/n, 08860 Castelldefels, Spain

M. García-Comas Instituto de Astrofísica de Andalucía, CSIC, Granada, Spain, maya@iaa.es

C.E. García-Dabó European South Observatory, Karl-Schwarzschild-Strasse 2, 85748 Garching, Germany

R.J. García López (on behalf of the MAGIC Collaboration) Instituto de Astrofísica de Canarias, 38205 La Laguna, Tenerife, Spain
and
Departamento de Astrofísica, Universidad de La Laguna, 38206 La Laguna, Tenerife, Spain, rgl@iac.es

E. García-Melendo Esteve Duran Observatory Foundation, Montseny 46, 08553 Seva, Spain, duranobs@astrogea.org

Domingo García-Senz Departament de Fisica i Enginyeria Nuclear (UPC), Jordi Girona 1-3 B5, 08034 Barcelona, Spain, domingo.garcia@upc.edu

Miguel García-Torres SVO/LAEX-CAB(INTA-CSIC), LAEFF-European Space Astronomy Center (ESAC), P.O. Box 78, 28691 Villanueva de la Cañada, Madrid, Spain, mgarciat@laeff.inta.es

C. Garland Castleton State College, Vermont, USA

F. Garzón Instituto de Astrofísica de Canarias (IAC), 38200 La Laguna, Tenerife, Spain
and
Departamento de Astrofísica, Universidad de La Laguna, Tenerife, Spain, fgl@iac.es

M. Gaug Instituto de Astrofísica de Canarias, La Laguna, Tenerife, Spain

Enrique Gaztañaga Institut de Ciencies de l'Espai (IEEC/CSIC), www.ice.cat, Barcelona, gazta@ice.cat

J. Genebriera Observatorio de Tacande, El Paso, La Palma, Islas Canarias, Spain

Ricardo Génova-Santos Instituto de Astrofisica de Canarias (IAC), C/Via Lactea s/n, 38200 La Laguna, Tenerife, Spain.

W. Gieren Departamento de Física, Astronomy Group, Universidad de Concepción, Casilla 160-C, Concepción, Chile

A. Gil de Paz Departamento de Astrofísica y CC de la Atmósfera, Facultad de C.C. Físicas, Universidad Complutense de Madrid (UCM), Avda. Complutense s/n, 28040 Madrid, Spain

Rodrigo Gil-Merino Instituto de Física de Cantabria (CSIC-UC), Avda. de Los Castros s/n, 39005 Santander, Spain, gilmerino@ifca.unican.es

Luis J. Goicoechea Departamento de Física Moderna, Universidad de Cantabria, Avda. de Los Castros s/n, 39005 Santander, Spain, goicol@unican.es

A. Golovin Main Astronomical Observatory of National Academy of Science of Ukraine, Kyiv, Ukraine

A. Gómez IDOM, Avda. Lehendakari Aguirre, 3, 48014 Bilbao, Spain.

C. Gómez IDOM, Avda. Lehendakari Aguirre, 3, 48014 Bilbao, Spain.

V. Gómez Dpt. d'Astronomia i Meteorologia, Universitat de Barcelona, ICC, Avda. Diagonal 647, 08028 Barcelona, Spain

A.I. Gómez de Castro Fac. de CC. Matemáticas, Universidad Complutense de Madrid, Plaza de Ciencias 3, Madrid 28040, Spain, aig@mat.ucm.es

M.A. Gómez-Flechoso Univ. Autónoma de Madrid, 28049 Cantoblanco, Madrid, Spain, mangeles.gomez@uam.es

F. Gómez-Reñasco Instituto de Astrofisica de Canarias (IAC), C/Via Lactea s/n, 38200 La Laguna, Tenerife, Spain.

C. González IAC, Spain, cgonzal@iac.es

J.J. González Instituto de Astronomía UNAM, México D.F, México, jesus@astroscu.unam.mx

R.M. González Delgado Instituto de Astrofíca de Andalucía (CSIC), P.O. Box 3004, 18080 Granada, Spain, rosa@iaa.es

J.M. González-Pérez Instituto de Astrofísica de Canarias (IAC), Via Láctea s/n, La Laguna, Tenerife, Spain

V. Gonzalez-Perez Institut de Ciències de l'Espai (CSIC/IEEC), F. de Ciencies, UAB, Torre C5 Par 2a, Bellaterra, 08193 Barcelona, Spain, gonzalez@ieec.uab.es

S. González Sánchez Departamento de Astrofísica y Ciencias de la Atmósfera, Facultad de Física, Universidad Complutense de Madrid, 28040 Madrid, Spain, caballero@astrax.fis.ucm.es

J.I. González-Serrano Instituto de Física de Cantabria, CSIC-Universidad de Cantabria, Avda de los Castros s/n, 39005 Santander, Spain, gserrano@ifca.unican.es
and
Departamento de Física Moderna, Universidad de Cantabria, Santander, Spain

E.A. González-Solares Institute of Astronomy, University of Cambridge, Madingley Road, Cambridge CB3 0HA, UK, eglez@ast.cam.ac.uk

J. Gorgas Departamento de Astrofísica y C.C. de la Atmósfera, Facultad de C.C. Físicas, Universidad Complutense de Madrid, Avda. Complutense s/n, 28040 Madrid, Spain

J. Gorosabel Instituto de Astrofísica de Andalucía (IAA-CSIC), C. Bajo de Huétor 50, 18080 Granada, Spain

K. Grainge Astrophysics Group, Cavendish Laboratory, University of Cambridge, Madingley Road, Cambridge CB3 0HE, UK

M. Griffin Department of Physics and Astronomy, Cardiff University, 5 The Parade, Cardiff CF24 3AA, UK

N. Gruel University of Florida, USA

M.A. Guerrero Instituto de Astrofísica de Andalucía (IAA-CSIC), C. Bajo de Huétor 50, 18080 Granada, Spain

Ana Guijarro Departamento de Física Teórica y del Cosmos, Universidad de Granada, Granada, Spain

Instituto Carlos I, Spain Centro Astronómico Hispano Alemán, Almería, Spain and
Calar Alto Observatory, Almeria, Spain, aguijarr@ugr.es

E.F. Guinan Department of Astronomy and Astrophysics, Villanova University, 800 Lancaster Avenue, Villanova, PA 19085, USA

José Carlos Guirado Universidad de Valencia, Dr. Moliner 50, 46100 Burjassot, Valencia, Spain, jose.c.guirado@alumni.uv.es, Jose.C.Guirado@uv.es

J.O. Gundersen Department of Physics, University of Miami, 1320 Campo Sano Drive, Coral Gables, FL 33146, USA

Carlos M. Gutiérrez Instituto de Astrofísica de Canarias, La Laguna, Spain

Raul Gutiérrez SVO/LAEX-CAB (INTA-CSIC), LAEFF, European Space Astronomy Center (ESAC), P.O. Box 78, 28691 Villanueva de la Cañada, Madrid, Spain, raul@laeff.inta.es, raul.gutierrez@laeff.inta.es

Raúl Gutiérrez-Sánchez SVO/LAEFF-CAB/INTA-CSIC, PO Box 78, 28691 Villanueva de la Cañada, Madrid, Spain, raul.gutierrez@laeff.inta.es

J. Gutiérrez-Soto Observatori Astronòmic de la Universitat de València, Ed. Instituts d'Investigació, Polígon La Coma, 46980 Paterna, València, Spain
and
GEPI, Observatoire de Paris, CNRS, Université Paris Diderot, Place Jules Janssen, 92195 Meudon Cedex, France

S. Guziy Nikolaev State University, Nikolskaya 24, 54030 Nikolaev, Ukraine

R. Guzmán Departamento de Astrofísica y C.C. de la Atmósfera, Facultad de C.C. Físicas, Universidad Complutense de Madrid, Avda. Complutense s/n, 28040 Madrid, Spain
and
Universidad de Florida, FL, USA

M. Halpern Department of Physics and Astronomy, University of British Columbia, 6224 Agricultural Road, Vancouver, BC, Canada V6T 1Z1

P.L. Hammersley IAC, Spain, plh@iac.es

L. Hanlon University College, Belfield, Dublin 4, Ireland

K.T. Hanna University of Florida, 211 Bryant Space Science Center, Gainesville, FL 32611, USA, hanna@astro.ufl.edu

P.C. Hargrave Department of Physics and Astronomy, Cardiff University, 5 The Parade, Cardiff CF24 3AA, UK

M. Hernán-Obispo Departamento de Astrofísica y CC. de la Atmósfera, Facultad de Física, Universidad Complutense de Madrid, Avda. Complutense s/n, 28040 Madrid, Spain, mho@astrax.fis.ucm.es

D. Herranz Instituto de Fisica de Cantabria (IFCA), CSIC-Univ. de Cantabria, Avda. los Castros s/n, 39005 Santander, Spain

Margarita Hernanz Institut de Ciències de l'Espai (CSIC-IEEC), Campus UAB, Facultat Ciències, Torre C5 par., 2a planta, 08193 Bellaterra, Spain, hernanz@ieec.uab.es

Artemio Herrero Instituto de Astrofísica de Canarias, C/Vía Láctea S/N, 38200 La Laguna, Tenerife, Spain, ahd@iac.es

J.M. Herreros Instituto de Astrofisica de Canarias (IAC), C/Via Lactea s/n, 38200 La Laguna, Tenerife, Spain

P.C. Hewett Institute of Astronomy, University of Cambridge, Madingley Road, Cambridge CB3 0HA, UK, eglez@ast.cam.ac.uk

S.R. Hildebrandt Instituto de Astrofisica de Canarias (IAC), C/Via Lactea s/n, 38200 La Laguna, Tenerife, Spain

M. Hobson Astrophysics Group, Cavendish Laboratory, University of Cambridge, Madingley Road, Cambridge CB3 0HE, UK

G. Holsclaw Laboratory for Atmospheric and Space Physics, University of Colorado, Boulder, USA

J. Holt Leiden Observatory, Leiden University, Leiden, The Netherlands

R. Hoyland Instituto de Astrofisica de Canarias (IAC), C/Via Lactea s/n, 38200 La Laguna, Tenerife, Spain

M. Huang National Astronomical Observatories, CAS, China

R. Hudec Astronomical Institute of the Czech Academy of Sciences, Ondrejov, Czech Republic

M. Huertas-Company LESIA, Observatoire de Paris, CNRS, UPMC, Universit Paris Diderot, 5 Place Jules Janssen, 92195 Meudon, France, marc.huertas@obspm.fr

R. Hueso Grupo de Ciencias Planetarias, Dpto Fisica Aplicada I, Escuela Superior Ingenieros, Universidad del Pais Vasco/Euskal Herriko Unibertsitatea, Bilbao, Spain, ricardo.hueso@ehu.es

D.H. Hughes Instituto Nacional de Astrofísica Óptica y Electroica, Aptdo. Postal 51 y 72000 Puebla, Mexico

A. Ibarra ESAC (ESA), Madrid, Spain, Aitor.Ibarra@sciops.esa.int

Y. Isasi Departament d'Astronomia i Meteorologia and IEEC-UB, Institut de Ciències del Cosmos de la Universitat de Barcelona, Martí i Franquès, 1, 08028 Barcelona, Spain, isasi@am.ub.es

J. Isern Institut d'Estudis Espacials de Catalunya (IEEC), c/Gran Capità 2–4 (Edifici Nexus), 08034 Barcelona, Spain
and
Institut de Ciències de l'Espai (CSIC-IEEC), Facultat Ciències, Universitat Autònoma de Barcelona, Torre C-5 Parell, 2ona planta, 08193 Bellaterra, Spain

J. Izquierdo Departamento de Astrofísica y C.C. de la Atmósfera, Facultad de C.C. Físicas, Universidad Complutense de Madrid, Avda. Complutense s/n, 28040 Madrid, Spain

D.L. Jauncey Australian Telescope National Facility, P.O. Box 76, Epping, NSW 2121, Australia, David.Jauncey@atnf.csiro.au

R. Jay Gabany Black Bird Observatory, NM, USA

M. Jelínek Instituto de Astrofísica de Andalucía (IAA-CSIC), C. Bajo de Huétor 50, 18080 Granada, Spain

F.J. Jiménez-Fernández Instituto de Astrofsica de Andaluca, CSIC, Camino Bajo de Hutor 50, 18008 Granada, Spain

F. Jiménez-Luján Instituto de Física de Cantabria (CSIC-Universidad de Cantabria), Avda. de los Castros s/n, 39005 Santander, Spain, jimenezf@ifca.unican.es
and
Departamento de Física Moderna, Universidad de Cantabria, Avda de los Castros s/n, 39005 Santander, Spain

Sergio Jiménez-Monferrer Universidad de Valencia, Dr. Moliner 50, 46100 Burjassot, Valencia, Spain, sergio.jimenez@alumni.uv.es

Jorge Jiménez-Vicente Dpto. de Física Teórica y del Cosmos, Universidad de Granada, Spain, jjimenez@ugr.es

D.L. Jones Jet Propulsion Laboratory, California Institute of Technology, 4800 Oak Grove Drive, Pasadena, CA 91109, USA, dayton.jones@jpl.nasa.gov

Carme Jordi Departament d'Astronomia i Meteorologia, ICC-IEEC, Universitat de Barcelona, c/Martí i Franquès 1, Avda. Diagonal 647, 08028 Barcelona, Spain, carme@am.ub.es, carme.jordi@ub.edu

J. Julian University of Florida, 211 Bryant Space Science Center, Gainesville, FL 32611, USA, julian@astro.ufl.edu

S.R. Kane NASA Exoplanet Science Institute, Caltech, MS 100-22, 770 SouthWilson Avenue, Pasadena, CA 91125, USA

D.A. Kann Thüringer Landessternwarte Tautenburg, Sternwarte 5, 07778 Tautenburg, Germany

N. Kappelmann Institut fur Astronomie und Astrophysic, Abteilung Astronomie (IAAT), Universitat Tubingen, Tubingen 72076, Germany

Elena Khomenko Instituto de Astrofísica de Canarias, C/Vía Láctea s/n, 38205 Tenerife, Spain
and
Main Astronomical Observatory, NAS, 03680 Kyiv, Ukraine, khomenko@iac.es

J. Klein Department of Physics and Astronomy, University of Pennsylvania, 209 South 33rd Street, Philadelphia, PA 19104, USA

S. Klose Thüringer Landessternwarte Tautenburg, Sternwarte 5, 07778 Tautenburg, Germany

J.H. Knapen Instituto de Astrofísica de Canarias, 38200 La Laguna, Spain, jhk@iac.es

J.P. Kneib LAM, CNRS-Université de Provence, 38 rue Frédéric Joliot-Curie, 13388 Marseille Cedex 13, France

M. Krumpe Astrophysikalisches Institut Potsdam, An der Sternwarte 16, 14482 Postdam, Germany

V. Krushevska Main Astronomical Observatory of National Academy of Science of Ukraine, Kyiv, Ukraine

P. Kubánek Instituto de Astrofísica de Andalucía (IAA-CSIC), C. Bajo de Huétor 50, 18080 Granada, Spain

R.-P. Kudritzki Institute of Astronomy, 2680 Woodlawn Drive, Honolulu, HI 96822, USA

Yu. Kuznyetsova Main Astronomical Observatory of National Academy of Science of Ukraine, Kyiv, Ukraine

C.G. Lacey Department of Physics, Institute for Computational Cosmology, University of Durham, South Road, Durham DH1 3LE, UK

J.F. Lahulla National Astronomical Observatory, Madrid, Spain, lahulla@oan.es

Dharam Vir Lal Max-Planck-Institut für Radioastronomie, Auf dem Hügel 69, 53121 Bonn, Germany, dharam@mpifr-bonn.mpg.de

U. Lammers ESA/European Space Astronomy Centre, Urbanizacion de Villafranca del Castillo, Avda. de los Castillos s/n, 28692 Villanueva de la Canada, Madrid, Spain

M.A. Lara-López Instituto de Astrofísica de Canarias, 38205 La Laguna, Spain, mall@iac.es

A. Lasenby Astrophysics Group, Cavendish Laboratory, University of Cambridge, Madingley Road, Cambridge CB3 0HE, UK

N.M. Lasso Cabrera University of Florida, 211 Bryant Space Science Center, Gainesville, FL 32611, USA, nestor@astro.ufl.edu

R. Lenzen Max Plank Institut für Astronomie (MPIA), Königstuhl 17, 69117 Heidelberg, Germany

Stéphane Leon Instituto de RadioAstronomía Milimétrica (IRAM), Avenida Divina Pastora, 7, 18012 Granada, Spain, leon@iram.es

J.-F. Lestrade Observatoire de Paris/LERMA, Rue de l'Observatoire 61, 75014 Paris, France, jean-francois.lestrade@obspm.fr

Javier Licandro Instituto de Astrofísica de Canarias, La Laguna, Spain, jlicandr@iac.es

L. Lindegren Lund Observatory, Box 43, 221 00 Lund, Sweden

J.F. Ling Astronomical Observatory R.M. Aller (USC), Santiago de Compostela, Spain, josefinaf.ling@usc.es

U. Lisenfeld Dept. Física Teórica y del Cosmos, Universidad de Granada, Spain, ute@ugr.es

Andrei P. Lobanov Max-Planck-Institut für Radioastronomie, Auf dem Hügel 69, 53121 Bonn, Germany, alobanov@mpifr-bonn.mpg.de

Mauro López SVO/LAEX-CAB (INTA-CSIC), LAEFF, European Space Astronomy Center (ESAC), P.O. Box 78, 28691 Villanueva de la Cañada, Madrid, Spain, mauro@laeff.inta.es

C. López Caraballo Instituto de Astrofisica de Canarias (IAC), C/Via Lactea s/n, 38200 La Laguna, Tenerife, Spain

M. López-Puertas Instituto de Astrofísica de Andalucía, CSIC, Granada, Spain, puertas@iaa.es

Carlos López San Juan Instituto de Astrofísica de Canarias, C.Vía Láctea s/n, 38200 La Laguna, Tenerife, Spain
and
Universidad de La Laguna, Tenerife, Spain, clsj@iac.es

J. López-Santiago Departamento de Astrofísica, Facultad de Ciencias Físicas, Universidad Complutense de Madrid, 28040 Madrid, Spain, jls@astrax.fis.ucm.es

Javier Lorenzo Alicante University, P.O. Box 99, 03080, Spain, javihd64315@gmail.com

X. Luri Departament d'Astronomia i Meteorologia and IEEC-UB, Institut de Ciències del Cosmos de la Universitat de Barcelona, C/Martí i Franquès 1, 08028 Barcelona, Spain, xluri@am.ub.es

K.-H. Mack INAF – Istituto di Radioastronomia, Bologna, Italy

B.F. Madore Observatories of the Carnegie Institution of Washington, Pasadena, CA, USA

B. Maffei Jodrell Bank Centre for Astrophysics, School of Physics and Astronomy, University of Manchester, Oxford Road, Manchester M13 9PL, UK

Steven R. Majewski Department of Astronomy, University of Virginia, USA

J. Maldonado Dpto. Física Teórica C-XI, Facultad de Ciencias, Universidad Autónoma de Madrid, Madrid, Spain, jesus.maldonado@uam.es

Luis Manuel Sarro Dpto. Inteligencia Artificial, ETSI Informática – UNED. C Juan del Rosal, 16–3, 28040. Madrid, Spain
and
Spanish Virtual Observatory Thematic network, Spain, lsb@dia.uned.es

J.M. Marcaide Universidad de Valencia, Dr. Moliner 50, 46100 Burjassot, Valencia, Spain, J.M.Marcaide@uv.es

A. Marín-Franch Instituto de Astrofísica de Canarias, C/Vía Láctea s/n, 38205 La Laguna, Tenerife, Spain, amarin@iac.es

E. Mármol-Queraltó Departamento de Astrofísica y C.C. de la Atmósfera, Facultad de C.C. Físicas, Universidad Complutense de Madrid, Avda. Complutense s/n, 28040 Madrid, Spain

G. Marsden Department of Physics and Astronomy, University of British Columbia, 6224 Agricultural Road, Vancouver, BC, Canada V6T 1Z1

C. Martayan GEPI, Observatoire de Paris, CNRS, Université Paris Diderot, Place Jules Janssen, 92195 Meudon Cedex, France
and
Royal Observatory of Belgium, 3 Avenue Circulaire, 1180 Brussels, Belgium

I. Martí-Vidal Universidad de Valencia, Dr. Moliner 50, 46100 Burjassot, Valencia, Spain, I.Marti-Vidal@uv.es
and
Max-Planck-Institut für Radioastronomie, Auf dem Hügel 69, 53121 Bonn, Germany, imartiv@mpifr-bonn.mpg.de

Eduardo L. Martín Instituto de Astrofísica de Canarias, Calle Vía Láctea s/n, 38200 La Laguna, Tenerife, Spain

P.G. Martin Canadian Institute for Theoretical Astrophysics, University of Toronto, 60 St. George Street, Toronto, ON, Canada M5S 3H8
and
Department of Astronomy and Astrophysics, University of Toronto, 50 St. George Street Toronto, ON, Canada M5S 3H4

J.M. Martín-Hernández Dpto. Astrofísica, Universidad Complutense de Madrid, Avda. Complutense s/n, 28040 Madrid, Spain, josemm@astrax.fis.ucm.es

M.L. Martín-Manjón Universidad Autónoma de Madrid, Madrid, Spain, mariluz.martin@uam.es

Vicent J. Martínez Observatori Astronòmic, Universitat de València, Paterna, Spain
and
Departament d'Astronomia i Astrofísica, Universitat de València, Apartat de Correus 22085, 46071 València, Spain

R.M. Martinez-Arnáiz Dpto. Astrofísica, Facultad de C.C. Físicas, Universidad Complutense de Madrid, 28040 Madrid, Spain

Vincent Martinez-Badenes Instituto de Astrofisica de Andalucia, IAA-CSIC, Spain, vicentm@iaa.es

David Martínez-Delgado Instituto de Astrofisica de Canarias (IAC), C/Vía Láctea s/n, 38205 La Laguna, Spain, ddelgado@iac.es
and
Max-Planck-Institute fur Astronomie, Heidelberg, Germany

Enrique Martínez-González Instituto de Fisica de Cantabria (IFCA), Universidad de Cantabria-CSIC, Avda. los Castros s/n, 39005 Santander, martinez@ifca.unican.es

I.G. Martínez-Pais Instituto de Astrofísica de Canarias (IAC), Vía Láctea s/n, 38205 La Laguna, Spain, igm@iac.es

V. Martínez Pillet Instituto de Astrofísica de Canarias, 38200 La Laguna, Tenerife, Spain, vmp@iac.es

Inma Martinez-Valpuesta Instituto de Astrofísica de Canarias, C/Via Láctea s/n, 38200 La Laguna, S/C Tenerife, Spain, imv@iac.es

Isabel Márquez IAA (CSIC), Apdo 3004, 18080 Granada, Spain, isabel@iaa.es

M.J. Márquez Universidad Nacional de Educación a Distancia, Spain

E. Masana Dpt. d'Astronomia i Meteorologia, IEEC-UB-ICC, C/Martí Franquès 1, Avda. Diagonal 647, 08028 Barcelona, Spain, emasana@am.ub.es

Josefa Masegosa IAA (CSIC), Apdo 3004, 18080 Granada, Spain, pepa@iaa.es

J.M. Mas-Hesse Centro de Astrobiología, Laboratorio de Astrofísica Espacial y Física Fundamental (CSIC–INTA), Apartado 78, 28691 Villanueva de la Cañada, Spain, mm@laeff.inta.es
and
Ministerio de Ciencia e Innovación, Spain
and
Centro de Astrobiología, Madrid, Spain

S. Mateos X-ray and Observational Astronomy Group, Department of Physics and Astronomy, University of Leicester, Leicester LE1 7RH, UK

Satoki Matsushita Academia Sinica, Institute for Astronomy and Astrophysics, Taiwan, satoki@asiaa.sinica.edu.tw

P. Mauskopf Department of Physics and Astronomy, Cardiff University, 5 The Parade, Cardiff CF24 3AA, UK

A.F. Maximov Special Astrophysical Observatory, Karachaevo-Cherkesia, Russia, max@sao.ru

B. McBreen University College, Belfield, Dublin 4, Ireland

W. McClintock Laboratory for Atmospheric and Space Physics, University of Colorado, Boulder, USA

R.G. McMahon (and the UKIDSS Collaboration) Institute of Astronomy, University of Cambridge, Madingley Road, Cambridge CB3 0HA, UK

A. Mediavilla Departamento de Ingenieria de COMunicaciones (DICOM), Laboratorios de I+D de Telecomunicaciones, Plaza de la Ciencia s/n, 39005 Santander, Spain

G. Melady University College, Belfield, Dublin 4, Ireland

Ignacio Mendigutía Laboratorio de Astrofísica Espacial y Física Fundamental (CAB/LAEFF/INTA), European Space Astronomy Centre, PO Box 78, 28691 Villanueva de la Cañada, Madrid, Spain, Ignacio.Mendigutia@laeff.inta.es

G. Micela INAF – Osservatorio Astronomico di Palermo, Piazza del Parlamento 1, 90134 Palermo, Italy, giusi@astropa.unipa.it

J.M. Miller (on behalf of a large collaboration team) Department of Astronomy, University of Michigan, 500 Church Street, Dennison 814, Ann Arbor, MI 48105, USA

G. Miniutti Laboratory APC, Paris, Italy

D. Miralles-Caballero DAMIR-IEM-CSIC, Serrano 121, 28006 Madrid, Spain, dmiralles@damir.iem.csic.es

F.X. Miret Observatori Astronòmic del Garraf, Barcelona, Catalunya, Spain

A. Molina Departamento de Fsica Aplicada, Facultad de Ciencias, Universidad de Granada, Avenida Severo Ochoa s/n, 18071 Granada, Spain
and
Instituto de Astrofsica de Andaluca, CSIC, Camino Bajo de Hutor 50, 18008 Granada, Spain

M. Mollá CIEMAT, Avda. Complutense 22, 28040 Madrid, Spain, mercedes.molla@ciemat.es

Ana Monreal-Ibero ESO, Karl-Schwarzschild-Strasse 2, 85748 Garching bei München, Germany, amonreal@eso.org

F.M. Montenegro-Montes INAF – Istituto di Radioastronomia, Via P. Gobetti 101, 40129 Bologna, Italy
and
Dpto. de Astrofísica, Universidad de La Laguna, La Laguna, Spain
and
Instituto de Astrofísica de Canarias, La Laguna, Spain, fmm@ira.inaf.it

D. Montes Departamento de Astrofísica y Ciencias de la Atmósfera, Facultad de Física, Universidad Complutense de Madrid, 28040 Madrid, Spain, dmg@astrax.fis.ucm.es

Benjamín Montesinos (on behalf of the Spanish Astronomical Society) CAB/LAEFF (CSIC–INTA), POB 78, 28691 Villanueva de la Cañada, Madrid, Spain, bmm@laeff.inta.es

Alcione Mora Departamento de Física Teórica C-XI, Universidad Autónoma de Madrid (UAM), 28049 Madrid, Spain, alcione.mora@uam.es

A. Moralejo Institut de Física d'Altes Energies, Bellaterra, Spain

Juan Antonio Morales Departamento de Astronomía y Astrofísica, Universidad de Valencia, 46100 Burjassot, Valencia, Spain, antonio.morales@uv.es

Juan Carlos Morales Institut d'Estudis Espacials de Catalunya (IEEC), Edif. Nexus, C/Gran Capità 2-4, 08034 Barcelona, Spain, morales@ieec.uab.es
and
Institut de Ciències de l'Espai (IEEC-CSIC), Spain, jcmorales@am.ub.es

Carmen Morales Durán LAEFF-INTA, ESAC, 28691 Villanueva de la Cañada, Madrid, Spain

F. Moreno Instituto de Astrofsica de Andaluca, CSIC, Camino Bajo de Hutor 50, 18008 Granada, Spain

E. Moreno Instituto de Astronomía, Universidad Nacional Autónoma de México, A.P. 70-264, 04510 Mexico D.F., Mexico, barbara@astroscu.unam.mx

D.J. Mortlock (and the UKIDSS Collaboration) Astrophysics Group, Blackett Laboratory, Imperial College London, Prince Consort Road, London, SW7 2AZ, UK

J. Moustakas Center for Cosmology and Particle Physics, New York University, New York, NY, USA

G. Murga IDOM, Avda. Lehendakari Aguirre, 3, 48014 Bilbao, Spain

V.M. Muñoz Marín Instituto de Astrofíca de Andalucía (CSIC), P.O. Box 3004, 18080 Granada, Spain, manuel@iaa.es

J.C. Muñoz-Mateos Departamento de Astrofísica y C.C. de la Atmósfera, Facultad de C.C. Físicas, Universidad Complutense de Madrid, Avda. Complutense s/n, 28040 Madrid, Spain, jcmunoz@astrax.fis.ucm.es

C.B. Netterfield Department of Physics, University of Toronto, 60 St. George Street, Toronto, ON, Canada M5S 1A7
and
Department of Astronomy and Astrophysics, University of Toronto, 50 St. George Street Toronto, ON, Canada M5S 3H4

D. Nieto Dpto. Física Atómica, UCM, Madrid, Spain

N.A. Nizhelskij Special Astroph. Obs. (SAO-RAS), Nizhnij Arkhyz, Karachai-Cirkassian Rep. 369167, Russia

Ignacio Negueruela Departamento de Física, Ingeniería de Sistemas y Teoría de la Señal, Escuela Politécnica Superior, Universidad of Alicante, Apdo.99, 03080 Alicante, Spain, ignacio@dfists.ua.es

D. Nurenberger European Southern Observatory (ESO), Casilla 19001, Santiago 19, Chile

R. Oliver Departament de Física, Universitat de les Illes Balears, 07122 Palma de Mallorca, Spain

L. Olmi Department of Physics, University of Puerto Rico, Rio Piedras Campus, Box 23343, Puerto Rico, USA
and
Instituto di Radioastronomia, Largo E. Fermi 5, 50125 Firence, Italy

W. O'Mullane ESA/European Space Astronomy Centre, Urbanizacion de Villafranca del Castillo, Avda. de los Castillos s/n, 28692 Villanueva de la Canada, Madrid, Spain

R. Oreiro Katholieke Universiteit Leuven, Instituut voor Sterrenkunde, Celestijnenlaan 200D BUS 2401, 3001 Leuven, Belgium, raquel@ster.kuleuven.be

Amelia Ortiz Gil Observatorio Astronómico de la Universidad de Valencia, Pol.La Coma s/n, 46980 Paterna, Spain, amelia.ortiz@uv.es

P. Orviz IFCA (CSIC-UC), Avda Los Castros, Santander, Spain, orviz@ifca.unican.es

H. Otí-Floranes Laboratorio de Astrofísica Espacial y Física Fundamental, CAB (CSIC–INTA), POB 78, 28691 Villanueva de la Cañada, Spain
and
Dpto. de Física Moderna, Facultad de Ciencias, Universidad de Cantabria, 39005 Santander, Spain, otih@laeff.inta.es

I. Oya Dpto.Física Atómica, UCM, Madrid, Spain, oya@gae.ucm.es

C. Packham University of Florida, 211 Bryant Space Science Center, Gainesville, FL 32611, USA, packham@astro.ufl.edu

M.J. Page Mullard Space Science Laboratory, University College London, Holmbury St. Mary, Dorking, Surrey RH5 6NT, UK

X. Palau Fundació Joan Oró, Travessera de les Corts 272, 08014 Barcelona, Spain

J. Pan IDOM, Avda. Lehendakari Aguirre, 3, 48014 Bilbao, Spain

J.C. Pandey The Indian Astronomical Observatory, Mt. Saraswati, Hanle, India

F. Panessa INAF, Rome, Italy

E. Pascale Department of Physics and Astronomy, Cardiff University, 5 The Parade, Cardiff CF24 3AA, UK

J.P. Pascual Departamento de Ingenieria de COMunicaciones (DICOM), Laboratorios de I+D de Telecomunicaciones, Plaza de la Ciencia s/n, 39005 Santander, Spain

S. Pascual Departamento de Astrofísica y C.C. de la Atmósfera, Facultad de C.C. Físicas, Universidad Complutense de Madrid, Avda. Complutense s/n, 28040 Madrid, Spain, sergiopr@fedoraproject.org, spr@astrax.fis.ucm.es

G. Patanchon Laboratoire APC, 10, rue Alice Domon et Léonie Duquet 75205 Paris, France

M. Patel (and the UKIDSS Collaboration) Astrophysics Group, Blackett Laboratory, Imperial College London, Prince Consort Road, London, SW7 2AZ, UK

E. Pavlenko Crimean Astrophysical Observatory, Nauchny, Ukraine

L. Pavlenko Crimean Astrophysical Observatory, Nauchny, Ukraine

Santos Pedraz Centro Astronómico Hispano Alemán, Jesús Durbán Remón 2, 04004 Almería, Spain, pedraz@caha.es

Reynier F. Peletier Kapteyn Astronomical Institute, University of Groningen, 9700 AV Groningen, The Netherlands, peletier@astro.rug.nl

R. Pelló Laboratoire d'Astrophysique de l'Observatoire Midi-Pyrénées, France

Jorge Peñarrubia Institute of Astronomy, Cambridge, UK

J. Peralta Departament d'Astronomia i Meteorologia, Universitat de Barcelona, C/Martí Franquès 1, Barcelona 08028, Spain
and
Grupo de Ciencias Planetarias, Dpto Física Aplicada I, UPV/EHU, E.T.S. Ingeniería, Bilbao, Spain

M. Pereira-Santaella Instituto de Estructura de la Materia, CSIC, 28006 Madrid, Spain
and
Steward Observatory, University of Arizona, Tucson, AZ 85721, USA

Isabel Pérez Kapteyn Astronomical Institute, University of Groningen, Postbus 800, Groningen 9700 AV, the Netherlands
and
Departamento de Física Teórica y del Cosmos, Campus de Fuentenueva, Universidad de Granada, 18071 Granada, Spain, isa@astro.rug.nl
and
Instituto Carlos I, Spain, isa@ugr.es

S. Pérez-Hoyos Grupo de Ciencias Planetarias, Dpto Física Aplicada I, UPV/EHU, E.T.S. Ingeniería, Alda. Urkijo s/n, 48013 Bilbao, Spain, santiago.perez@ehu.es

J. Pérez-Gallego University of Florida, USA

Ana Pérez García IAC, La Laguna, Spain

A.M. Pérez García Instituto de Astrofísica de Canarias, 38200 La Laguna, Spain, apg@iac.es

P.G. Pérez-González Departamento de Astrofísica y C.C. de la Atmósfera, Facultad de C.C. Físicas, Universidad Complutense de Madrid (UCM), Avda. Complutense s/n, 28040 Madrid, Spain

Ricardo Pérez Martínez ESAC/INSA, Madrid, Spain, ricardo.perez@sciops.esa.int

D. Pérez-Ramírez Facultad de Ciencias Experimentales, U. de Jaén, Campus Las Lagunillas, 23071 Jaén, Spain

Paolo Persi Istituto di Astrofisica Spaziale e Fisica Cosmica, Roma, Italy, paolo.persi@iasf-roma.inaf.it

L. Piccirillo Jodrell Bank Centre for Astrophysics, School of Physics and Astronomy, University of Manchester, Oxford Road, Manchester M13 9PL, UK

B. Pichardo Instituto de Astronomía, Universidad Nacional Autónoma de México, A.P. 70-264, 04510 Mexico D.F., Mexico, barbara@astroscu.unam.mx

G. Pietrzynski Departamento de Física, Astronomy Group, Universidad de Concepción, Casilla 160-C, Concepción, Chile

Carmen Pilar Padilla-Torres Instituto de Astrofísica de Canarias, La Laguna, Spain, cppt@iac.es

Göran L. Pilbratt Research and Scientific Support Department, European Space Agency, ESTEC/SRE-SA, Keplerlaan 1, 2201 AZ Noordwijk, The Netherlands, gpilbratt@rssd.esa.int

I. Pillitteri Dip. di Scienze Fisiche e Astronomiche, Sez. di Astronomia, Università di Palermo, Piazza del Parlamento 1, 90134 Palermo, Italy, pilli@astropa.unipa.it

D.J. Pisano National Radio Astronomy Observatory, Green Bank, USA

G. Pisano Jodrell Bank Centre for Astrophysics, School of Physics and Astronomy, University of Manchester, Oxford Road, Manchester M13 9PL, UK

Susana Planelles Departament d'Astronomia i Astrofísica, Universitat de València, 46100 Burjassot, Valencia, Spain, susana.planelles@uv.es

Michael Pohlen School of Physics and Astronomy, Cardiff University, Cardiff CF24 3AA, Wales, UK, Michael.Pohlen@astro.cf.ac.uk

M. Pović Instituto de Astrofísica de Canarias, 38205 La Laguna, Spain, mpovic@iac.es

R.A. Preston Jet Propulsion Laboratory, California Institute of Technology, 4800 Oak Grove Drive, Pasadena, CA 91109, USA, rap@sgra.jpl.nasa.gov

Mercedes Prieto Instituto de Astrofísica de Canarias (IAC), C.Vía Láctea s/n, La Laguna, Tenerife, Spain
and
Universidad de La Laguna, Tenerife, Spain, mpm@iac.es

K.G. Puschmann Instituto de Astrofísica de Canarias, 38200 La Laguna, Tenerife, Spain, kgp@iac.es

S.N. Raines University of Florida, 211 Bryant Space Science Center, Gainesville, FL 32611, USA, raines@astro.ufl.edu

Rafael Rebolo Instituto de Astrofísica de Canarias, La Laguna, Spain
and
Consejo Superior de Investigaciones Científicas, Spain, rrl@iac.es

F. Reale INAF – Osservatorio Astronomico di Palermo, Piazza del Parlamento 1, 90134 Palermo, Italy, reale@astropa.unipa.it
and
Dip. di Scienze Fisiche e Astronomiche, Sez. di Astronomia, Università di Palermo, Piazza del Parlamento 1, 90134 Palermo, Italy

R. Rebolo Instituto de Astrofisica de Canarias (IAC), C/Via Lactea s/n, 38200 La Laguna, Tenerife, Spain

M. Rex Department of Physics and Astronomy, University of Pennsylvania, 209 South 33rd Street, Philadelphia, PA 19104, USA

J.E. Reynolds Australian Telescope National Facility, P.O. Box 76, Epping, NSW 2121, Australia, John.Reynolds@atnf.csiro.au

S. Rial Departament de Física, Universitat de les Illes Balears, 07122 Palma de Mallorca, Spain, samuel.rial@uib.es

B. Riaz Instituto de Astrofísica de Canarias, Calle Vía Láctea s/n, 38200 La Laguna, Tenerife, Spain

Ignasi Ribas Institut d'Estudis Espacials de Catalunya (IEEC), c/Gran Capità 2–4 (Edifici Nexus), 08034 Barcelona, Spain
and
Institut de Ciències de l'Espai (CSIC-IEEC), Facultat de Ciències, Universitat Autònoma de Barcelona, Campus UAB, Torre C-5 – parell, 2a planta, 08193 Bellaterra, Spain, iribas@ieec.uab.es

Salvador J. Ribas Consorci del Montsec, Pl. Major 1, 25691 Ager (Lleida), Spain
and
Departament d'Astronomia i Meteorologia, Universitat de Barcelona, ICC, Mart i Franques 1, Avda. Diagonal 647, 08028 Barcelona, Spain, sribas@am.ub.es

J. Rico (for the MAGIC Collaboration) Institut de Física d'Altes Energies, Bellaterra, Spain
and
Institució Catalana de Recerca i Estudis Avançats, Barcelona, Spain

G. Rieke University of Arizona, USA

G.H. Rieke Instituto de Estructura de la Materia, CSIC, 28006 Madrid, Spain
and
Steward Observatory, University of Arizona, Tucson, AZ 85721, USA

D. Risquez Centro de Astrobiología – LAEFF (CSIC–INTA), Apartado 78, 28691 Villanueva de la Cañada, Madrid, Spain, omc-support@laeff.inta.es

Hans-Walter Rix Max-Planck-Institute fur Astronomie, Heidelberg, Germany

Ricardo Rizzo LAEFF-CAB/INTA-CSIC, Apdo 78, 28691 Villanueva de la Cañada, Madrid, Spain, ricardo@laeff.inta.es

A.C. Robin Institut UTINAM, CNRS-UMR6213, Observatoire de Besanon, France, annie@obs-besancon.fr

J.J. Rodes Universitat d'Alacant, Apartat de correus 99, E03080 Alacant, Spain, rodes@dfists.ua.es

M. Rodgers Optical Research Associates, 3280 East Foothill Boulevard Suite 300, Pasadena, CA 91107, USA

F. Rodler Instituto de Astrofísica de Canarias, Calle Vía Láctea s/n, 38200 La Laguna, Tenerife, Spain

J.R. Rodón IFCA (CSIC-UC), Avda Los Castros, Santander, Spain, rodon@ifca.unican.es

C. Rodrigo SVO, LAEX-CAB (INTA-CSIC), LAEFF, European Space Astronomy Center (ESAC), P.O. Box 78, 28691 Villanueva de la Cañada, Madrid, Spain

L. Rodriguez Departamento de Astrofísica y C.C. de la Atmósfera, Facultad de C.C. Físicas, Universidad Complutense de Madrid, Avda. Complutense s/n, 28040 Madrid, Spain

P. Rodríguez-Gil Isaac Newton Group of Telescopes, La Palma, Spain

J. Rodríguez Gómez Instituto de Astrofísica de Andalucía (IAA-CSIC), P.O. Box 3004, 18080 Granada, Spain

Inés Rodríguez Hidalgo Instituto de Astrofísica de Canarias, C/ Vía Láctea s/n, 38200 La Laguna, Tenerife, España, irh@iac.es

C. Rodríguez-López Laboratoire d'Astrophysique Toulouse-Tarbes, Université de Toulouse-CNRS, 14 Av. Edouard Belin, 31400 Toulouse, France
and
Universidade de Vigo, Campus Marcosende-Lagoas, 36310 Vigo, Spain, crodrigu@ast.obs-mip.fr

A.Y. Rodríguez Marrero Institut de Ciències de l'Espai (IEEC-CSIC), Campus UAB, Torre C5, 2a planta, 08193 Barcelona, Spain, arodrig@ieec.uab.es

J.F. Rojas Grupo de Ciencias Planetarias, Dpto Fisica Aplicada I, Escuela Universitaria Ingenieria Tecnica Industrial, Universidad del Pais Vasco/Euskal Herriko Unibertsitatea, Bilbao, Spain, josef.rojas@ehu.es

D. Rouan LESIA, Observatoire de Paris, CNRS, UPMC, Universit Paris Diderot, 5 Place Jules Janssen, 92195 Meudon, France

José Alberto Rubiño-Martín Instituto de Astrofísica de Canarias (IAC), C/Vía Láctea s/n, 38200 La Laguna, Tenerife, Spain, jalberto@iac.es

A. Ruiz Instituto de Física de Cantabria (CSIC-UC), Santander 39005, Spain, ruizca@ifca.unican.es

José Enrique Ruiz Instituto de Astrofísica de Andalucia, CSIC, Camino Bajo de Huétor 50, Apdo 3004, 18080 Granada, Spain
and
Red temática Observatorio Virtual Español, Spain, jer@iaa.es

B. Ruiz-Granados Dpto. Física Teórica y del Cosmos, Edif. Mecenas, planta baja, Campus Fuentenueva, Universidad de Granada, 18071 Granada, Spain
and
Instituto de Física Teórica y Computacional Carlos I, Granada, Spain, bearg@ugr.es

R.G.M. Rutten Isaac Newton Group of Telescopes, La Palma, Spain

Enn Saar Tartu Observatoorium, Tõravere, 61602 Estonia

L. Sabau-Graziati Instituto Nacional de Técnica Aerospacial (INTA), Ctra. de Ajalvir km. 4, 28750 Torrejón de Ardoz, Madrid, Spain

M. Sachkov Institute of Astronomy, Russian Academy of Sciences, Pyatnitskaya 48, 119017 Moscow, Russia

Diego Sáez Departamento de Astronomía y Astrofísica, Universidad de Valencia, 46100-Burjassot, Valencia, Spain, diego.saez@uv.es

P. San Martín Centro de Biología Molecular Severo Ochoa, C/Nicolás Cabrera, 1, Campus de Cantoblanco, 28049 Madrid, Spain, psanmartin@cbm.uam.es

Néstor Sánchez Instituto de Astrofísica de Andalucía, CSIC, Granada, Spain, nestor@iaa.es

O. Sánchez Dpto. Matemática Aplicada, Univ. de Granada, Granada, Spain, ossanchez@ugr.es

S. Sánchez Centro Astronómico Hispano Alemán CAHA, 18008 Granada, Spain, sanchez@caha.es

E. Sánchez-Blanco Instituto de Astrofísica de Andalucía (IAA-CSIC), P.O. Box 3004, 18080 Granada, Spain

P. Sánchez-Blázquez Centre For Astrophysics, University of Central Lancashire, PR1 2HE Preston, UK, psanchez-blazquez@uclan.ac.uk

A. Sánchez-Lavega Grupo de Ciencias Planetarias, Dpto Fisica Aplicada I, Escuela Superior Ingenieros, Universidad del Pais Vasco/Euskal Herriko Unibertsitatea, E.T.S. Ingeniería, Alda. Urkijo s/n, 48013 Bilbao, Spain, agustin.sanchez@ehu.es

Miguel Sánchez-Portal Herschel Science Centre, ESAC/SRE-SDH/INSA, Apdo. 78, 28691 Villanueva de la Cañada, Madrid, Spain, miguel.sanchez@sciops.esa.int

M. Santander-García Isaac Newton Group of Telescopes, La Palma, Spain, msantander@ing.iac.es

R. Sanquirce IDOM, Avda. Lehendakari Aguirre, 3, 48014 Bilbao, Spain

Luis M. Sarro Dpto. Inteligencia Artificial, ETSI Informática, Universidad Nacional de Educación a Distancia, C\Juan del Rosal 16–3, 28040 Madrid, Spain, lsb@dia.uned.es

R. Saunders Astrophysics Group, Cavendish Laboratory, University of Cambridge, Madingley Road, Cambridge CB3 0HE, UK

H.R. Schmitt Remote Sensing Division, Naval Research Laboratory, Washington, DC 20375, USA
and
Interferometrics, Inc., Herdon, VA 20171, USA

S. Schulze Thüringer Landessternwarte Tautenburg, Sternwarte 5, 07778 Tautenburg, Germany

A. Schwope Astrophysikalisches Institut Potsdam, An der Sternwarte 16, 14482 Postdam, Germany

D. Scott Department of Physics and Astronomy, University of British Columbia, 6224 Agricultural Road, Vancouver, BC, Canada V6T 1Z1

P. Scott Astrophysics Group, Cavendish Laboratory, University of Cambridge, Madingley Road, Cambridge CB3 0HE, UK

C. Semisch Department of Physics and Astronomy, University of Pennsylvania, 209 South 33rd Street, Philadelphia, PA 19104, USA

A. Sergeev Terskol Branch of Institute of Astronomy RAS, Russia

A. Serrano Department of Computer Architecture and Technology, Computer Science and Artificial Intelligence, Rey Juan Carlos University, High Technical School of Computer Engineering, 28933 Móstoles, Madrid, Spain, angel.serrano@urjc.es

T. Shahbaz Instituto de Astrofísica de Canarias (IAC), Via Láctea s/n, La Laguna, Tenerife, Spain

Vyacheslav N. Shalyapin Institute for Radiophysics and Electronics, National Academy of Sciences of Ukraine, 12 Proskura St., Kharkov 61085, Ukraine, vshal@ire.kharkov.ua

R.G. Sharp (and the UKIDSS Collaboration) Anglo-Australian Observatory, P.O. Box 296, Epping, NSW 1710, Australia

B. Shustov Institute of Astronomy, Russian Academy of Sciences, Pyatnitskaya 48, 119017 Moscow, Russia

Sergio Simón Observatoire Astronomique, Université de Genève, Switzerland, Sergio.Simon-Diaz@obs.unige.ch

Alicia M. Sintes Departament de Física, Universitat de les Illes Balears, Cra. Valldemossa Km. 7.5, 07122 Palma de Mallorca, Spain, alicia.sintes@uib.es, sintes@aei.mpg.de
and
Max-Planck-Institut für Gravitationsphysik (Albert-Einstein-Institut), Am Mühlenberg 1, 14476 Golm, Germany

D. Sluse Lab. d'Astroph., Ecole Pol. Fédérale de Lausanne (EPFL) Obs., 1290 Sauverny, Switzerland

V.V. Sokolov Special Astroph. Obs. (SAO-RAS), Nizhnij Arkhyz, Karachai-Cirkassian Rep. 369167, Russia

Enrique Solano SVO/LAEX-CAB (INTA-CSIC), LAEFF, European Space Astronomy Center (ESAC), P.O. Box 78, 28691 Villanueva de la Cañada, Madrid, Spain, esm@laeff.inta.es, enrique.solano@laeff.inta.es

J. Soler Dpto. Matemática Aplicada, Univ. de Granada, Granada, Spain, jsoler@ugr.es

E. Sonbas Special Astroph. Obs. (SAO-RAS), Nizhnij Arkhyz, Karachai-Cirkassian Rep. 369167, Russia

P. Spinelli Universitäts-Sternwarte München, Scheinerstr. 1, 81679 München, Deutschland

J.R. Stauffer Spitzer Science Center, Caltech, Pasadena, CA 91125, USA, stauffer@ipac.caltech.edu

G.P. Stiller Institut für Meteorologie und Klimaforschung, University and Forschungszentrum Karlsruhe, Germany, gabriele.stiller@imk.fzk.de

T. Storchi-Bergmann Instituto de Física, Universidade Federal do Rio Grande do Sul, C.P. 15001, 91501-970 Porto Alegre, Brazil

Moisès Suades Institut de Ciències de l'Espai (CSIC-IEEC), Campus UAB, Facultat Ciències, Torre C5 par., 2a planta, 08193 Bellaterra, Spain, suades@ieec.uab.es

J. Sulentic Department of Astronomy, University of Alabama, Tuscaloosa, USA, jsulenti@bama.ua.edu

J. Suso Observatori Astronòmic de la Universitat de València, Ed. Instituts d'Investigació, Polígon La Coma, 46980 Paterna, València, Spain

V.S. Tamazian Astronomical Observatory R.M. Aller (USC), Santiago de Compostela, Spain, vakhtang.tamazian@usc.es

D. Tapiador ESAC (ESA), Madrid, Spain, Daniel.Tapiador@sciops.esa.int

R. Tata University of Central Florida, Physics Department, 4000 Central Florida Blvd, Orlando, USA

E. Toloba Departamento de Astrofísica y C.C. de la Atmósfera, Facultad de C.C. Físicas, Universidad Complutense de Madrid, Avda. Complutense s/n, 28040 Madrid, Spain

J. Torra Institut d'Estudis Espacials de Catalunya (IEEC), Gran Capità 2–4 (Edifici Nexus), 08034 Barcelona, Spain
and
Departament d'Astronomia i Meteorologia, Institut de Ciències del Cosmos (UB-IEEC), Universitat de Barcelona, Av.Diagonal 647, 08028 Barcelona, Spain
and
Ministerio de Ciencia e Innovación, Spain

Ignacio Trujillo Instituto de Astrofisica de Canarias, La Laguna, Spain, ddelgado@iac.es

M. Tucci Instituto de Astrofisica de Canarias (IAC), C/Via Lactea s/n, 38200 La Laguna, Tenerife, Spain

S. Tulloch Isaac Newton Group of Telescopes, La Palma, Spain

A. Ulla Universidade de Vigo, Campus Lagoas-Marcosende, 36310 Vigo, Spain, ulla@uvigo.es

Luisa Valdivielso Instituto de Astrofísica de Canarias, Calle Vía Láctea s/n, 38200 La Laguna, Tenerife, Spain, lval@iac.es

Juan Angel Vaquerizo Gallego Laboratorio de Astrofísica Espacial y Física Fundamental (LAEFF-INTA), Apartado 78, 28691 Villanueva de la Cañada, Madrid, Spain, jvaquerizog@laeff.inta.es

F. Varosi University of Florida, 211 Bryant Space Science Center, Gainesville, FL 32611, USA, varosi@astro.ufl.edu

Almudena Velasco SVO/LAEX-CAB(INTA-CSIC), LAEFF-European Space Astronomy Center (ESAC), P.O. Box 78, 28691 Villanueva de la Cãnada, Madrid, Spain, avelasco@laeff.inta.es

B.P. Venemans (and the UKIDSS Collaboration) Institute of Astronomy, University of Cambridge, Madingley Road, Cambridge CB3 0HA, UK

Lourdes Verdes-Montenegro IAA-CSIC, Camino Bajo de Huétor 50, 18008 Granada, Spain, lourdes@iaa.es

P. Vielva Instituto de Fisica de Cantabria (IFCA), CSIC-Univ. de Cantabria, Avda. los Castros s/n, 39005 Santander, Spain

E. Villa Departamento de Ingenieria de COMunicaciones (DICOM), Laboratorios de I+D de Telecomunicaciones, Plaza de la Ciencia s/n, 39005 Santander, Spain

V. Villar Departamento de Astrofísica y C.C. de la Atmósfera, Facultad de C.C. Físicas, Universidad Complutense de Madrid, Avda. Complutense s/n, 28040 Madrid, Spain

A. Vizcargüenaga IDOM, Avda. Lehendakari Aguirre, 3, 48014 Bilbao, Spain

H. Voss University of Barcelona, ICC-IEEC, Martí i Franquès 1, 08028 Barcelona, Spain

S.J. Warren (and the UKIDSS Collaboration) Astrophysics Group, Blackett Laboratory, Imperial College London, Prince Consort Road, London, SW7 2AZ, UK

R.A. Watson Jodrell Bank Centre for Astrophysics, School of Physics and Astronomy, University of Manchester, Oxford Road, Manchester M13 9PL, UK

Part I
Plenary Sessions

New Insights into X-ray Binaries

J. Casares

Abstract X-ray binaries are excellent laboratories to study collapsed objects. On the one hand, transient X-ray binaries contain the best examples of stellar-mass black holes while persistent X-ray binaries mostly harbor accreting neutron stars. The determination of stellar masses in persistent X-ray binaries is usually hampered by the overwhelming luminosity of the X-ray heated accretion disc. However, the discovery of high-excitation emission lines from the irradiated companion star has opened new routes in the study of compact objects. This paper presents novel techniques which exploits these irradiated lines and summarizes the dynamical masses obtained for the two populations of collapsed stars: neutron stars and black holes.

1 Introduction

X-ray binaries (XRBs hereafter) are interacting binaries where X-rays arise from the accretion of matter onto a neutron star (NS) or a black hole (BH). Accretion processes are found in other astrophysical environments such as cataclysmic variables (i.e. interacting binaries with accreting white dwarfs), T Tauri stars, protoplanetary discs, etc. but the unique property of XRBs is the presence of central compact objects that are the remnants of collapsed, massive stars. Therefore, they provide the best laboratories to study their properties in detail, such as masses, spin or NS equation of state. This paper is not meant to give a thorough review of XRBs but instead it will focus on three selected topics with implications for our knowledge of the mass spectrum of collapsed stars. These are:

1. *Evidence for BHs in XRBs:* A summary of dynamical masses is presented.
2. *The Bowen Project:* A new technique to trace the orbit of companion (donor) stars in persistent XRBs.

J. Casares
Instituto de Astrofísica de Canarias, E-38200 La Laguna, Tenerife, Spain
e-mail: jcv@iac.es

J.M. Diego et al. (eds.), *Highlights of Spanish Astrophysics V*, Astrophysics and Space Science Proceedings, DOI 10.1007/978-3-642-11250-8_1,

3. *Echo Tomography:* Reprocessed light from the donor star can constrain the binary inclination, the parameter with largest impact in the mass error.

Excellent reviews on other aspects of XRBs can be found in [10] and [24].

2 Black Holes in X-ray Binaries

The mass distribution of Black Holes (BHs hereafter) has strong impact in several areas of Astrophysics, in particular SNe models, the evolution of massive stars, chemical enrichment of the galaxy, jet formation etc.. Stellar evolution theories predict $\sim 10^9$ BHs remnants in the Galaxy [2] but only BHs in interacting binaries can be easily detected through X-ray radiation triggered by accretion. This is the reason why the history of BH discoveries has run in parallel with the development of X-ray astronomy.

In particular, Soft X-ray transients (SXTs hereafter) provide the best systems to find stellar-mass BHs, since $\sim$75% of these transients likely harbor a BH. SXTs are a subclass of X-ray binaries with low-mass donor stars (typically of K-M spectral types) which exhibit episodic outbursts due to mass transfer instabilities in the accretion disc [23]. During outburst, the X-ray luminosity increases by factors up to 10^6–10^7 and, therefore, they are easily spotted by X-ray satellites. Unfortunately, the companion star is overwhelmed by the X-ray heated disc at all wavelengths, precluding its detection. However, after a few months of activity, X-rays switch off, the reprocessed light drops off several magnitudes into quiescence and the companion starts to dominate the optical flux. This provides a special opportunity for spectroscopic detection, perform dynamical studies and derive stellar masses. Figure 1 shows the cumulative histogram of BH SXTs discovered since 1966, when Cen X-2 was first detected during a rocket flight. A linear increase at a rate of $\sim 1.7\,\mathrm{yr}^{-1}$ is apparent since the late 1980's, when a new fleet of X-ray satellites with higher sensitivity and All-Sky-Monitor capabilities became operative.

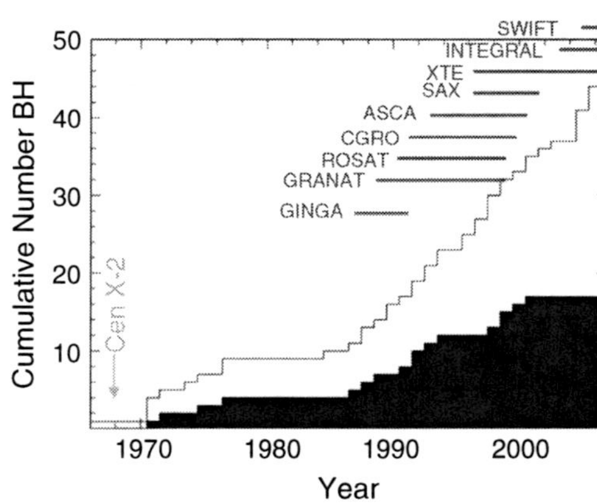

Fig. 1 Cumulative distribution of BH SXTs discovered during the era of X-ray astronomy. The black histogram indicates dynamical BHs

The best evidence for the presence of BHs is dynamical, i.e. a compact object whose mass exceeds the maximum allowed for stable neutron stars or $\sim 3\,M_\odot$ [36]. And this is relatively easy to prove through the observed radial velocity curve of the companion star during quiescence. The orbital period $P_{\rm orb}$ and velocity semi-amplitude $K_{\rm C}$ combine in the standard mass function equation $f(M)$

$$f(M) = \frac{P_{\rm orb} K_{\rm C}^3}{2\pi G} = \frac{M_{\rm X} \sin^3 i}{(1+q)^2} \;, \qquad (1)$$

which relates the mass of the compact object $M_{\rm X}$ and the companion star $M_{\rm C}$ through the orbital inclination i and mass ratio $q = M_{\rm C}/M_{\rm X}$. It is easy to show that $f(M)$ yields a secure lower limit to the BH mass $M_{\rm X}$. This experiment typically requires resolving powers $\geq$1,500 and can be performed with targets brighter than $R \sim 23$ using current instruments on 10 m class telescopes.

But getting actual BH masses rather than lower limits is not so straightforward. The reason being that, by their very nature, BHs do not burst nor pulse and hence one cannot trace their orbital motion. We are facing a single-line spectroscopic binary where the extra observables (q and i) must be extracted from the optical star. And this can be accomplished with two further experiments:

- *Resolving the rotational broadening* of the donor's photospheric lines. Because the donor star is filling the Roche lobe and synchronized, the rotational broadening $V \sin i$ scales with $K_{\rm C}$ as a function of q [42]. Therefore, by measuring $V \sin i$ one can determine q directly.
- *Fitting synthetic models to the ellipsoidal modulation.* The changing visibility of the tidally distorted companion star generates a double-humped light curve (the so-called ellipsoidal modulation) whose amplitude is a strong function of i and q. For extreme mass ratios $q < 0.2$, typical of BH binaries, the shape of the light curve is weakly sensitive to q and hence i can be easily determined [38].

By combining the mass function with constraints on q and i one gets a full dynamical solution and hence the BH mass with minimum assumptions. Further details on this prescription and possible systematics involved can be found in several reviews such as [3].

BHs have also been found in a handful of High Mass X-ray Binaries (HMXBs hereafter), i.e. XRBs with early-type massive donor stars. However, here we find several limitations which complicate the analysis. A key factor is $M_{\rm C}$, which for a HMXB is a large number and has a wide range of uncertainty. The optical star is likely to be undermassive for its spectral type as a result of mass transfer and binary evolution [35]. Furthermore, mass transfer is usually produced through winds rather than Roche lobe overflow and this has a two-fold effect. On one side, wind emission can contaminate the radial velocities of the donor star. On the other, since the optical star does not fill its Roche lobe, q and i values derived through $V \sin i$ and ellipsoidal fits may be overestimated. These caveats can only be side-lined in eclipsing binaries such as X-7 in M33, which so far is the only case where this has been possible. In addition to the eclipse duration, the distance provides an extra restriction which lead to tight constraints in the space parameter. In particular, the radius of the donor, the Roche lobe filling factor and the inclination are accurately determined and yield a very precise BH mass [33].

Table 1 presents an updated list of confirmed BHs based on dynamical arguments, with their best mass estimates. We currently have 21 BHs, with orbital periods between 33.5 days and 4.1 h. The great majority are SXTs (17) while 4 are persistent HMXBs: Cyg X-1 plus the 3 extragalactic binaries LMC X-1, LMC X-3 and M33 X-7. The case of GX 339-4 deserves special mention because it is the only SXT where the presence of a BH was proven during the outburst phase. This was possible thanks to the detection of fluorescent lines arising from the irradiated companion (see Sect. 3). GRS 1915+105 is also noteworthy, not only because of its long orbital period and large mass function but also because IR spectroscopy was essential to overcome the >25 magnitudes of optical extinction and reveal the radial velocity curve of the companion star [20]. However, it should be noted that recent photometry reports a slightly shorter orbital period and evidence for irradiated light

Table 1 Dynamical BHs

System	P_{orb} (days)	$f(M)$ ($M_\odot$)	Donor Spect. Type	Classification	M_x [a] ($M_\odot$)
GRS 1915+105[b]	33.5	9.5 ± 3.0	K/M III	LMXB/Transient	14 ± 4
V404 Cyg	6.471	6.09 ± 0.04	K0 IV	"	12 ± 2
Cyg X-1	5.600	0.244 ± 0.005	09.7 Iab	HMXB/Persistent	10 ± 3
LMC X-1[c]	3.909	0.143 ± 0.007	07 III	"	10.3 ± 1.3
M33 X-7	3.453	0.46 ± 0.08	07-8 III	"	15.7 ± 1.5
XTE J1819-254	2.816	3.13 ± 0.13	B9 III	IMXB/Transient	7.1 ± 0.3
GRO J1655-40	2.620	2.73 ± 0.09	F3/5 IV	"	6.3 ± 0.3
BW Cir[d]	2.545	5.73 ± 0.29	G5 IV	LMXB/Transient	> 7.0
GX 339-4[e]	1.754	5.8 ± 0.5	–	"	> 6.0
LMC X-3	1.704	2.3 ± 0.3	B3 V	HMXB/Persistent	7.6 ± 1.3
XTE J1550-564	1.542	6.86 ± 0.71	G8/K8 IV	LMXB/Transient	9.6 ± 1.2
4U 1543-475	1.125	0.25 ± 0.01	A2 V	IMXB/Transient	9.4 ± 1.0
H1705-250	0.520	4.86 ± 0.13	K3/7 V	LMXB/Transient	6 ± 2
GS 1124-684	0.433	3.01 ± 0.15	K3/5 V	"	7.0 ± 0.6
XTE J1859+226[f]	0.382	7.4 ± 1.1	–	"	
GS2000+250	0.345	5.01 ± 0.12	K3/7 V	"	7.5 ± 0.3
A0620-003	0.325	2.72 ± 0.06	K4 V	"	11 ± 2
XTE J1650-500	0.321	2.73 ± 0.56	K4 V	"	
GRS 1009-45	0.283	3.17 ± 0.12	K7/M0 V	"	5.2 ± 0.6
GRO J0422+32	0.212	1.19 ± 0.02	M2 V	"	4 ± 1
XTE J1118+480	0.171	6.3 ± 0.2	K5/M0 V	"	6.8 ± 0.4

[a] Masses compiled by [10] and [32].
[b] New photometric period of 30.8 ± 0.2 days reported by [30].
[c] Updated after [34].
[d] Updated after [7].
[e] Updated after [27].
[f] Period is uncertain. See [43].

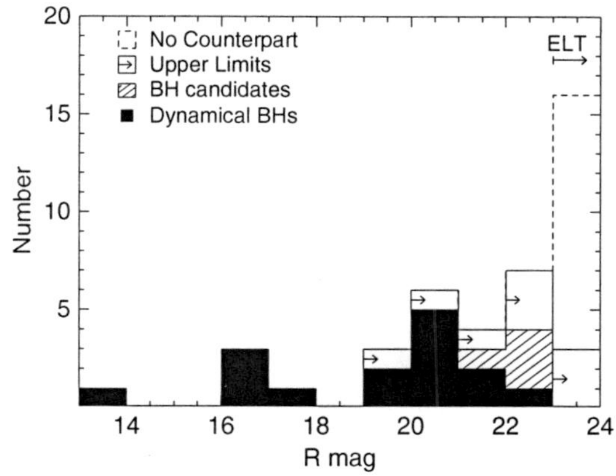

Fig. 2 Magnitude distribution of BH SXTs in quiescence. The black histogram indicates dynamical BHs while the rest are BH candidates. Targets without optical counterpart are likely fainter than $R > 23$

curves [30]. The combination of these two effects will likely decrease the mass function and BH mass. XTE J1859+226 also needs revisiting because its orbital period is uncertain [43]. In summary we have 16 BH masses ranging between 4 and 16 $M_\odot$ with ~5–30% errors. These can be compared with theoretical distributions of stellar remnants such as [19]. The model includes binary interaction under Case C mass transfer (i.e. Common Envelope evolution after core helium ignition), wind mass-loss in the Wolf–Rayet phase and SN Ib explosion. The computation predicts a continuum distribution of remnants with a mass cut at 12 $M_\odot$ which is difficult to reconcile with some of the observed masses. However, the model entails many theoretical uncertainties which dominate the final mass spectrum such as the Common Envelope efficiency, the wind mass-loss rate or the progenitor's mass cut. Clearly more SXT discoveries and lower uncertainties in BH masses are required before these issues can be addressed and the form of the distribution is used to constrain BH formation models and XRB evolution.

In addition to dynamical BHs, there are 27 other SXTs with similar X-ray spectral and timing properties during outburst[1]. Unfortunately, these BH *candidates* become too faint in quiescence for dynamical studies or even lack accurate astrometry. This is illustrated in Fig. 2 which shows the magnitude distribution of the 44 currently known BH transients. Dynamical studies are only possible with the current largest telescopes for sources brighter than $R \leq 23$. Not shown in the figure is the heavily reddened GRO 1915+105 which was studied in the NIR. The figure depicts the bright tail of a *dormant* population of galactic BH SXTs which several works have estimated in a few thousand systems ([37] and included references). Improving the statistics of dynamical BHs requires not only a new generation of ELT telescopes to tackle fainter targets but also new strategies aimed at unveiling new *hibernant* SXTs before they go into outburst. Quiescent BH SXts typically have $EW(H_\alpha)$ ~20–50 Å and hence they should show up in deep H_α surveys such

[1] This number has been updated after [24] with new detections reported in several Astronomical Telegrams.

as IPHAS [17]. However, clever diagnostics need to be defined to clear out other populations of H_α emitters such as cataclysmic variables or T Tauris (see [15]).

3 The Bowen Project

Aside from transient XRBs, there are $\sim$150 persistent XRBs in the Galaxy, the great majority hosting neutron stars (NS hereafter) accreting at the Eddington limit. They are considered the progenitors of Binary Millisecond Pulsars (BMPs hereafter) because is the sustained accretion during their long active lives that spins the NS up to millisecond periods. The discovery of millisecond pulses in 8 transient XRBs and coherent oscillations during X-ray bursts in 13 persistent XRBs gave strong support to this *recycle* pulsar scenario. And burst oscillations were detected in addition to persistent pulses in the transient XRBs SAX J1808-3658 [9] and XTE J1814-338 [40] with identical frequencies. This confirmed that burst oscillations are indeed modulated with the spin of the NS. The interest of these discoveries stands in the fact that one can use the orbital Doppler shift of pulses/oscillations to trace the NS orbit and obtain the X-ray mass function.

On the other hand, optical emission in persistent XRBs is triggered by reprocessing of the intense X-ray radiation in different binary sites, mainly the accretion disc. The companion star is $\sim$1,000 times fainter than the irradiated disc at optical-IR wavelengths and hence completely undetectable. This has systematically plagued attempts to determine system parameters and, in most cases, only the orbital period is known. Fortunately, there are methods which can exploit the effects of irradiation and X-ray variability. New prospects were opened by the discovery of sharp high excitation emission lines arising from the irradiated face of the companion star in Sco X-1 [39]. The most prominent are found in the core of the Bowen feature, a blend of CIII/NIII lines which are mainly powered by fluorescence. These lines trace the motion of the companion star and provided the first dynamical information on this protypical LMXB (see Fig. 3). Since then, sharp Bowen lines from companion stars have been discovered in 7 other persistent LMXBs and 3 transients during outburst: Aql X-1, GX 339-4 and the BMP XTE J1814-338. These transient studies beautifully demonstrate the power of this technique in systems which otherwise cannot be studied in quiescence because either they are too faint (case of GX 339-4 and XTE J1814-338) or are contaminated by a bright interloper (Aql X-1). In particular, the case of GX 339-4 is remarkable because the Bowen study provides the first solid evidence for the presence of a BH in this classic transient.

The radial velocity curves of the Bowen lines are biased because they arise from the irradiated face of the star instead of its center of mass. Therefore, a *K-correction* needs to be applied in order to obtain the true velocity semi-amplitude K_C from the observed velocity K_{em}. The *K-correction* parameterizes the displacement of the center of light with respect to the donor's center of mass through the mass ratio and disc flaring angle α. The latter dictates the size of the disc shadow projected over the irradiated donor [25]. Extra information on q and α is thus required to

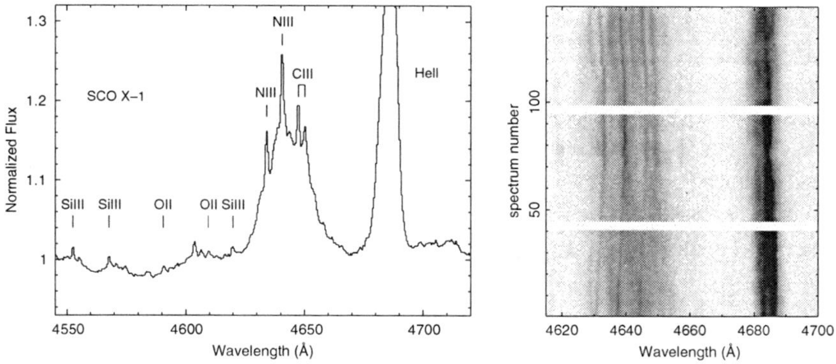

Fig. 3 Detecting companion stars in persistent XRBs. *Left*: the main high excitation emission lines due to irradiation of the donor star in Sco X-1. Adapted from [39]. Right: Trail spectrum showing the radial velocity motion of the Bowen CIII/NIII lines as a function of time After [39]

Table 2 NS masses obtained using the Bowen Technique

System	P_{orb} *(hr)*	Mag	Type	K_{em} (km/s)	M_{NS} [a] ($M_\odot$)	Reference
Sco X-1	18.9	$B = 12.2$	Persistent	87 ± 1	>0.2	[39]
LMC X-2	8.1	$B = 18$	"	351 ± 28	>1.2	[13]
X1822-371	5.6	$B = 15.8$	"	300 ± 15	1.6–2.3	[4]
V926 Sco (X1735-444)	4.7	$B = 17.9$	"	226 ± 22	>0.5	[6]
GX9+9 (X1728-16)	4.2	$B = 16.8$	"	230 ± 35	>0.3	[12]
GR Mus	3.9	$B = 19.1$	"	245 ± 30	1.2–2.6	[1]
V801 Ara (X1636-536)	3.8	$B = 18.2$	"	277 ± 22	>0.8	[6]
EXO 0748-676	3.8	$B = 16.9$	"	310 ± 10	1.1–2.6	[29]
Aql X-1	19	$V \sim 22$	Transient	247 ± 8	>1.6	[11]
GX 339-4	42.1	$V > 21$	"	317 ± 10	>6.0	[21][27]
XTE J1814-338[b]	4.2	$V = 23.3$	"	345 ± 19	>1.0	[8]

[a] After K-correction and constraints to the inclination, mass ratio or NS velocity (when available)
[b] Preliminary results

get the real K_C. Furthermore, useful limits to the NS mass can be set if the binary inclination is well constrained through eclipses.

Table 2 summarizes the NS masses obtained through the Bowen technique during several campaigns at the WHT, AAT and VLT. The list of persistent systems is almost a complete sample of Galactic LMXBs brighter than $B \simeq 19$. In the cases of Aql X-1 and X1822-371 the evidence of NS more massive than canonical is very persuasive. The latter is a particularly favorable binary because it is eclipsing and the NS is a pulsar. Then its radial velocity curve is known through the study of orbital pulse delays. Good constraints on the NS velocity are also available for V801 Ara through the detection of pulse oscillations during a superburst [6]. Tight limits to the inclination and mass ratio are also available for the eclipsing EXO 0748-676 [29] and the dipper GR Mus [1]. In the remaining cases the NS mass is not well constrained due to large uncertainties in the inclination and/or mass ratio. However, it is important to stress that these are the first dynamical constraints in persistent LMXBs since their discovery, 40 years ago. Other techniques (such as the Echo Tomography) need to be exploited to further refine these limits and derive more accurate NS masses. Previous reviews presenting results of the Bowen project can be found in [5] and [14].

4 Echo Tomography

Echo Tomography uses time delays between X-ray and UV/optical variability as a function of orbital phase to map the reprocessing sites in a binary [31]. The optical variability can be modeled by the convolution of the X-ray light curve with a *transfer function* which depends on the binary geometry. The *transfer function* encodes information on the most fundamental parameters such as the binary inclination, star separation and mass ratio. And in particular, the component associated with the companion star is most sensitive to these parameters so detecting echoed emission from the donor offers the best opportunity to constrain them. There has been several attempts at detecting correlated optical and X-ray variability using white light or broad band filters (e.g. [22, 41]). These works have detected delays which are mostly consistent with reprocessing in the outer disc implying that the disc is the dominant source of continuum reprocessed light.

Exploiting emission-line reprocessing rather than broad-band photometry has two potential benefits: (a) it amplifies the response of the donor's contribution by suppressing most of the background continuum light (dominated by the disc); (b) since the emission line reprocessing time is instantaneous, the response is sharper (i.e. only smeared by geometry). Through the Bowen project we know that high energy radiation is very efficiently reprocessed by the donor's atmospheres into Bowen fluorescence lines. Therefore, we decided to search for optical echoes of X-ray variability using ULTRACAM [16] equipped with a special set of narrow band filters, centered at the Bowen blend and a red continuum. The latter is essential to subtract the continuum light and hence amplify the reprocessed signal from the companion. During an RXTE/WHT campaign on Sco X-1 correlated variability was detected at phase $\simeq$0.5 i.e. superior conjunction of the companion star, when the heated face presents its maximum visibility [26]. Time delays of 14–16 s are measured after the continuum light is subtracted from the Bowen light curves (see Fig. 4). These delays are consistent with the light traveltime between the NS and the companion star and hence provide the first evidence of reprocessing in the companion of Sco X-1. However, one needs to detect several optical echoes as a function of orbital phase in order to constrain i and q and derive masses.

In a second campaign we observed the burster X1636-536 simultaneously with RXTE and VLT+ULTRACAM. Three X-ray bursts and their corresponding optical echoes were recorded at orbital phases 0.55, 0.20 and 0.83 and these are shown in the left panel of Fig. 5. The optical bursts clearly lag X-ray burst and are also

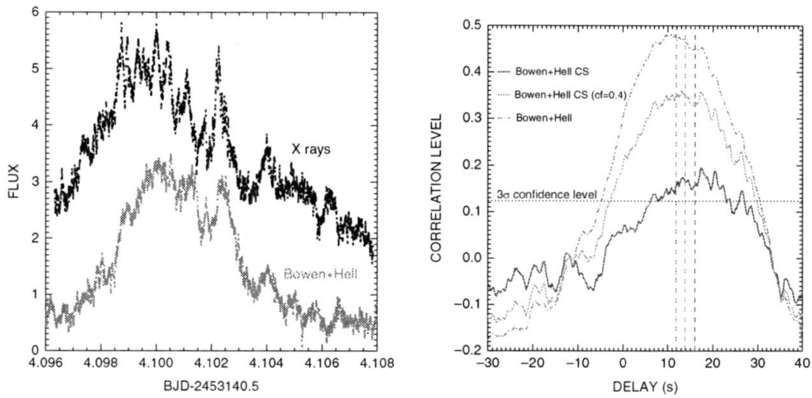

Fig. 4 Echo Tomography experiment in Sco X-1. *Left*: large amplitude X-ray variability and correlated optical light curve observed at orbital phase 0.5. Sco X-1 happened to be in the flaring branch state. *Right*: Cross-correlation functions between X-rays and optical light curves observed in the continuum (*top*), Bowen+HeII window (*middle*) and Bowen+HeII after continuum subtraction (*bottom*). After [26]

smeared, indicating an extended reprocessing site. Delay times are in the range 2–3 sec showing little evidence for orbital variability. However, these delays drift when several amounts of continuum light (parameterized by the factor *cf*) are subtracted from the Bowen+HeII light. And for $cf \simeq 0.8$–0.95 the 3 delays become consistent with reprocessing in the companion for $M_{NS} = 1.4\,M_{\odot}$, $q = 0.3$, $\alpha = 12^{\circ}$ and $i = 36$–60°, as derived through radial velocities of the Bowen lines [6]. This is illustrated in the right panel of Fig. 5. Note that, in particular, delays observed at phase ∼0.5 are especially sensitive to the inclination angle. The main difficulty which hinders us from constraining the inclination is the unknown amount of continuum substraction. In principle, there must be an optimum *cf* factor which results in a perfect subtraction. However, this is not easy to find because the continuum filter is placed ∼1,500 Å away from the Bowen lines due to the optical layout of ULTRACAM. New high-speed spectrophotometry devices such as ULTRASPEC will provide pure emission line light curves for echo mapping experiments. These are likely to yield accurate inclinations and, when combined with dynamical information from the Bowen lines and X-ray mass functions, the first accurate NS masses in persistent XRBs.

5 Conclusions

In the past 20 years the field of X-ray binaries has experienced significant progress with the discovery of 17 new BHs and 8 transient BMPs in LMXBs. Dynamical masses are available for 16 BHs but better statistics and improved errors are required before using the observed distribution to constrain XRB evolution and supernova

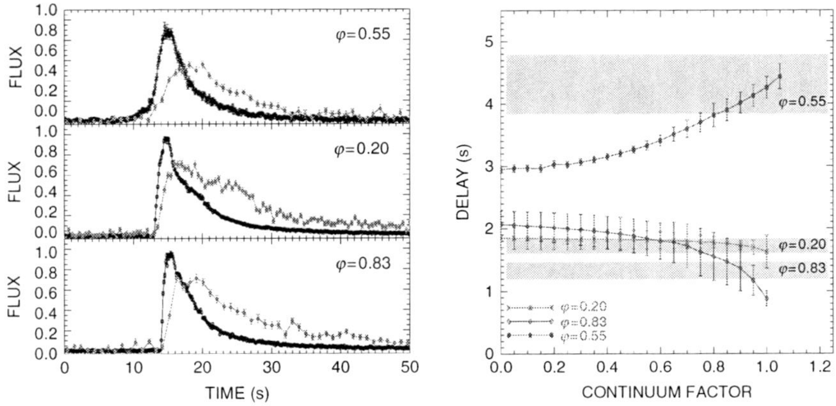

Fig. 5 Echo Tomography of X1636-536. *Left*: the three X-ray bursts detected and their optical (Bowen+HeII) counterparts. *Right*: delay times between the X-ray and Bowen+HeII burst light curves as a function of continuum subtraction factor. Shaded regions correspond to delays expected for reprocessing in the companion at each orbital phase. They are computed for $M_{NS} = 1.4\dot{M}_{\odot}$, $q = 0.3$, $i = 36–60^{\circ}$ and $\alpha = 12^{\circ}$. After [28]

models. Exploiting deep Hα surveys of the Galactic plane, such as IPHAS, may unveil a significant fraction of a large expected population of quiescent XRBs.

The discovery of fluorescence emission from the companion star has opened the door to derive NS masses in persistent and new transient XRBs. This is possible thanks to: (a) dynamical information from irradiated donors through high-resolution spectroscopy of the Bowen blend; (b) echo-mapping reprocessing sites through simultaneous Bowen-line/X-ray lightcurves. These techniques, together with results from burst oscillations and transient BMPs, will likely provide the first accurate NS masses in XRBs in the near future and perhaps confirm the existence of massive NS. Thanks to these new techniques, which have proven their worth, the future is bright as new instruments and telescopes will allow to push ahead our sample of BHs and NS masses. High-speed and high-resolution instruments, such as OSIRIS at GTC, RSS at SALT and ULTRASPEC, will play a crucial role in this goal.

Acknowledgements I would like to acknowledge helpful comments from my colleagues D. Steeghs, R. Cornelisse and T. Muñoz-Darias. I'm also grateful for support from the Spanish MCYT grant AYA2007-66887.

References

1. Barnes, A.D., Casares, J., Cornelisse, R., Charles, P.A., Steeghs, D., Hynes, R.I., O'Brien, K., MNRAS 380, 1182 (2007)
2. Brown, G.E., Bethe, H.A., ApJ 423, 659 (1994)
3. Casares, J., in *Binary Stars: Selected Topics on Observations and Physical Processes*, eds., F.C., Lazaro, & M.J., Arevalo, LNP 563, p. 277 (2001)

4. Casares, J., Steeghs, D., Hynes, R.I., Charles, P.A, O'Brien, K., ApJ 590, 1041 (2003)
5. Casares, J., Steeghs, D., Hynes, R.I., Charles, P.A., Cornelisse, R., O'Brien, K., 2004, Rev Mex AA 20, 21 (2004)
6. Casares, J., Cornelisse, R., Steeghs, D., Charles, P.A., Hynes, R.I., O'Brien, K., Strohmayer, T.E., MNRAS 375, 1463 (2006)
7. Casares, J., et al., ApJS, in press (2009)
8. Casares, J., et al., in preparation (2009)
9. Chakrabarty, D., Morgan, E.H., Muno, M.P., Galloway, D.K., Wijnands, R., van der Klis, M., Markwardt, C.B., Nature 424, 42 (2003)
10. Charles, P.A., Coe, M.J., in *Compact Stellar X-ray Sources*, Lewin, W.H.G., & van der Klis, M., eds., Cambridge Astrophysics Series No. 39, Cambridge: Cambridge University Press, p. 215 (2006)
11. Cornelisse, R., Casares, J., Steeghs, D., Barnes, A.D., Hynes, R.I., O'Brien, K., MNRAS 375, 1463 (2007)
12. Cornelisse, R., Steeghs, D., Casares, J., Charles, P.A., Barnes, A.D., Hynes, R.I., O'Brien, K., MNRAS 380, 1219 (2007)
13. Cornelisse, R., Steeghs, D., Casares, J., Charles, P.A., Shih, I.C., Hynes, R.I., O'Brien, K., MNRAS 381, 194 (2007)
14. Cornelisse, R., Casares, J., Muñoz-Darias, T., Steeghs, D., Charles, P.A., Hynes, R.I., O'Brien, K., Barnes, A., in *A Population Explosion: The Nature & Evolution of X-ray Binaries in Diverse Environments*, AIP Conf. Proc., Vol. 1010, p. 148 (2008)
15. Corral Santana, J.M., this volume (2009)
16. Dhillon, V.S., et al., MNRAS 378, 825 (2007)
17. Drew, J.E., et al., MNRAS 362, 753 (2005)
18. Friedman, J.L., Ipser, J.R., ApJ 314, 594 (1987)
19. Fryer, C.L., Kalogera, V., ApJ 554, 548 (2001)
20. Greiner, J., Cuby, J.G., McCaughrean, M.J., Nature 414, 522 (2001)
21. Hynes, R.I., Steeghs, D., Casares, J., Charles, P.A., O'Brien, K., ApJ 583, L95 (2003)
22. Hynes, R.I., in *Correlated X-ray and Optical Variability in X-ray Binaries*, ed., Hameury, J.M., & Lasota, J.P., ASP Conf. Ser., Astronomical Society of the Pacific, San Francisco, Vol. 330, p. 237 (2005)
23. King, A.R., Phys. Rev. 311, 337 (1999)
24. McClintock, J.E., Remillard, R.A., in *Compact Stellar X-ray Sources*, Lewin, W.H.G., & van der Klis, M., eds., Cambridge Astrophys. Ser. No. 39, Cambridge: Cambridge University Press, p. 157 (2006)
25. Muñoz-Darias, T., Casares, J., Martínez-Pais, I.G., ApJ 635, 502 (2005)
26. Muñoz-Darias, T., Martínez-Pais, I.G., Casares, J., Dhillon, V.S., Marsh, T.R., Cornelisse, R., Steeghs, D., Charles, P.A., MNRAS 379, 1673 (2007)
27. Muñoz-Darias, T., Casares, J., Martínez-Pais, I.G., MNRAS 385, 2205 (2008)
28. Muñoz-Darias, T., et al., in *High Time Resolution Astrophysics: The Universe at Sub-Second Timescales*, AIP Conf. Proc., Vol. 984, p. 15 (2008)
29. Muñoz-Darias, T., et al., in preparation (2009)
30. Neil, E.T., Bailyn, C.D., Cobb, B.E., ApJ 657, 409 (2007)
31. O'Brien, K., Horne, K., Hynes, R.I., Chen, W., Haswell, C.A., Still, M.D., MNRAS 334, 426 (2002)
32. Orosz, J.A., in *A Massive Star Odyssey, from Main Sequence to Supernova*, van der Hucht, K.A., Herrero, A., & Esteban, C., eds., Proc. IAU Symp. No. 212, San Francisco: Astronomical Society of the Pacific, p. 365 (2003)
33. Orosz, J.A., et al., Nature 449, 872 (2007)
34. Orosz, J.A., et al., arXiv:0810.3447 (2008)
35. Rappaport, S.A., Joss, P.C., in *Accretion-driven X-ray Sources*, Lewin, W.H.G., & van den Heuve, E.P.J., eds., Cambridge: Cambridge University Press, p. 33 (1983)
36. Rhoades, C.E., Ruffini, R., Phys. Rev. Lett. 32, 324 (1974)
37. Romani, R.W., A&A 333, 583 (1998)

38. Shahbaz, T., Naylor, T., Charles, P.A., MNRAS 268, 756 (1994)
39. Steeghs, D., Casares, J., ApJ 568, 273 (2002)
40. Strohmayer, T.E., Markwardt, C.B., Swank, J.H., Zand, J.J.M.in't, ApJ 596, 67 (2003)
41. van Paradijs, J., van der Klis, M., van Amerongen, S., et al., A&A 234, 181 (1990)
42. Wade, R.A., Horne, K., ApJ 324, 411 (1998)
43. Zurita, C., et al., MNRAS 334, 999 (2002)

OSIRIS: Final Characterization and Scientific Capabilities

Jordi Cepa

Abstract OSIRIS, the optical Day One instrument for the GTC, will shortly be shipped to La Palma to start the commissioning at the telescope. Some results of the final laboratory characterization of the instrument are shown, together with the upgrades that are planned to be operational after Day One. Several large programs using the OSIRIS Tunable Filters are presented as well, to demonstrate the scientific capabilities of this characteristic OSIRIS observing mode.

1 Introduction

1.1 Brief History

OSIRIS (Optical System for Imaging and low Resolution Integrated Spectroscopy), was supported by the GTC Scientific Advisory Committee as the optical Day One instrument, in March 1999, after an international Announcement of Opportunity (AoO). The final concept of the instrument was fixed in July 2000, while the manpower required was available in Autumn 2000, and the total budget needed was secured in November 2001. After a Preliminary Design Review (PDR) held in April 2001 by a pannel of international experts, who issued a report where only minor technical amendments were suggested, OSIRIS entered the final design and fabrication phases.

The instrument was installed at the Nasmyth rotator of the Assembly, Integration, and Verification (AIV) laboratory of the Instituto de Astrofísica de Canarias (IAC) in May 2007 (Fig. 1), to start laboratory tests. After Factory acceptance in October 2008, OSIRIS will be shipped to La Palma in November 2008 for the on–site

J. Cepa
Departamento de Astrofísica, Facultad de Física, Universidad de La Laguna, Instituto de Astrofísica de Canarias, E–38200 La Laguna, Tenerife, Spain
e-mail: jcn@iac.es

J.M. Diego et al. (eds.), *Highlights of Spanish Astrophysics V*, Astrophysics and Space Science Proceedings, DOI 10.1007/978-3-642-11250-8_2,

Fig. 1 OSIRIS instrument mounted at the Nasmyth rotator of the AIV room at the IAC for laboratory tests

Commissioning at the GTC, in preparation for the scheduled Day One operation in March 2009.

1.2 Institutions and Budget

OSIRIS has been designed and builded by the IAC and by the Instituto de Astronomía of the Universidad Nacional Autónoma de México (IA–UNAM). The IA–UNAM was responsible for the optical design and the fabrication of some camera lenses.

The project was funded by the Spanish Ministry of Science and Technology, by GRANTECAN S.A., and by the IAC.

1.3 The Challenge

OSIRIS is the first instrument designed and builded in Spain for a telescope larger than 4 m. This represented in itself a challenge both from a technological and a managerial points of view, specially when the following main general requirements were imposed by the OSIRIS Instrument Definition Team:

- Field of view of at least 8 arcmin in diameter (goal $8' \times 8'$).
- Excellent image quality ($\sim$ 1 pixel) to fully exploit the excellent site and GTC optics.
- Red optimized but blue sensitive (down to 365 nm) optics.

- Small pupil (80 mm ⊘) to accommodate sensible sized Tunable Filters (TF).
- For installation either at the Nasmyth or Cassegrain foci to ease GTC operation.
- Overheads due to instrument configuration changes limited by detector readout time.
- Able to accommodate the number of masks, grisms and filters required to operate in service mode without the need of changing these dispersive elements during the night.
- Maximum spectral resolution $\lambda/\Delta\lambda$ of 5,000 (goal 8,000).

Since its initial concept, maximum priority was given to the use of Tunable Filters for narrow band imaging from 365 to 1,000 nm. This is the main driver of the instrument, and its main distinctive characteristic, amongst similar optical camera–spectrographs for large telescopes.

2 OSIRIS Characteristics

The main general characteristics of OSIRIS, their observing modes, and the dispersive elements that can be accommodated in the instrument are summarized in Tables 1, 2, and 3, respectively.

3 User Information and Pipelines

More information about the instrument, including exposure time calculators for broad band, tunable imaging and spectroscopy, and mask designer tools, can be found at www.iac.es/project/OSIRIS. More details on the mask designer software can be found in [3]. To train future GTC observers in the use of the tunable filters and specific MOS features, several workshops have been organized in La Palma, Granada and Mérida (México).

Table 1 Summary of the main characteristics of OSIRIS

Parameter	Value
FOV	$8.5' \times 8.7'$
Plate scale	$0.125''$/pix
Detector	2 MAT 4 k×2 k ($8''$ gap)
Broad band	ugriz & TF order sorters
Narrow band	2 Tunable Filters covering from 365 to 1,000 nm, with FWHM tunable from 1.2 to $\sim$ 4 nm depending on λ
Spectral resolutions	$\lambda/\Delta\lambda = 300, 500, 1,000, 2,000, 2,500$, and $5,000$ ($0.6''$ slit width)
MOS (masks)	$\sim$ 40 targets using classical slits or several hundreds of targets using multiplexing modes

Table 2 OSIRIS observing modes. Tunable imaging allows synchronizing the wavelength tuned with charge shuffling on detector for better continuum subtraction, while MOS mode allows Nod & Shuffle for excellent sky subtraction. Fast modes are achieved by combining charge shuffling or frame–transfer with windows on detector

Mode	Submodes
Imaging	Broad band
	Narrow band using TF (8′ ⊘)
Spectroscopy	Long slit (8.7′)
	Multiple Object (MOS)
Fast modes	Photometry
	Spectroscopy

Table 3 OSIRIS elements that the instrument can accommodate

Type	Number
Optical elements	2 TF
	24 filters
	6 grisms
Masks	13 masks

There will be two pipelines available for observers. One fully automated that work only on site using GTC specific image formats and recipes, and another interactive, standalone, based on Pyraf and running on Linux-based computers. The users will be provided with FITS format raw and reduced date using the on site pipeline.

4 Characterization Tests

The laboratory tests have demonstrated that the stringent general requirements described in Sect. 1.3 are fulfilled. For example, 80% of the polichromatic encircled energy is confined within ~1 pixel, which is equivalent to a resolution of 0.15 arcsec. Also, the optical distortion is lower than 2%, as required.

The image movement, due to the combination of gravitational flexures and instrument rotation, is controlled by moving the collimator in two axis via an open loop. The residuals are ~ 1 pix in the spatial direction and ~ 0.15 pix in the spectral direction, within the specs of 1.0 and 0.5 pixels, respectively. However, work is still ongoing to reduce the image movement in the spatial direction to 0.5 pix.

4.1 Instrument Transmission

The instrument transmission versus wavelength is shown in Fig. 2, and includes the collimator and the camera, but does not include the telescope, filters or detector. It is excellent in the red, and very competitive in the blue, below 400 nm. In spite

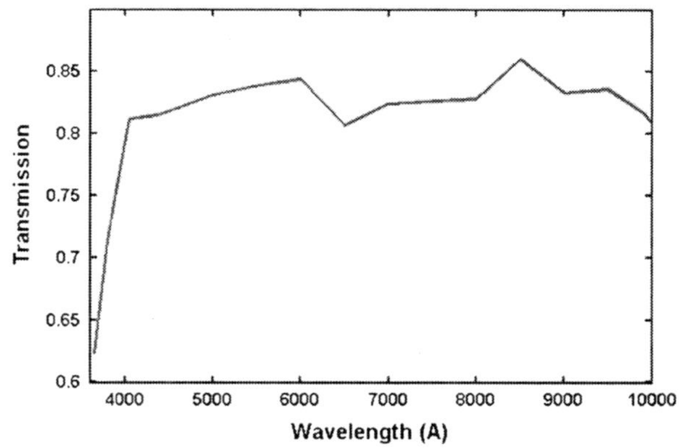

Fig. 2 OSIRIS transmission versus wavelength. Includes all OSIRIS optics but does not include telescope, detector or filters/grisms

Table 4 Mean times, in seconds, for changing instrument configuration of different instruments for 8–10 m–class telescopes. The information has been retrieved from on–line manuals available in the www of the different instruments. It is important to note that all elements (filters, grisms, TF and masks) can be changed simultaneously to save time

Telescope	Instrument	Mask	Grism	Filter
GTC	OSIRIS	20	6	3
VLT	VIMOS	210	90	180
GEMINI	GMOS	120	90	20
SUBARU	FOCAS	120	90	90

of being a blue sensitive instrument, the transmission in the red is comparable of higher than that of instruments such as DEIMOS (Table 6), specifically optimized in the red at the price of low blue performance. See [2] for details.

4.2 *Overheads*

The instrument overheads due to configuration changes are summarized in Table 4. The performance is far better than that of similar instruments in 8–10 m–class telescopes. Also, the 4 wheels holding TF, grisms and filters can be moved simultaneously and together with the mask loader. Hence the slowest element drives the final time for configuration changes: either the mask loader or the detector readout (that takes from 10 to 40 sec depending on the speed and the binning). Specially in service mode, when different observing programs requiring different configurations are scheduled during the night, it can save a substantial amount of observing time.

4.3 Optical Elements

The different broad band and order sorter filters have been characterized in the laboratory by measuring the transmission profile versus wavelength in different parts of the filter optical area, and at different tilt angles. The behavior is excellent and within specs (Figs. 3 and 4).

4.4 Tunable Filters

Tunable filters are etalons of very low resolution, with typical gap spacings of $\sim 2\,\mu$m, whose plate parallelism and distance between plates are controlled with an accuracy of $\sim$ 1 nm. They allow tuning any wavelength within their corresponding wavelength range (365–670 nm for the blue TF, 650–1,000 nm for the red TF) with a variety of FWHM available (Fig. 5). The tuning accuracy both in wavelength and FWHM is better than 0.1 nm. The tuning time range between 1 ms and 0.1 s depending on the gap change required. This fast tuning speed allows fast photometry with frequency switching between exposures.

The TF calibration involves checking plate parallelism, and establishing wavelength calibration, i.e. the equivalence between gap spacing in 16-bit counts and wavelength/order. Checking parallelism is a procedure that can be done in day time, although it is not expected to vary with time or even after switching off and on again the TF controller. Wavelength calibration is a procedure that takes about 90 s, and that can be done in day time using the A&G GTC unit or during the night. This procedure should be done every night, and checking it during the night might be required, depending on temperature changes.

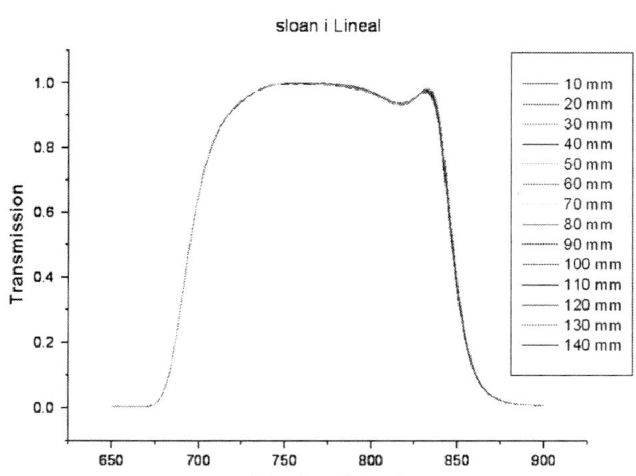

Fig. 3 Transmission of the i′ filter versus wavelength for different areas of the filter showing the excellent uniformity of the response

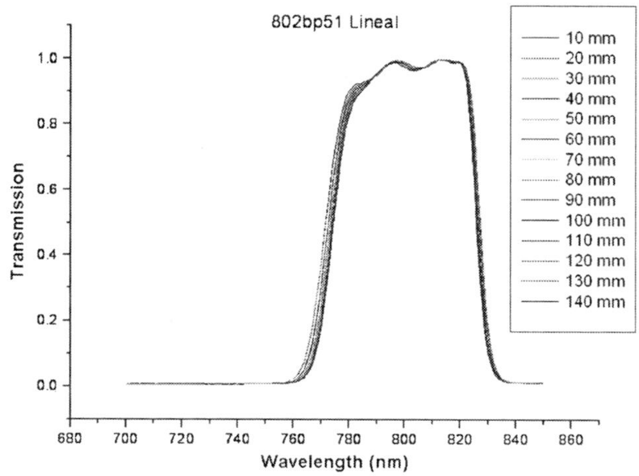

Fig. 4 Transmission of one of the order sorter filters versus wavelength for different areas of the filter showing the excellent uniformity of the response

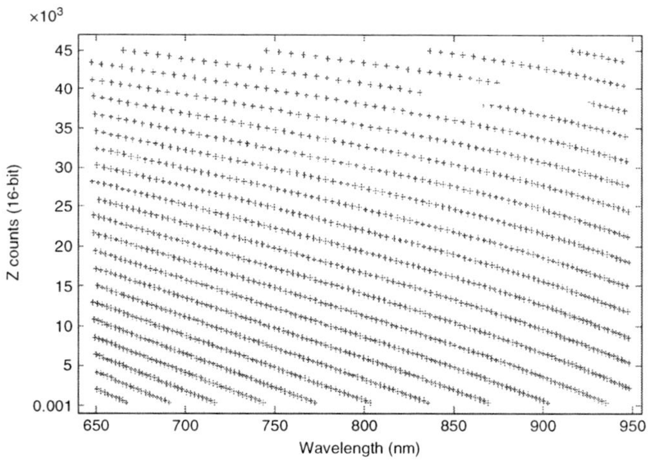

Fig. 5 Red TF calibration: Etalon gap in counts of 16-bit versus wavelength. Each set of points defining a straight line represent a different order. For each wavelength the different orders define the different FWHMs that can be tuned at this wavelength

5 OSIRIS Evolution and Context

5.1 Instrument Evolution

Sometimes the instrument capabilities, or even the basic instrument concept, face reality of budget, schedule or feasibility constraints, driving to a reduction of the instrument capabilities or its performance. This has not been the case of OSIRIS.

Table 5 Evolution of OSIRIS characteristics over time. They have not been reduced but increased

Feature	Letter of Intend	Announcement of Opportunity	Day One
Date	February 1998	December 1998	March 2009
Field of View	8′ ⊘	8′ ⊘	$8.5' \times 8.7'$
Maximum R	5,000	5,000 (goal 8,000)	5,000 (goal 10,000)
Observing modes	Broad band TF Long slit MOS	Broad band TF Long slit MOS Fast photometry	Broad band TF Long slit MOS Fast photometry Fast spectroscopy
Image quality	–	$< 0.4''$	$< 0.2''$
Number of masks	–	> 6	13
Numer of filters	–	–	24
Number of grisms	–	–	6

On the contrary, OSIRIS capabilities and performance have been increased over time (Table 5), and the instrument to be delivered for Day One has more observing modes and capabilities than initially promised.

5.2 A Comparison

OSIRIS has been designed and optimized for imaging using Tunable Filters. However, its capabilities as spectrograph make it competitive with VIMOS at the VLT or DEIMOS at Keck (Table 6). OSIRIS has a MOS field and spectral resolution similar to DEIMOS, albeit with smaller spectral coverage. However, DEIMOS is not sensitive below 400 nm and its efficiency below 500 nm is smaller than that of OSIRIS. Also, although VIMOS field is quite large, its spectral resolution is limited to about 2,000, due to spectral stability limitations induced by instrument flexures. As a consequence, OSIRIS has advantage over DEIMOS for its blue sensitivity and over VIMOS for its higher resolution.

6 Future Upgrades

There are currently several OSIRIS upgrades under development or planned:

- *Integral Field Units*: This mode will be implemented by using 100 μm diameter OH doped fibers, thus with high UV and red transmissions, with microlenses of $0.6''$ diameter at both ends. Then, a square array of fibers in the sky is rearranged to form a linear array in the focal plane as input for the spectrograph. Two IFU are planned: a compact array of 12×12 arcsec^2, and a sparse array of 45×45 arcsec^2. These IFUs will be mounted in the mask loader and can use any of the grisms

Table 6 Comparison of optical imaging–spectrographs for 8–10 m–class telescopes. Data have been retrieved from the instrument www pages or reference publications. OSIRIS has higher resolution than VIMOS and higher UV-blue sensibility than DEIMOS

Feature	VIMOS	DEIMOS	OSIRIS	FORS	GMOS	FOCAS
FoV (arcmin^2)	$4 \times (7 \times 8)$	16.7×5.0	8.5×8.7	6.8×6.8	5.5×5.5	6 ⊘
Max. R	$\sim 2,200$	$\sim 5,000$	5,000	2,800	3,600	1,600
Blue transmission	Yes	No	Yes	Yes	Yes	Yes
Max. transmission		0.85	0.87	0.78	0.81	0.80
Masks	15	11	13	10	13	10
Filters	6	7	24	22	8	14
Grisms/gratings	6	2	6	6	3	5
IFU	Yes	No	Planned	No	Yes	No

available, thus yielding up to the maximum resolution that the instrument can achieve ($R = 5000$).

- *Higher resolution*: Additional VPH–based grisms for $R = 5,000$ in the spectral range 400–500 nm, and $R = 10,000$ in the red are planned. It is important to point out that this resolution in the blue is not currently available in any of the spectrographs for 8–10 m–class telescopes.
- *3D spectroscopy*: This mode is implemented by changing one of the TF for a higher resolution etalon. The etalon, already purchased by IA–UNAM and currently under characterization, has $R = 10,000$, and a spectral range of operation from 650 to 900 nm.

7 OSIRIS Core Team Surveys

There are several surveys in which the OSIRIS Core Team is engaged, together with other collaborators. All of them are based in the spectral tomography either by using the TF or the order sorters. In what follows, several surveys using the TF tomography technique are briefly summarized.

7.1 TF Tomography

In the TF tomography technique, several images at the same pointing on the sky are taken using different TF tunings. Then, a data cube with the wavelength or redshift as the third dimension is obtained. For each emission line, a perfectly defined volume of the Universe in redshift range and limiting flux is scanned.

Using this technique, three different surveys will be tackled, funded by a coordinated project of the Spanish *Plan Nacional de Astronomía y Astrofísica*:

- *OTELO*: (OSIRIS Tunable Emission Line Object) survey using the red TF to detect low and high redshift emitters including Lyα up to redshift 7.

- *HORUS*: (Hydrogen and Oxygen Recombination line Unified Survey) using the blue TF to search for Lyα emitters from redshifts 2 up to 4.
- *GLACE*: (GaLAxy Cluster Emission line survey) using the TF for observing emitters in known galaxy clusters at redshifts 0.4 up to 0.9. The different optical lines to be scanned will allow deriving star formation rates and metallicities, and studying AGNs and optical cooling flows.

7.2 OTELO

In OTELO survey, the TF tomography will be applied to scan through two areas relatively free of OH sky lines, at 815, and 925 nm approximately. Table 7 summarizes the main survey characteristics comparing them with the most conspicuous and deep narrow band survey to date (SUBARU Deep Field). OTELO will be the deepest emission line survey, yielding redshifts with spectroscopic accuracy and deblending Hα from [NII] for low redshift emitters.

The scientific applications of OTELO include studying star formation rates, metallicity evolution [7], AGNs [10], distant QSO [4], Lyα emitters [5], the stellar component [9], and galaxy color evolution [1].

OTELO survey will provide a large database of about 40,000 emission line objects at different redshifts from 0.24 to 7.0 (Table 8).

It is very important to note that the minimum equivalent width (EW) of 15 Å stated in Table 7 corresponds to the minimum detectable flux. For brighter objects, such as emission line ellipticals or S0, the minimum detectable EW is of $\sim$ 0.3 Å.

Table 7 OTELO survey characteristics compared with SUBARU Deep Field narrow band surveys (from [8])

Characteristics	SUBARU	OTELO
Flux limit at 5σ	6×10^{-18} erg cm^{-2} s^{-1}	10^{-18} erg cm^{-2} s^{-1}
Minimum EW	15 Å	2 Å
Area	0.25 sq.deg.	0.10 sq.deg.
Redshift accuracy	10^{-1}–10^{-2}	10^{-3}–10^{-4}
Cosmic statistics	Single field	Different fields
Deblend Hα from [NII]	No	Yes

Table 8 Expected OTELO census of galaxies at different emission lines, including emission line ellipticals. Assuming no evolution and a concordance Cosmology $H_0 = 65$ km/s Mpc^{-1}, $\Omega_{m0} = 0.3$, and $\Omega_{\Lambda 0} = 0.7$

Morphology	Max. z	Number
E/S0	0.84	$\sim 10^3$
Sa–b–c–d–Im	1.50	3×10^4
Sy	1.50	7×10^3
BCD	0.84	10^3
Lyα	7.0	10^3

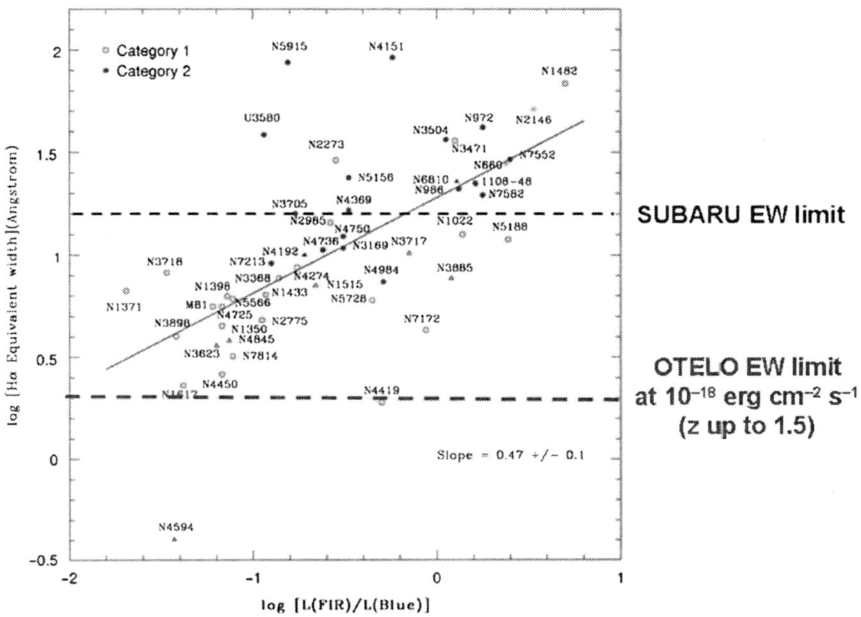

Fig. 6 OTELO *EW* limit will allow observing for first time in emission line surveys, all spirals up to redshift 1.5, and emission line ellipticals and S0 up to a redshift 0.84. Figure adapted from [6]

This implies that, *for first time in this kind of surveys*, all spirals up to $z = 1.5$, and most emission line ellipticals and S0 up to $z = 0.84$ can be detected (Fig. 6).

7.3 Lyα Emitters

The combination of HORUS and OTELO will render an important view on Lyα emitters (LAEs). HORUS is expected to gather LAEs, and Lyα blobs (LABs) at redshifts ranging from 2 to 4, while OTELO would provide LAEs at redshift 6, and 7, the latter representing the most distant LAEs known to date (Table 9). Also, some of the most conspicuous optical lines will be observed with NIR spectrographs to determine SFR and metallicites for the lowest redshift LAEs and LABs (Table 9). This database will allow studying the evolution of the LAEs luminosity function, and constraining the reionization epoch.

8 Summary

OSIRIS is an instrument of a wide field of view, with high red transmission and UV–blue sensitivity, very small overheads for changes of instrument configuration, and optimized for the use of Tunable Filters.

Table 9 Lyα emitters that will be collected by HORUS and OTELO TF tomography surveys. The rest frame optical emission lines that can be observed in the NIR are indicated

z	Lyα (nm)	[OII] (μm)	Hβ,[OIII] (μm)	Hα (μm)	Age (Gyr)	Project
2.1	377	1.2	1.6	2.0	3.2	HORUS
2.5	426	1.3	1.8	2.3	2.8	HORUS
3.1	499	1.5	2.0	–	2.2	HORUS
3.8	584	1.8	2.4	–	1.7	HORUS
5.7	815	–	–	–	1.1	OTELO
6.6	925	–	–	–	0.9	OTELO
7.0	980	–	–	–	0.8	OTELO

Although narrow band imaging using the TF in the blue and red is a unique mode in 8–10 m–class telescopes, OSIRIS has other special modes such as MOS Nod & Shuffle, fast photometry (with frequency switching using the TF), and fast spectroscopy. Its field of view and spectral resolution place OSIRIS in a competitive place with respect to VIMOS and DEIMOS.

Several large format surveys using the TF tomography technique will allow obtaining the deepest emission line surveys to date, that will allow studying galaxy formation and evolution including the farthest known LAEs (up to $z = 7$) and normal spirals and emission line ellipticals (up to $z = 1.5$).

OSIRIS will be shipped to La Palma in November 2008 to start on–the–sky tests at the GTC, in preparation for starting Day One operation in March 2009.

8.1 More Information

More information about the instrument including exposure time calculators can be found at www.iac.es/project/OSIRIS.

Acknowledgements OSIRIS instrument has been funded by the Spanish *Plan Nacional de Astronomía y Astrofísica* of the Ministry of Science and Technology under grants AYA2000–0333–P4–02, AYA2002–12070–E, and AYA2005–04149, GRANTECAN S.A., and the IAC.

References

1. Cepa, J., et al., A&A, in press (2008)
2. Cobos, F., González, J.J., Rasilla, J.L., Cepa, J., RMA&A 29, 139 (2007)
3. González–Serrano, J.I., et al., Exp. Astron. 18, 65 (2004)
4. González–Serrano, J.I., et al., RMA&A 24, 245 (2005)
5. González–Serrano, J.I., et al., RMA&A 24, 247 (2005)
6. Hameed, S., Devereux, N., AJ 129, 2597 (2005)
7. Lara–López, M.A., et al., A&A submitted (2008)
8. Ly, C., et al., ApJ 657, 738 (2007)
9. Pérez–García, A.M., et al., RMA&A 29, 155 (2007)
10. Povic, M., et al., ApJ, submitted (2008)

Gravitational Lenses: An Update

Emilio E. Falco

Abstract The impact of gravitational lenses on our knowledge of the Universe is inversely proportional to their scarcity. In the weak-field limit, lensing studies are based on well-established physics and thus offer a direct, simple approach to address many pressing problems of astrophysics and cosmology. Examples of these are the significance of dark matter and the density, age and size of the Universe. I describe examples of these applications. I also present new developments in cosmological applications of gravitational lenses, regarding estimates of the Hubble constant using strong lensing of quasars. I describe our recent measurements of time delays for the images of SDSS J1004+4112, and discuss prospects for the future utilizing synoptic telescopes, planned, under construction, and beginning operations.

1 Introduction

Gravitational lens systems (hereafter GLS) consist of a source and the lens proper, which deflects light from the source and forms distorted images; Fig. 1 is a sketch of a typical configuration. Given the right geometry, GLS form multiple images of a source, such as those shown in Fig. 2. Here, I consider only the gravitational weak-field limit, where deflections are always $\ll 1$ radian, or a few to several seconds of arc. In this limit, the deflection is achromatic (with exceptions, e.g. due to extinction or from the variation of the sizes of sources with wavelength). For the typical GLS, the images are unchanging with the wavelengths of observations. For example, lensed multiple images of quasars have very similar flux ratios in different wavebands (Fig. 2) and their spectra are very similar to one another, thus yielding a single redshift (an example is shown in Fig. 3) that in each case corresponds to the source quasar. The examples in these and in Figs. 4 and 5 show cases of *strong* gravitational lensing with different morphologies; the typical scale for separation

E. E. Falco
F. L. Whipple Observatory, Smithsonian Institution, P.O. Box 6369, Amado, AZ 85645, USA
e-mail: falco@cfa.harvard.edu

J.M. Diego et al. (eds.), *Highlights of Spanish Astrophysics V*, Astrophysics and Space Science Proceedings, DOI 10.1007/978-3-642-11250-8_3,

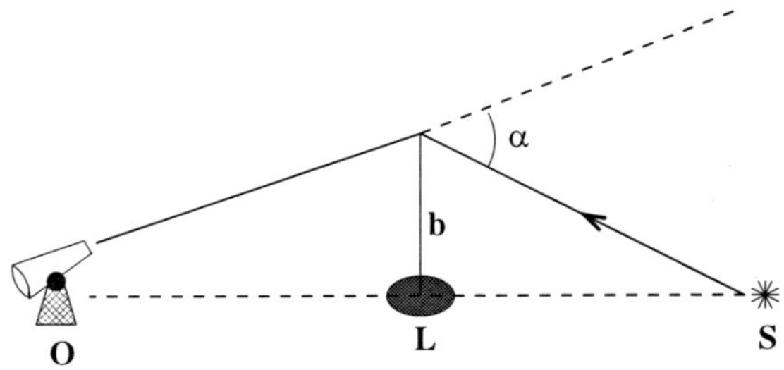

Fig. 1 The sketch shows light emitted by a source (S), deflected by a lens (L) and detected by an observer (O). The impact parameter is b; the corresponding deflection angle is α ($\ll$ 1 radian, greatly exaggerated in this sketch)

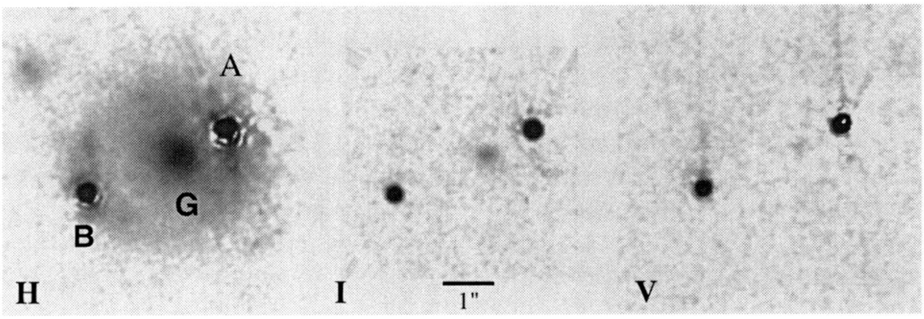

Fig. 2 HST H, I and V filter views of HE1104-1805, a double system. The lens galaxy is G, the images A and B. North (East) is at the top (*left*). Here, we fitted a photometric model, subtracted the best-fit models for the quasar images, and re-added them as gaussians with the same width as the original PSFs. This procedure removes artifacts due to the diffraction pattern of the HST PSFs

between distinct multiple images of a quasar is a few to several arcsec and the lens is most frequently a massive elliptical galaxy. Larger separations such as for SDSSJ1029+2623 (the largest at 22.5 arcsec for the separation of images A and B) are caused by clusters of galaxies. The label *strong* is used to distinguish from *weak* lensing, where there is no multiple imaging and lenses merely distort the images.

In a different occurrence of strong lensing, the effect yields distorted (at times multiple) images of distant galaxies, varying from *arclets* (slightly distorted) to *giant arcs* (greatly sheared), produced by intervening clusters of galaxies. There are two additional classes of gravitational lensing: *weak* lensing, where the strength of the lensing is insufficient to form multiple images and the effect is only measurable statistically and *microlensing* where unresolved (at the micro-arcsec level) sub-images are formed and the detectable results are large, rapid changes in detected fluxes.

The gravitational potential fluctuations of weak lenses cause modest distortions in the shapes of background sources. By measuring such distortions, we can determine the amplitude of density fluctuations as a function of cosmic distance. Tomographic surveys that include the radial cosmic coordinate yield 3D information; otherwise, one determines 2D variations, where the radial coordinate is

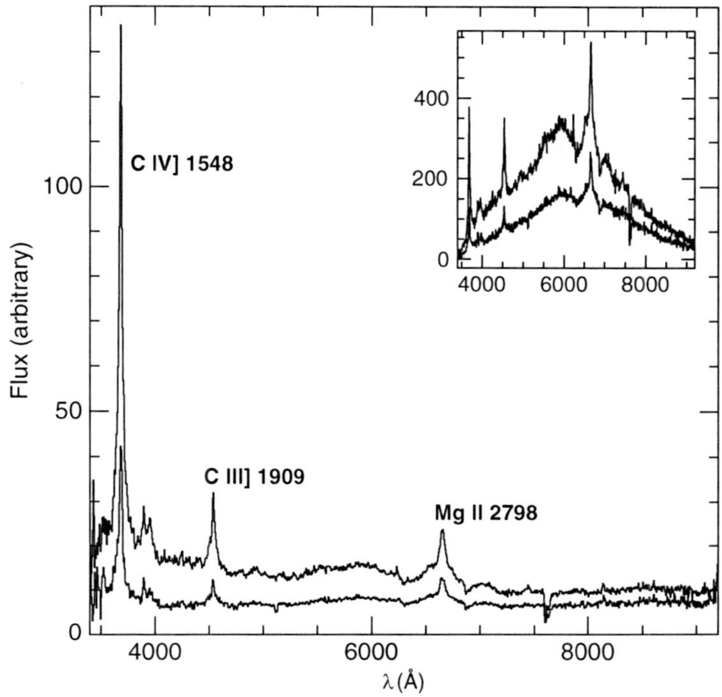

Fig. 3 Spectra of images B and C of SDSS J1029+2623 obtained with the LRIS-ADC spectrograph on the Keck I telescope [15]. Quasar emission lines redshifted to $z_s = 2.197$ are indicated by vertical *dotted lines*. The ratio of the spectra is shown at the *bottom*

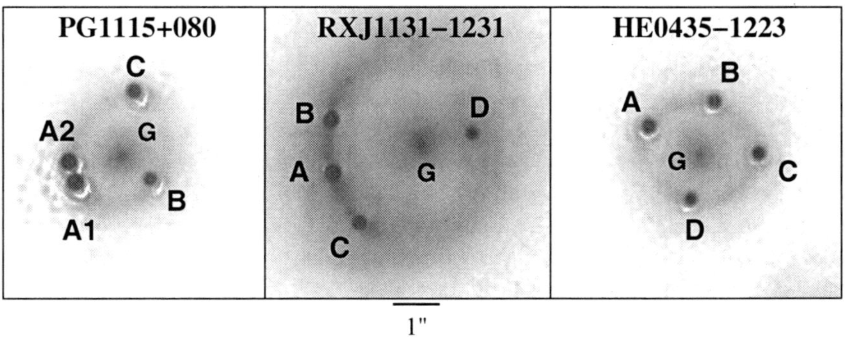

Fig. 4 CASTLES (*see the text*) HST NICMOS H-band images of three quadruple GLS. The different configurations in the three panels arise from the positioning of the source relative to the caustic curves generated by the lens (e.g. [8]). We used the same procedure as in Fig. 2 to remove PSF artifacts

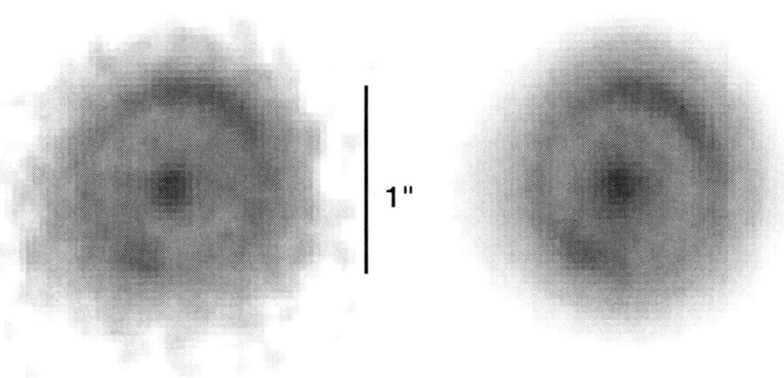

Fig. 5 *Left panel*: CASTLES HST NICMOS image of the complete Einstein-ring BLS B1938+666. *Right panel*: a photometric model including an elliptical lens galaxy and a lensed extended source to produce the ring

integrated out. The current weak-lensing measurements are degenerate: measurements may be consistent with small (large) fluctuations in a high (low) density universe. A recent weak-lensing determination is $\Omega_m \approx 0.3$ with a 10% error [25].

Microlensing forms multiple images, but with very small separations compared to strong lensing, $10^{-6}-10^{-3}$ arcsec. Such separations are usually unresolvable. But one can easily detect the corresponding large increases in the total brightness of a source that is microlensed. Many interesting results arose from the pioneering work of several projects, the Massive Compact Halo Object (MACHO), Expérience de Recherche d'Objets Sombres (EROS) and Optical Gravitational Lens Experiment (OGLE) collaborations and the world-wide collaboration Probing Lensing Anomalies NETwork (PLANET). These groups monitored stars in the LMC and attempted to find rapid brightness (thus, lens magnification) variations due to compact dark matter in the galactic halo (e.g. [1]). MACHO concluded that about 20% of the Milky Way halo is in compact objects of about 0.5 solar masses. Such a high mass content exceeds the total mass of the known stars. EROS found fewer microlensing events than MACHO, and set an upper limit of 25% for the compact dark matter content of the halo. Thus, only a small fraction of the dark matter in the halo of the Milky Way appears to consist of MACHOs: dark matter is of a still undiscovered type.

In the following, I concentrate on applications of observations of strong GLS to the study of cosmology. I first summarize CASTLES (CfA-Arizona Space Telescope Lens Survey) and then discuss our monitoring over the past 4.5 years of the SDSS J1004+4112 lens system and our results so far. I conclude with the future of surveys that are or will in the next several years be poised to make substantial contributions to these studies. I assume a concordance cosmology.

2 CASTLES: A Lensed Quasar Sample

As the characteristic size of galaxy-scale lenses is $\Delta\theta \sim 1$ arcsec, precision photometric studies of the lensing galaxies and of lensed quasar host galaxies are only practical with HST. HST also provides the best possible refinement of astrometry at optical and infrared wavelengths for components of GLS.

The CASTLES project (CfA/Arizona Space Telescope LEns Survey[1]) is a non-proprietary survey of known galaxy-mass GLS using the Hubble Space Telescope (HST) with a fixed set of filters: H, I and V. A fundamental goal of CASTLES is to obtain accurate astrometric measurements and photometry of lens galaxies and lensed images to refine lens models, particularly for systems where a time delay may provide a direct measurement of H_0. Other significant CASTLES goals are direct estimates of the mass-to-light ratio M/L of lens galaxies up to $z \sim 1$, a comparison of the dark matter and stellar light distributions in the lens galaxies, measurements of the properties of the interstellar medium in distant galaxies using differential extinction between the lensed images, identification of as yet undetected lens galaxies in known multiple-image systems, and understanding the environments of lens galaxies. CASTLES observations yield lens models that allow us to use gravitational lenses as cosmological tools.

CASTLES currently includes $\sim$ 100 small-separation ($\Delta\theta \leq 15$ arcsec) GLS. The GLS in CASTLES were found as a product of optical quasar surveys, radio lens surveys and serendipity. In all cases, there is a dominant lens galaxy which may be a member of a group or small cluster. The heterogeneity of the overall sample is important for some questions (e.g. the separation distribution), but relatively unimportant for others (e.g. the evolution of the lens galaxies). We observed our targets in the near infrared (principally the H band, but in a few cases, J and K) with the NICMOS camera NIC2. The infrared observations are complemented by WFPC2 imaging in the optical I and V bands to obtain uniform multi-color photometry of the systems. In recent HST cycles, we continued to use NIC2 and we also used ACS/WFC (Advanced Camera for Surveys) with the wide-field camera and V and I filters. The ACS failure in January 2007 prevented continuation of the survey until possibly the next HST cycle.

2.1 CASTLES Follow-Up Results

2.1.1 Dark Matter

Significant progress has recently been achieved in studies of early-type (elliptical) galaxies. The homogeneity of these galaxies results in a strong constraint on formation models. First, the population exhibits very uniform colors both locally and at $z \sim 1$. Second, early-type galaxies follow a tight correlation among their central

---

[1] see www.cfa.harvard.edu/castles

velocity dispersion, effective (half-light) radius, and surface brightness known as the fundamental plane (FP). The scatter in the FP, which is closely related to the scatter in mass-to-light ratio, is locally small, does not evolve significantly with redshift, and shows little dependence on the environment of the galaxies. In current hierarchical models of galaxy formation, the mergers of late-type (spiral) galaxies create elliptical galaxies. Support for the model is provided by the observation that high-redshift clusters exhibit both larger merger rates and smaller fractions of elliptical galaxies, compared to their present-day counterparts. Semi-analytic CDM models predict that early-type galaxies in the field should contain recently-formed ($z_f < 1$) stellar populations, while those in clusters should have significantly older stellar populations. Local studies have difficulty separating the effects of age and metallicity (e.g. [24]), but such degeneracies can be broken by measuring the evolution of mass-to-light ratios with redshift.

We recently completed an analysis of the evolution of 28 mass-selected early-type field galaxies spanning the redshift range $0 < z < 1$ [21]. We measured an evolution rate for the mass-to-light ratio in the rest-frame B band of $d\,\log(M/L)B/dz = -0.54 \pm 0.09$, consistent with other recent determinations. However, our study shows that the stellar populations of early-type field galaxies formed at $z_f > 1.8$ and argues against significant episodes of star formation at $z < 1$.

2.1.2 Extinction

An understanding of the interstellar medium through extinction laws is required for models of galaxy evolution, to establish a global history of star formation. Extinction also affects the light curves of γ-ray bursts for example; deriving the extinction law from afterglows requires theoretical assumptions about the intrinsic spectrum of the burst. Precision measurements of extinction curves are generally limited to the Galaxy and the Magellanic Clouds (Small, SMC, and Large, LMC), because at greater distances it is impossible to obtain the precise photometry or spectroscopy of individual stars needed for accurate extinction law measurements. In the SMC and LMC, the UV extinction curves can deviate significantly from the Galactic models, most obviously in having a far weaker 2,175 Å feature. Physically, the extinction law depends on the mean size and composition of the dust grains along the line of sight, so it should not be surprising that it varies with the environment. With the increasing need for extinction corrections at increasingly higher redshifts, it is clear we need more quantitative measurements of dust properties at similar redshifts. GLS allow estimates of the extinction properties of high redshift galaxies.

In most of the known lens galaxies we see 2 or 4 images of a background quasar produced by the deflection of light by a foreground lens galaxy. When light from each image traverses the lens galaxy, it is extincted by the dust at that position. As the dust distribution is generally not uniform, each image suffers a different amount of extinction; the observational signature is that the flux ratios of the images depend on wavelength. We demonstrated a method using these properties in [2], where we determined differential extinction in 23 gravitational lens galaxies over the range

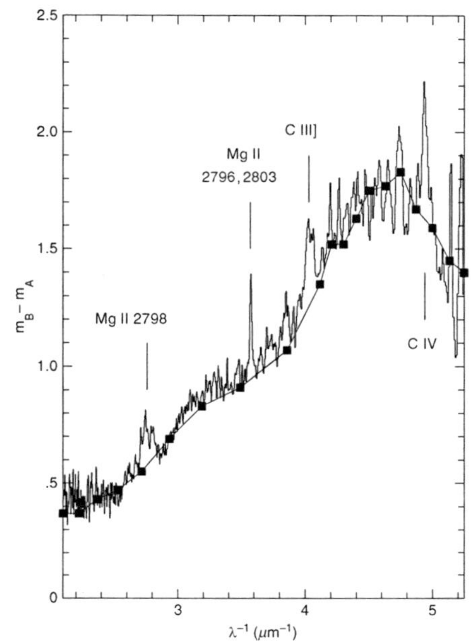

Fig. 6 The differential extinction curve for SBS0909+532. The continuous line is the $m_{B_u} - m_{A_u}$ magnitude difference curve obtained from the spectra of A and B. The abscissa is the inverse wavelength at the lens galaxy rest frame (the standard approach to display extinction properties)

$0 < z_l < 1$. Of the 23 systems we analyzed, 16 have spectral differences consistent with differential extinction. The extinction is patchy and shows no correlation with impact parameter. The directly measured extinction distributions are consistent with the mean extinction estimated by comparing the statistics of quasar and radio GLS surveys, thereby confirming the need for extinction corrections when using the statistics of lensed quasars to estimate the cosmological model.

Recently, we showed that GLS can be used to measure extinction curves at intermediate redshifts with high accuracy [11]. In that study, we derived the extinction curve of a distant galaxy ($z = 0.83$) by comparing two images of a gravitationally lensed quasar that are differentially reddened by the lens galaxy. We observed the double-image GLS SBS 0909+532 with the 2D spectrograph INTEGRAL–WYFFOS. From the spectra in our integral-field data, we derived the differential extinction curve between the two images of the quasar (Fig. 6). Ours is the first determination of an extragalactic extinction curve with confidence and quality similar to those derived for galaxies in the Local Group. The presence of a significant 2,175 Å feature (bump) in the extinction curve is surprising, for it has been considered weak or non existent outside the Milky Way. The average Milky Way extinction curve also fits well the SBS 0909+532 extinction curve with $R_v = 2.1 \pm 0.09$. The dust redshift estimated using as reference the zero redshift extinction curve is $z = 0.88 \pm 0.02$, in good agreement with the spectroscopic redshift of the galaxy.

In [12] we estimated the dust extinction laws in two intermediate-redshift galaxies. The dust in the lens galaxy of LBQS1009-0252, which has an estimated lens redshift of $z_l \simeq 0.88$, appears to be similar to that of the SMC with no significant feature at 2,175 Å. Only if the lens galaxy is at a redshift of $z_l \simeq 0.3$, completely inconsistent with the galaxy colors, luminosity or location on the fundamental plane, can the data be fit with a normal Galactic extinction curve. The dust in the $z_l = 0.68$ lens galaxy for B 0218+357, whose reddened image lies behind a molecular cloud, requires a very flat ultraviolet extinction curve with (formally) $R_V = 12 \pm 2$. Both lens systems seem to have unusual extinction curves by Galactic standards.

3 Time Delay Measurements

Estimating the Hubble constant H_0 has been a challenge since the 1920s. The classical approach is to build a cosmic distance ladder. One uses nearby celestial objects, for which distances can be measured relatively easily, to calibrate the distances to objects farther away. In that fashion, we can bootstrap to distances to far-off objects that move predominantly with the Hubble flow (i.e. for which the cosmic expansion velocity is much larger than their peculiar velocities). Unfortunately, we have a limited understanding of the underlying physics of many of the objects that are used to construct the distance ladder. Therefore, the empirical corrections that need to be applied to observations can hide biases that limit the reliability of the results.

GLS yield accurate, independent estimates of H_0 that bypass the distance ladder. The concept, as first proposed by [19], is to monitor multiple images created by a strong lens such as those in Fig. 4. The travel times associated with the different images formed by such lenses differ from the single travel time in the absence of a lens. The differences in the geometrical pathlength for each image and in the gravitational potential experienced by deflected light rays account for the differences in these travel times, or time delays Δt. Delays range from days to years, and are inversely proportional to H_0. One can conduct lensing measurements on relatively nearby sources ($z < 2$, say), which allow one to determine a value for H_0 that depends only weakly on other cosmological parameters such as Ω_m and Ω_Λ (e.g. [23]). The determination of H_0 by measuring time delays does require that the mass distribution of the lens be determined accurately.

Small perturbations to the gravitational potential from other galaxies near the lens must also be taken into account. Observational efforts have often focused on measuring time delays, but have neglected the systematics of the mass distributions. Consequently, measurements of H_0 inferred from different lens systems have been inconsistent. The quadruple system CLASS B1608+656 (Fig. 7) is particularly interesting because observers have measured all possible time delays [3]. These range from $\sim$ 30 to $\sim$ 77 days with uncertainties between 2 and 5%. Combining time-delay and mass estimates yields an estimate of H_0 of 75 km/s/Mpc with an error of about 10%. In recent years, additional *clean* lens systems (where the mass distribution of the lens is well constrained) have been analyzed in detail and the corresponding estimates of H_0 seem to be converging (see e.g. [10, 17]). That is encouraging for further improvements in the technique and for observational programs to add to the sample of measured time delays.

Currently, there are 17 GLS with 41 measured time delays (see Table 1 of [14]). The relative uncertainties in the delays range between $\sim$ 1 and 37%. The range

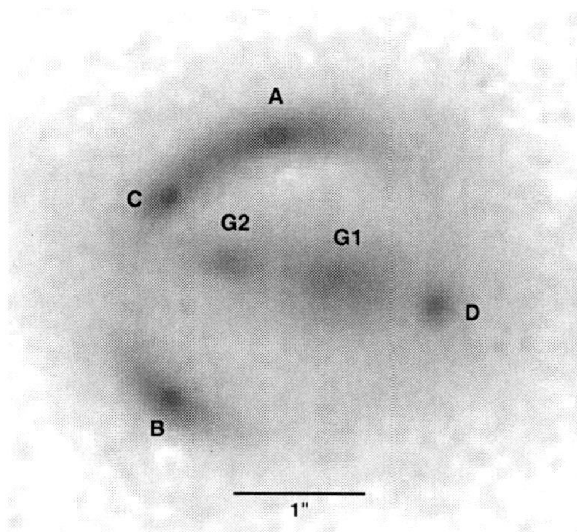

Fig. 7 HST NICMOS H-band image of the quadruple system B1608+656. The same procedure as in Fig. 3 removed PSF artifacts. The lens consists of a pair of elliptical galaxies, G1 and G2. Note the partial Einstein ring joining images A and C

reflects the difficulties of measurements that require prolonged monitoring of weak signals. For example, because of the large separation of the images in the first known strong lens, Q0957+561, and the asymmetry in the image configuration, the time delay Δt for the 2 images is large, $\approx$ 1.1 year. Thus, long campaigns to measure Δt were conducted, in particular that of Schild for well over a decade (e.g. [22]). The feeble variations of the quasar contributed to a long but now faded controversy on the true value of the delay (e.g. [18]). The complications in the modeling of the mass distribution of this system are such that the estimates of H_0 agree with newer ones, but the uncertainty of 25% detracts from its usefulness as a test of cosmological models [7].

We need time delay uncertainties of 1% so that these errors are smaller than mass modeling errors. That is achievable because once a delay is identified, intensive monitoring with observations phased by the delays will reduce the errors.

For the application of time-delay measurements to estimates of H_0, a reliable determination of the mass profile of the lens is essential. Therefore, *simple* systems with a single galaxy are preferable, rather than multiple deflectors such as in Q0957+561. Mass perturbations near the lines of sight of GLS are unavoidable, but a sufficiently large sample of GLS will yield significant numbers of systems where such perturbations are minimized. Oguri [14] derived a statistical procedure based on two simple measures for each lens galaxy and lensed image pair: the degree of asymmetry and opening angle of each image pair, relative to the center of each lens. He showed that based on the extant sample of time delays, the results for H_0 agree with the HST Kep project estimate within about 10% errors. The limitations are the small size of the sample and the assumed Gaussian distribution of measured time delays. Both of these will improve with larger samples.

For the past few years, we have monitored several GLS in an attempt to estimate time delays and identify microlensing when it occurs (with J. Fohlmeister, J. Wambsganss and C. Kochanek). Among our targets is a wide-separation system, SDSS J1004+4112 [5]. The system is set apart from *classic* quadruples because of

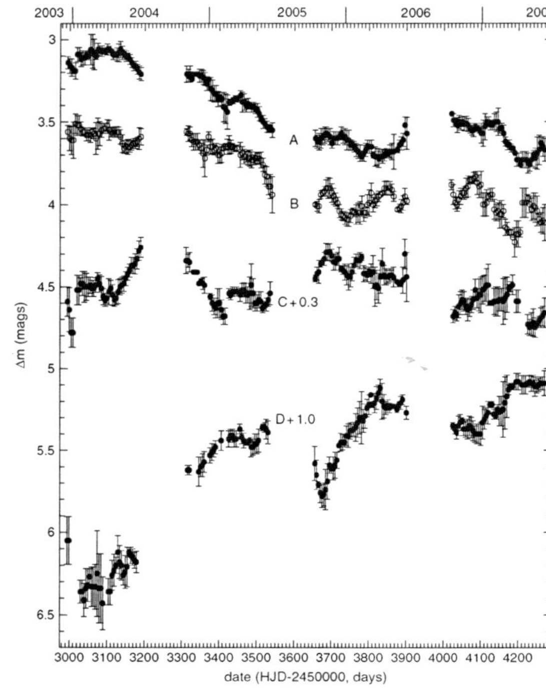

Fig. 8 Light curves for images A–D in SDSS J1004+4112 in the Sloan r band [4]

its $\sim$ 15 arcsec image separation and because it actually contains a 5th faint image [6]. We observed SDSS J1004+4112 with HST NICMOS and ACS in 2004. We obtained Sloan r-band lightcurves for the 4 brightest images between December 2003 and June 2007 and we are continuing the monitoring. We were able to determine the time delays for images A, B and C; see Fig. 8 for the lightcurves. We found that A leads B by $\Delta t_{BA} = 40.6 \pm 1.8$ days, and that image C leads image A by $\Delta \tau_{CA} = 821.6 \pm 2.1$ days. For the last independent delay, we find a lower limit such that image D lags image A by $\Delta \tau_{AD} > 1250$ days.

The presence of microlensing in SDSS J1004+4112 was first pointed out in spectra of the images [20]. In addition to the intrinsic variations of the source quasar in SDSS J1004+4112 that we saw in images A–D (Fig. 8), we confirmed that the images undergo microlensing at the $\sim$ 0.15 mag level in the Sloan r band. Based on our microlensing estimates for images A and B, we estimate an accretion disk size at a rest wavelength of 2,300 Å of $10^{14.8 \pm 0.3}$ cm.

4 Conclusions

Our lightcurve measurements for the images A–D of SDSS J1004+4112 yield an estimate of the accretion disk size of the lensed quasar. Unfortunately, the complexity of the lens, a cluster of galaxies, precludes a useful estimate of H_0. In spite of that, based on the delays and by assuming a value for H_0, we will be able to refine mass models for the lens cluster. The long delays allowed us to fill in the seasonal gaps and assemble a continuous, densely sampled light curve spanning 5.7 years whose variability implies a structure function with a logarithmic slope of $\gamma = 0.35 \pm 0.02$. As C is the leading image, sharp features in the C light curve can be intensively studied 2.3 years later in the A/B pair, allowing detailed reverberation mapping studies of a quasar at minimal cost.

One simple step is always required for strong lensing to achieve its full potential as a tool: redshifts need to be measured with 10-m class telescopes (e.g. [13]). At present, half of the current sample consists of GLS with measured source and lens redshifts. A modest investment in observing time will yield a very significant scientific return by allowing complete analyses of GLS. A step that requires a larger commitment of telescope time is the monitoring of GLS. Such monitoring continues at various sites under many projects.

The current, continually growing sample of $\approx$ 100 strong lenses must grow significantly larger, at least to the level of a few hundred GLS. When that level is reached, selection biases (e.g. radio, optical, serendipitous discovery) should become well understood and come under our control. Wide and deep surveys are now being conducted with ground-based telescopes, which will observe $\sim 10^7$ galaxies and measure their weak-lensing distortions on $\sim 1°$ angular scales. They should also discover many new cases of strong lensing. Some of those surveys are able to determine the distances of the galaxies based on photometric redshifts, and thus enable tomographic studies.

Further advances can be expected from new space and ground-based telescopes with large effective apertures and wide fields of view that are under design. These will conduct gravitational lensing studies among others. The 2-m class orbital SuperNova Acceleration Probe (SNAP) is designed to detected $\sim 2,000$ supernovae up to $z \sim 1.7$, and to provide the best weak-lensing data set possible. SNAP should also discover $\sim 10^5$ galaxy-sized strong lenses and advance our understanding of galaxy mass distributions.

The proposed Large Synoptic Survey Telescope (LSST) and the now-starting Panoramic Survey Telescope and Rapid Response System (Pan-STARRS) surveys will take a complementary approach. These projects will survey about 1/2 of the sky from the ground. They should generate $\sim 10^9$ galaxy images, and may be able to look at the largest-scale mass structures in the local universe. Potentially, LSST and PanSTARRS will also find thousands of new strong lenses and will provide lightcurves for the new systems.

Acknowledgements I acknowledge support from the Smithsonian Institution and from HST grants in support of CASTLES. I would also like to thank the CASTLES team, Janine Fohlmeister and Joachim Wambsganss for always interesting discussions. I am very grateful to the SOC, as well as Xavier Barcons and Luis Goicoechea for the invitation to present a talk at SEA 2008 in Santander.

References

1. Evans, N. W., in *Gravitational lensing: a unique tool for cosmology*, ASP Conference Series, Valls-Gabaud, D., & Kneib, J.-P., eds., astro-ph/0304252 (2003)
2. Falco, E.E., Impey, C.D., Kochanek, C.S., Lehár, J., McLeod, B.A., Rix H.-W., Keeton, C.R., Muñoz, J.A., Peng, C.Y., ApJ 523, 617 (1999)
3. Fassnacht, C.D., Xanthopoulos, E., Koopmans, L.V.E., Rusin, D., ApJ 581, 823 (2002)
4. Fohlmeister, J., Kochanek, C.S., Falco, E.E., Morgan, C.W., Wambsganss, J., ApJ 676, 761 (2008)
5. Inada, N., et al., Nature 426, 810 (2003)
6. Inada, N., et al., PASJ 57, L7 (2005)
7. Keeton, C.R., Falco, E.E., Impey, C.D., Kochanek, C.S., Lehár, J., McLeod, B., Rix, H.-W., Muñoz, J.A., Peng, C.Y., ApJ 542, 74 (2000)
8. Kochanek, C.S., ArXiv Astrophysics e-prints, astro-ph/0407232 (2004)
9. Kochanek, C.S., Keeton, C.R., McLeod, B.A., ApJ 547, 50 (2001)
10. Kochanek, C.S., Morgan, N.D., Falco, E.E., McLeod, B.A., Winn, J.N., Dembicky, J., Ketzeback, B., ApJ 640, 47 (2006)
11. Motta, V., Mediavilla, E., Muñoz, J.A., Falco, E., Kochanek, C.S., Arribas, S., García-Lorenzo, B., Oscoz, A., Serra-Ricart, M., ApJ 574, 719 (2002)
12. Muñoz, J.A., Falco, E., Kochanek, C.S., McLeod, B.A., Mediavilla, E., ApJ 605, 614 (2004)
13. Ofek, E.O., Maoz, D., Rix, H.-W., Kochanek, C.S., Falco, E.E., ApJ 641, 70 (2006)
14. Oguri, M., ApJ 660, 1 (2007)
15. Oguri, M., Ofek, E.O., Inada, N., Morokuma, T., Falco, E.E., Kochanek, C.S., Kayo, I., Broadhurst, T., Richards, G.T., ApJ 676, L1 (2008)
16. Peng, C.Y., Ph.D. Thesis, Univ. of Arizona (2004)
17. Poindexter, S., Morgan, N., Kochanek, C.S., Falco, E.E., ApJ 660, 146 (2007)
18. Press, W., Rybicki, G., Hewitt, J., ApJ 385, 416 (1992)
19. Refsdal, S., MNRAS 128, 295 (1964)
20. Richards, G.T., et al., ApJ 610, 679 (2004)
21. Rusin, D., Kochanek, C.S., Falco, E.E., Keeton, C.R., McLeod, B.A., Impey, C.D., Lehár, J., Muñoz, J.A., Peng, C.Y., Rix, H.-W., ApJ 587, 143 (2003)
22. Schild, R., Thomson, D.J., AJ 109, 1970 (1997)
23. Schneider, P., Ehlers, J., Falco, E.E., *Gravitational Lenses*, Springer (1992)
24. Trager, S.C., Faber, S.M., Worthey, G., González, J.J., AJ 119, 1645 (2000)
25. Wittman, D., Margoniner, V.E., Tyson, J.A., Cohen, J.G., Becker, A.C., Dell'Antonio, I.P., ApJ 597, 218 (2003)

First Scientific Results from the ALHAMBRA: Survey

A. Fernández-Soto

Abstract The *Advanced, Large, Homogeneous Area, Medium-Band Redshift Astronomical* (ALHAMBRA)–Survey is mapping eight different areas in the Northern sky, totalling 4 square degrees, aiming at obtaining a photometric redshift catalogue of over 600,000 galaxies with a median redshift $\overline{z} \approx 0.7$. This sample will be used to measure cosmic evolution at large, including the processes of galaxy formation and differentiation, large-scale structure, and the history of star formation. The photometric redshift depth, completeness, and accuracy are better than in any previous similar effort, reaching $\delta z \approx 0.015(1+z)$ for 90% of the objects with $AB(I) < 24$. We present in this conference the present status of the project, including the observations, data analysis, and the first preliminary scientific results obtained with a small fraction of the total survey.

1 Introduction

Cosmic Evolution has become one of the main drivers of modern cosmology. Some cosmologists, perhaps more theoretically oriented, lean towards the knowledge of the cosmological parameters as the *Holy Grail* of our science. However, it is certainly true that before understanding the detailed properties of the spacetime frame in which galaxies grow and evolve, we may as well devote our efforts to the understanding of the processes in which this evolution is based. Even further back in the chain of knowledge, there are many physical properties related to galaxy evolution for which not only we lack a complete understanding–we have hardly started to measure them to an acceptable degree of accuracy.

A. Fernández-Soto (on behalf of the ALHAMBRA Core Team)
Instituto de Fsica de Cantabria (CSIC-UC), Av. de los Castros s/n, E-39005, Santander (SPAIN)
e-mail: fsoto@ifca.unican.es

The ALHAMBRA Core Team: M. Moles (PI), J. A. L. Aguerri, E. Alfaro, N. Benítez, T. Broadhurst, J. Cabrera-Caño, F. J. Castander, J. Cepa, M. Cerviño, D. Cristóbal-Hornillos, R. M. González Delgado, L. Infante , I. Márquez, V. J. Martínez, J. Masegosa, A. del Olmo, J. Perea F. Prada, J. M. Quintana, and S. F. Sánchez

J.M. Diego et al. (eds.), *Highlights of Spanish Astrophysics V*, Astrophysics and Space Science Proceedings, DOI 10.1007/978-3-642-11250-8_4,

Amongst the most basic such processes and properties we can include the origin of the Hubble sequence, the different evolution of galaxies as a function of the environment where they reside, or the intertwined properties of galaxies and their active nuclei, when the latter are present. The rate of star formation in different galaxies is a basic tool to understand many of the former, and simultaneously, by itself, part of the same riddle. Surely the best way to tackle these issues is the production of large-scale databases, where many thousands of galaxies can be studied in detail. The sheer force of numbers is necessary if one wants to be able to divide such sample as a function of redshift, age, type, luminosity, or environment. Only over the last decade some large scale, pivotal, programs have been able to gather such large samples [1, 9]. They have been based on large, dedicated efforts and have offered, for the first time, the possibility to analyze samples of the order of 10^5 galaxies.

1.1 Spectroscopy and Photometric Redshifts

In order to extract the maximal information from an individual galaxy the traditional tool of the observational cosmologist has been spectroscopy. Each spectrum, once analyzed and depending on its properties, yields data about the redshift of the galaxy, its present (and past) star formation, content, and dynamical properties. This wealth of information is reached, of course, at a price: spectroscopy is expensive in terms of telescope time and, for the faintest objects that are easily detected in our images, plain unfeasible.

A different approach to extracting information from a galaxy is based on photometry alone. We could say that photometry through a single filter is the 0-th order approach, and (assuming the redshift is known) it gives us a luminosity. Two filters, adequately selected, yield information about luminosity and spectral type (mixed with star formation history, dust content, metallicity,...). A set of broad-band filters, as was suggested as early as in [3], can by itself give us information on the properties of a galaxy *and* its redshift. The use of photometric redshift techniques has been part of the standard cosmological lore since the mid nineties, mostly thanks to their use in the analysis of the Hubble Deep Fields [4, 11, 12, 14, 18].

Over the last few years, in particular, the possibility to use many-band surveys has proved that photometric techniques can allow not only for the measurement of a redshift, but also for the measurement of other galactic properties, in what is slowly becoming a convergence of photometry through many filters, and low-resolution spectroscopy [21, 22].

2 ALHAMBRA: Origin and Design

The ALHAMBRA–Survey will produce accurate photometric redshifts for a large number of objects, enough to track cosmic evolution, i.e., the change with z of the content and properties of the Universe, a kind of *Cosmic Tomography*. ALHAMBRA is imaging eight different fields, for a total 4 square degrees (Table 1), with 20

Table 1 The ALHAMBRA–Survey Fields

Field name	RA(J2000)	DEC(J2000)	100 μm	$E(B-V)$	l	b
ALHAMBRA-1	00 29 46.0	+05 25 28	0.83	0.017	113	−57
ALHAMBRA-2/DEEP2	02 28 32.0	+00 47 00	1.48	0.031	166	−53
ALHAMBRA-3/SDSS	09 16 20.0	+46 02 20	0.67	0.015	174	+44
ALHAMBRA-4/COSMOS	10 00 28.6	+02 12 21	0.91	0.018	236	+42
ALHAMBRA-5/HDF-N	12 35 00.0	+61 57 00	0.63	0.011	125	+55
ALHAMBRA-6/GROTH	14 16 38.0	+52 25 05	0.49	0.007	95	+60
ALHAMBRA-7/ELAIS-N1	16 12 10.0	+54 30 00	0.45	0.005	84	+45
ALHAMBRA-8/SDSS	23 45 50.0	+15 34 50	1.18	0.027	99	−44

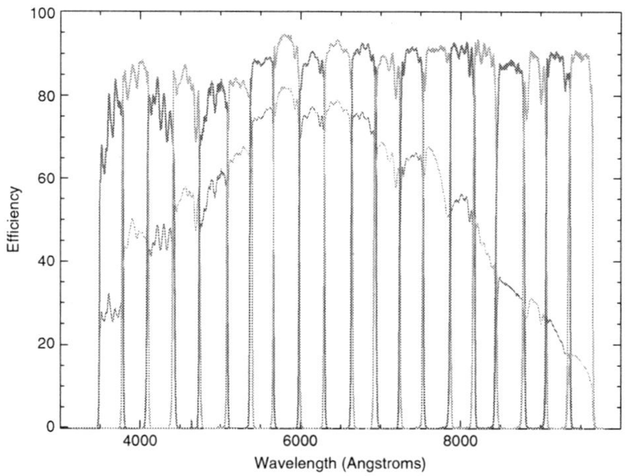

Fig. 1 Transmission curves measured for one of the optical filter sets, combined with the joint effect of the CCD detector, the atmosphere and the telescope

contiguous, equal-width, medium-band filters covering from 3,500 to 9,700 Å (see Fig. 1), plus the standard JHK_s near-infrared bands. The optical photometric system has been designed to maximize the number of objects with accurate classification by Spectral Energy Distribution type and redshift, and to be sensitive to moderate emission features in the spectrum [5].

The observations are being carried out with the Calar Alto 3.5 m telescope using the wide field cameras in the optical, LAICA, and in the NIR, OMEGA-2000. The magnitude limit, for a total of 100 ksec integration time per pointing, is $AB \geq 25$ mag (for an unresolved object, $S/N = 5$) in the optical filters from the blue to 8,300 Å, and ranges from $AB = 24.7$ to 23.4 for the redder ones. The limit in the NIR, for a total of 15 ks exposure time per pointing, is $K_s = 20$ mag, $H = 21$ mag, $J = 22$ mag (see Fig. 2) for the measured detection limits in one of the fields). We expect to obtain accurate redshift values, $\Delta z/(1+z) \leq 0.03$ for about 6.6×10^5 galaxies with $I \leq 25$ (60% completeness level), and $z_{med} = 0.74$. In Table 2 we compare the properties of the ALHAMBRA–Survey with other, similar,

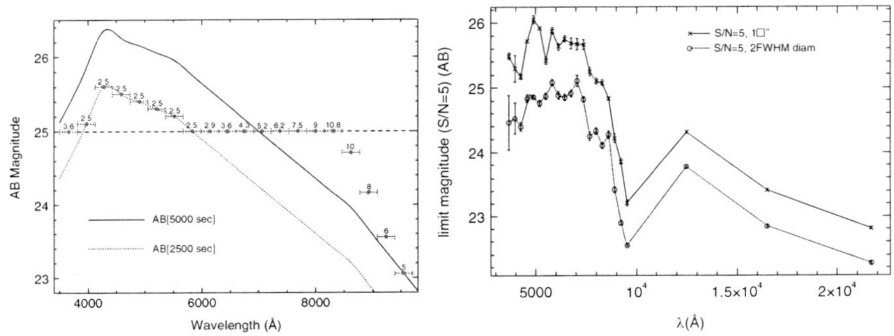

Fig. 2 *Left*: Expected magnitude limits as defined in our project for the visible ALHAMBRA filter set. For each filter, the exposure time in ks is also listed *Right*: The limit magnitude, at $S/N = 5$ in each filter, measured both in a 1" square aperture and in an aperture of $2 \times$ FWHM diameter

Table 2 Main characteristics of wide field (≥0.5 square degrees) spectroscopic surveys

Survey	Area (sqdeg)	Spectral range (Å)	z (median)	$N_{objects}$
SDSS/DR4	4783	3,800–9,200	0.1	6.4×10^5
LCRS	700	3,350–6,750	0.1	2.6×10^4
2dFGRS	1,500	3,700–8,000	0.11	2.2×10^5
VIRMOS	12	5,500–9,500	0.76	1.5×10^5
DEEP2	3.5	6,500–9,100	1.0	4.5×10^3
CfA/SRSS	13,000	4,300–6,900	0.017	2×10^4
ALHAMBRA-60	4	3,500–9,700 (+JHK)	0.74	$6.6\ (3.0)\times10^5$
ALHAMBRA-90	4	3,500–9,700 (+JHK)	0.63	$3.5\ (1.0)\times10^5$

endeavors in cosmic cartography. At the time of writing this contribution, the level of completion of the observations is ≈65%—slightly higher in the NIR, because of the earlier availability of OMEGA–2000 for our project.

This accuracy in redshift determination, together with the homogeneity of the selection function, will allow for the study of the large scale structure evolution with redshift, the galaxy luminosity function and its evolution, the identification of clusters of galaxies, and many other studies, without the need for any further follow-up. It will also provide exciting targets for detailed studies with 10 m class telescopes. Given its area, spectral coverage and its depth, apart from those main goals, the ALHAMBRA–Survey will also produce valuable data for galactic studies.

2.1 Comparison with Other Surveys

By its own nature, the type of many-filter photometric surveys that ALHAMBRA represents lies halfway in between classical photometric, large-area surveys, and narrow-area spectroscopic surveys. In recent years the advent of high multiplexing spectrographs has allowed for an extension of spectroscopic surveys, that can reach a significant completeness over the sky down to a relatively deep magnitude. Surveys such as the 2dFGRS or the SDSS include several times 10^5 galaxies, over a few thousand square degrees, reaching out to redshift $z \approx 0.2$. On the other extreme, deep surveys like DEEP2 reach redshifts $z \approx 1$ within areas of the order of 1 square degree.

2.2 Expected Redshift Accuracy and Number of Detections

In [5] we present the calculations that led to the definition of the ALHAMBRA filter system. The main driver, as has been discussed, is the optimization of the number of galaxies with precise redshifts. Using detailed simulations, described in full detail in [15], we have calculated the number of objects that the survey will detect to a fixed accuracy in the measured redshift value, for the adopted configuration.

Amongst the different ways to present the figure of merit of a planned survey, we show in Fig. 3 the total number of objects detected to a given redshift accuracy. It can be seen that the total number of objects with $\Delta z/(1+z) \leq 0.03$ is over 7×10^5, and it is over 5×10^5 for the higher accuracy of 0.015. Our simulations show that the survey will be complete at the 90% level with $\Delta z/(1+z) \leq 0.03$ (0.015) down to $I \approx 23.5(21.8)$, and to the 60% level till $I \approx 25.2(24.3)$. In the redshift-complete samples the number of objects expected are $3.5 \times 10^5(1.0 \times 10^5)$ and $6.6 \times 10^5(3.5 \times 10^5)$ respectively. All the calculations were done using the package BPZ [4].

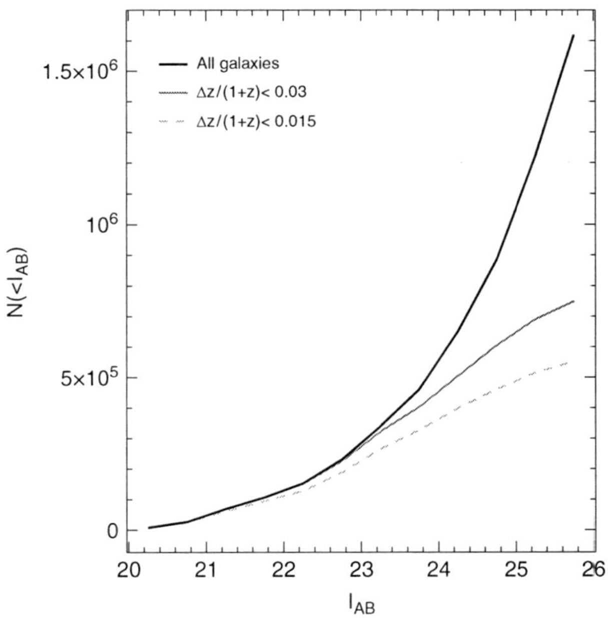

Fig. 3 The total number of galaxies detected in the ALHAMBRA–Survey (*thick line*). The thin continuous (*dashed*) line gives the total number of galaxies with $\Delta z/(1+z) \leq 0.03$ (0.015)

3 Present Status

At the time of this conference, the first basic milestones of the project have already been reached and met with success. In particular, in August 2006 the first pointing (within the field ALHAMBRA-8) was completed, having been observed to the expected depth through all 20+3 filters. At the end of 2008 a total area equivalent to one square degree (over several separate pointings and fields) has been completed, and the analysis is on-going.

The basic reduction steps have been performed using scripts based on well tested packages–basically IRAF and SExtractor [6, 20]. Cleaning and mosaicking of the images uses the XDIMSUM package [19], and the full images are combined with SWARP [7]. The astrometric solutions are based on the USNO-B1.0 catalogue [16], reaching typical astrometric precisions of $\approx$0.1 arcsec over a full field.

The photometry of the NIR images has been linked to the 2MASS survey [10]. Typical rms dispersions are below 0.03 magnitudes in all three filters. In the visible range we need to define a complete photometric system for the ALHAMBRA filters, and this is performed based on stars common to the SDSS survey and our data. The process is described in [15] and will be further developed in [2], its basis being the selection of stellar templates within large libraries for $\approx$10 bright stars per pointing using common SDSS and ALHAMBRA photometry, and the determination of expected fluxes from the spectral models anchored to the SDSS photometry. With this technique, that may if necessary be supplemented with spectrophotometry of some of the stars in each pointing, we reach an rms accuracy of $\approx 0.03-0.04$ magnitudes. The completeness limits in magnitude that are reached have been shown in Fig. 2 (right).

The percentage of the total area to be surveyed that has been covered to date is approximately 70% in the near infrared and 60% in the visible range–it must be remarked that OMEGA2000, the Calar Alto NIR imager, was available for our observations from the very beginning, whereas the optical imager LAICA was still under commissioning. We expect to complete our observations in 2010.

3.1 Photometric Redshifts and Spectroscopic Comparisons

In order to calculate redshifts for the galaxies and AGNs in our fields we will combine several different methods. We have ample experience in the application of photometric redshift techniques (see e.g. [4, 11]. These methods are based on the fitting of redshifted template spectra to the photometric data, and have been proved to provide the expected accuracy. We will improve the typical template set ([8], augmented with starburst spectra from [13]). We are also developing a new method [17] that fits linear combinations of different spectra, including line-dominated types, and is more able to detect the emission lines and use them to anchor the redshift measurements.

In any case we will need a spectroscopic sample, that will be used as calibration. Most of our fields overlap with SDSS data, and hence include some galaxies with

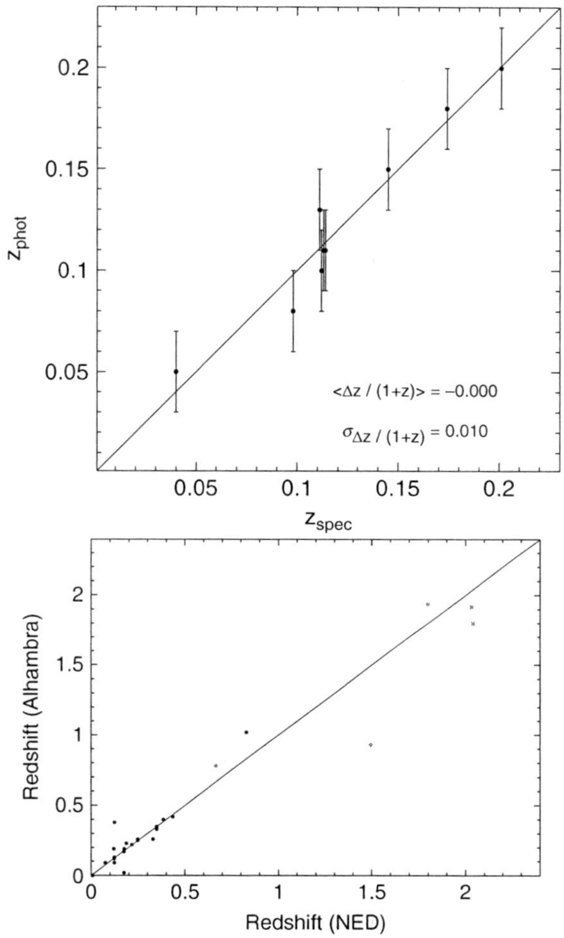

Fig. 4 Early comparison of photometric and spectroscopic redshifts in a CCD of the ALHAMBRA-8 field (*top*, using SDSS) and COSMOS field (*bottom*, using zCOSMOS). In the latter the red points correspond to AGN-dominated sources

spectroscopic redshifts, but only in small numbers (approximately 100 per field) and only out to $z \approx 0.2$. A few fields, as is the case of COSMOS and HDF-N, are richer in terms of number density of spectroscopic redshifts available. We will need, however, a spectroscopic sample in every field because the comparison with spectra is used as calibration not only for the photometric redshifts, but also for the photometry itself.

Although the photometric redshift studies are still only preliminary, we show in Fig. 4 the comparison for one CCD in the ALHAMBRA-8 field, only for SDSS galaxies, and in one of the CCDs in the COSMOS field.

4 Early Scientific Analysis

Our strategy has been from the beginning that of refraining from starting the data analysis until the data treatment pipelines and analysis tools are definitive. However, some basic analysis can be performed with the first set of complete pointings, and this work is in fact important for the definition and testing of the different tasks involved in the survey. We show in Fig. 5 a color image of the first complete pointing in the ALHAMBRA-8 field. The region covers an area of 15 arcmin side and contains approximately 10,000 objects to $AB \approx 25$.

Just to illustrate the depth that can be reached with the ALHAMBRA images, we show in Fig. 6 a small (1.3×1.8 arcmin) section of the ALHAMBRA-8 field. Here two bright red objects are found. Each of the ten panels corresponds to one of the ALHAMBRA filters running from F520 to F799, with central wavelengths that span the range between 5,200 and 7,990 Å. The reddest object (center left) has a strong break between 6,130 and 6,440 Å ($AB(6130) - AB(6440) > 1.5$ mag), and a second one between 7,060 and 7,370 Å ($AB(7060) - AB(7370) \sim 1$ mag), which identify it as either a moderate ($z \approx 1$) redshift galaxy with an old/reddened population, or a high-redshift object at $z \approx 4$.

Fig. 5 The first complete pointing of the ALHAMBRA-8 field in a square region of 15 arcmin side. This color image has been created making use of data from 14 out of the 23 filters

Fig. 6 A region of 1.3×1.8 arcmin is shown in this strip containing 10 images corresponding to 10 optical filters spanning from 5,200 to 7,990 Å. Two bright red objects having $AB(7990) - AB(5200) > 3$ can be easily identified

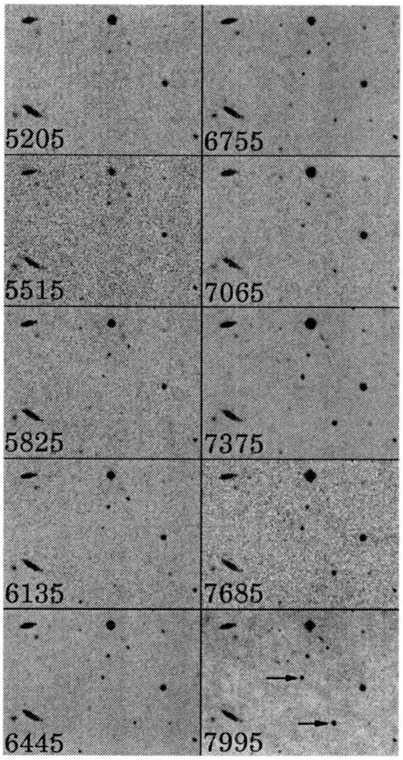

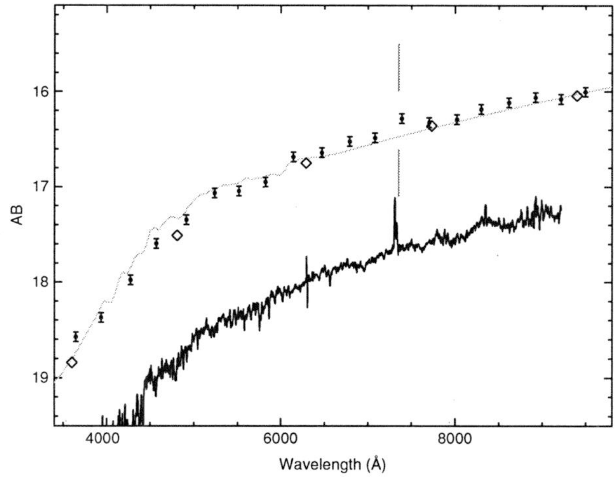

Fig. 7 SDSS spectrum (*line*) and photometry (*diamonds*), and ALHAMBRA photometry for one of the galaxies in our field with SDSS spectroscopy

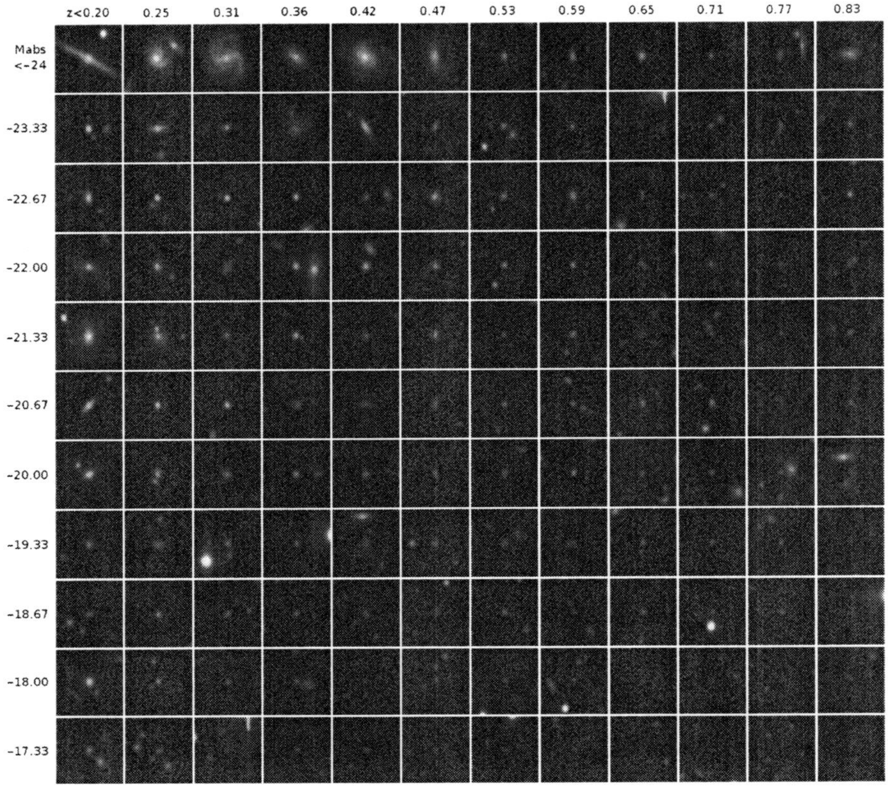

Fig. 8 A selection of ALHAMBRA galaxies, sorted according to their redshift and luminosity

A full, calibrated ALHAMBRA *spectrum* can be seen in Fig. 7, together with the corresponding SDSS spectrum and photometry.

Acknowledgements We acknowledge the decisive support given by the ALHAMBRA Extended Team to the project (see http://alhambra.iaa.csic.es:8080/ for the details regarding the project implementation and organization). We also wish to acknowledge the Calar Alto Director and staff for their strong support and warm assistance for a fruitful observation. The Ministerio de Ciencia e Innovación (formerly Ministerio de Educación y Ciencia) is acknowledged for its support through the grants AYA2002-12685-E and AYA2004-20014-E, and project AYA2006-14056.

References

1. Adelman-McCarthy, J.K., et al., ApJS 175, 297 (2008)
2. Alfaro, E., et al., in preparation (2009)
3. Baum, W.A., 1962, in IAU Symp. 15, 390 (1962)
4. Benítez, N., ApJ 536, 571 (2000)
5. Benítez, N., et al., ApJ 692, 5 (2009)

6. Bertin, E., Arnouts, S., A&AS 117, 393 (1996)
7. Bertin, E., et al., Astronomical Data Analysis Software and Systems XI 281, 228 (2002)
8. Coleman, G.D., Wu, C.C., Weedman, D.W., ApJS 43, 393 (1980)
9. Colless, M., et al., MNRAS 328, 1039 (2006)
10. Cutri, R.M., et al., *The IRSA 2MASSAll-Sky Point Source Catalog* (2003)
11. Fernández-Soto, A., Lanzetta, K.M., Yahil, A., ApJ 513, 34 (1999)
12. Gwyn, S.D.J., Hartwick, F.D.A., ApJ 468, L77 (1996)
13. Kinney, A.L., et al., ApJS 86, 5 (1993)
14. Lanzetta, K.M., Yahil, A., Fernández-Soto, A., Nature 381, 759 (1996)
15. Moles, M., et al., AJ 136, 1325 (2008)
16. Monet, D.G., et al., AJ 125, 984 (2003)
17. Sánchez, S.F., et al., in preparation (2009)
18. Sawicki, M.J., Lin, H., Yee, H.K.C., AJ 113, 1 (1997)
19. Stanford, S.A., Eisenhardt, P.R.M., Dickinson, M., ApJ 450, 512 (1995)
20. Valdes, F.G., in *Automated Data Analysis in Astronomy*, 309 (2002)
21. Wolf, C., et al., A&A 365, 681 (2001)
22. Wolf, C., et al., A&A 421, 913 (2004)

Magnetic Fingerprints of Solar and Stellar Oscillations

Elena Khomenko

Abstract Waves connect all the layers of the Sun, from its interior to the upper atmosphere. It is becoming clear now the important role of magnetic field on the wave propagation. Magnetic field modifies propagation speed of waves, thus affecting the conclusions of helioseismological studies. It can change the direction of the wave propagation, help channeling them straight up to the corona, extending the dynamic and magnetic couplings between all the layers. Modern instruments provide measurements of complex patterns of oscillations in solar active regions and of tiny effects such as temporal oscillations of the magnetic field. The physics of oscillations in a variety of magnetic structures of the Sun is similar to that of pulsations of stars that posses strong magnetic fields, such as roAp stars. All these arguments point toward a need of systematic self-consistent modeling of waves in magnetic structures that is able to take into account the complexity of the magnetic field configurations. In this paper, we describe simulations of this kind, summarize our recent findings and bring together results from the theory and observations.

1 Introduction

Any turbulent medium, as the interior of the Sun or stars, generates sound. The basics of the theory excitation of sound waves by the turbulent flow were developed by Lighthill in 1952 [34]. Since then, a vast amount of theoretical and numerical efforts has been dedicated to specify the properties of the spectrum of sound waves generated in a stratified stellar convection zone, by improving the description of the turbulent energy spectrum of the convective elements e.g. [1, 13, 14, 38, 40, 52, 53]. Without going into details of these works, the present knowledge can be summa-

E. Khomenko
Instituto de Astrofísica de Canarias, 38205, C/ Vía Láctea, s/n, Tenerife, Spain
and
Main Astronomical Observatory, NAS, 03680 Kyiv, Ukraine
e-mail: khomenko@iac.es

J.M. Diego et al. (eds.), *Highlights of Spanish Astrophysics V*, Astrophysics and Space Science Proceedings, DOI 10.1007/978-3-642-11250-8_5,

rized in the following way. The efficiency of the energy conversion from convective to acoustic is proportional to a high power of the Mach number of the convective motions ($M^{15/2}$ [13]). Most of the energy going into the p-modes, f-modes and propagating acoustic waves is emitted by convective eddies of size $h \sim M^{3/2}H$ (where H is the value of the pressure scale height), at frequencies close to the acoustic cut-off frequency $\omega \sim \omega_c$ and wavelengths similar to H. This defines the frequency dependence of the oscillation spectrum observed in the Sun and stars [13]. Since the Mach number is largest at the top part of the convection zone, the peak of the acoustic energy generation is located immediately below the photosphere [40, 53]. Recent numerical simulations of magneto-convection have shown that the magnetic field modifies the spectrum of waves [18]. Apart from the power suppression in regions with enhanced magnetic field, these simulations suggest an increase of high-frequency power (above 5 mHz) for intermediate magnetic field strengths (of the order of 300–600 G) caused by changes of the spatial–temporal spectrum of turbulent convection in a magnetic field.

Waves generated in the convection zone resonate inside a cavity formed by the stellar interior and the photosphere and are used by helio- and astroseismology to derive its properties [9, 57]. The information contained in the frequencies of the trapped wave modes is used by the classical helioseismology. A relatively newer branch called local helioseismology uses the information contained in the velocity amplitudes of the propagating waves measured in a region of interest on the solar surface [10]. By inversion of these measurements, variations of the wave speed and velocities of mass flows can be recovered below the visible solar surface. The inversion results have been obtained for quiet Sun regions as well as for magnetic active regions including sunspots. It is known that sunspots possess strong magnetic fields with a complicated structure in the visible layers of the Sun where the Doppler measurements used by local helioseismology are taken [50]. Consequently, such magnetic field can cause important effects on helioseismic waves, beyond the perturbation theories employed for helioseismic data analysis [2,12,29]. Recent numerical and analytical results demonstrate that the observed time-distance helioseismology signals in sunspot regions correspond to fast MHD waves [24, 37].

An estimation of the acoustic energy flux generated in the solar convection zone by the turbulent motions suggests that the it can be as large as e.g. $F_{\rm A} = 5 \times 10^7$ ergs/cm^2/s [38]. This is more than sufficient to maintain a hot chromosphere. It made the theory of acoustic heating of the upper atmosphere very attractive. However, soon after being proposed, the theory of acoustic heating encountered several major difficulties. It was found that both, low- and high-frequency waves are radiatively damped in the photosphere, reaching the upper layers with significantly less power [56]. An additional difficulty comes from the fact that the measured high-frequency acoustic fluxes in the photosphere and chromosphere are uncertain and non-conclusive [19, 59]. Low-frequency acoustic waves are reflected in the photosphere due to the effects of the cut-off frequency ($\sim$ 4 mHz) before reaching chromospheric heights. Due to their long wavelengths they have too large shock formation distances. Despite that, the five-min waves with enough energy were detected in the chromosphere and corona of the Sun mainly above solar facular and network areas [5, 31, 35, 44, 58]. Several explanations of these waves involving the magnetic field have been recently proposed [22, 45]. The wave energy can reach the upper layers not necessarily in the form of acoustic waves, but also in the form of other wave types, like magneto-acoustic waves, Alfvén waves or a family of waves propagating in thin magnetic flux tubes (see a recent example in e.g. [17]). All of them are related to the magnetic field structure. Osterbrock in 1961 [41] was the first to incorporate magnetic field into the theory of wave heating and to point out its importance on the propagation and refraction characteristics of the fast and slow MHD waves.

The above examples are only a few where the influence of the magnetic field on the wave properties is demonstrated to be important. Magnetic field not only changes the acoustic excitation rate and produces new wave modes. It also modifies the wave propagation paths and the direction of the energy propagation, it produces wave refraction and changes the reflection characteristics at the near-surface layers. Magnetic field defines the wave propagation speeds and changes the acoustic cut-off frequency. It can also change the wave dissipation rates and provides an additional energy source. Magnetic field connects all the atmospheric layers facilitating the channeling of waves from the lower to the upper atmosphere. This makes the magnetic field an important ingredient the theories of the wave propagation in the atmospheres of the Sun and stars.

Apart from the problems set by the local helioseismology in magnetic regions and the wave heating theory, of pure physical interest is the interpretation of oscillations observed in different magnetic structures in terms of MHD waves. For example, the wave dynamics seen in high-resolution DOT movies of a sunspot region [48] demonstrate that phenomena such as chromospheric umbral flashes and running penumbral waves are closely related. What type of waves are responsible for them? Solar small-scale and large-scale magnetic structures have distinct magnetic and thermal properties and support different wave types. The observed frequency spectrum of waves in these structures is not the same (see the introduction in [23]). Numerical simulations of waves in non-trivial magnetic structures (e.g. [4, 15, 16, 20, 23, 47]) have shown the complex pattern formed by waves of various types that can propagate simultaneously in various directions. Different wave modes can be detected in observations depending on the magnetic field configuration and the height where acoustic speed, c_S, and Alfvén speed, v_A, are equal relative to the height of formation of the spectral lines used in observations.

During the last years we applied efforts to develop a numerical code aimed at calculating the non-linear wave propagation inside magnetic fields in 2 and 3 dimensions. Using this code we focused our analysis on several problems described above, namely: magneto-acoustic wave propagation and refraction in sunspots and flux tubes; channeling the five-minute photospheric oscillations into the solar outer atmosphere through small-scale magnetic flux tubes; influence of the magnetic field on local helioseismology measurements in active regions. The results of these studies are published in [20, 22–24]. In the rest of the paper we briefly summarize the results and conclusions of these works. In addition, the last section gives our recent contribution to the problem of the interpretation of observations of waves in magnetic roAp stars.

2 Waves in Sunspots

Sunspots show a very complex dynamics as revealed by high-resolution observations. The umbral flashes observed in the chromosphere are thought to be acoustic shock waves [48] propagating along the nearly vertical magnetic field lines. The visible perturbation then expands quasi-circularly out to the penumbra producing the running penumbral waves [54]. This visible pattern can be interpreted as slow (acoustic) wave propagating along the inclined magnetic field lines in the penumbra [3]. The power distribution at different frequencies in active regions is rather complex as well. The recent observations of HINODE [39] confirmed the conclusions done with the ground-based observations (see e.g. [55]) that the dominating frequency of oscillations within a sunspot depends on the position in the umbra or penumbra, as well as on the height in the atmosphere. In the chromosphere the greatest power is observed in the umbra at 5–6 mHz and a sharp transition between umbra and penumbra where the dominating frequency is 3 mHz. In the umbral photosphere the oscillation power is generally suppressed compared to the quiet photosphere, except for the excess of low-frequency power at 1–2 mHz at the umbra-penumbra boundary. An unknown question is what drives the oscillations observed in sunspots? Are these waves due to a resonant response of the sunspot flux tube to the external driving by p-modes? Are there sources of oscillations inside the magnetized regions?

In the simulations described below we aim at identifying the types of waves modes observed in different layers of a sunspot atmosphere. We supposed a source of high-frequency (100 mHz) monochromatic waves located at photospheric level inside the umbra where the acoustic speed is slightly larger than the Alfvén speed. The waves are assumed to be linear. The initial unperturbed magnetostatic sunspot model was evaluated following the strategy described in [42] with a maximum magnetic field strength of 2.2 kG and characteristic radius of 6 Mm. The simulation domain is 2-dimensional, extending 0.86 Mm in the vertical and 3.5 Mm in the horizontal directions.

A snapshot of the simulations is given in Fig. 1. The source generates a set of fast magneto-acoustic waves propagating upwards. After the perturbation reaches the height where $c_S = v_A$ the mode transformation takes place. The fast (acoustic) mode propagating vertically below $c_S = v_A$ level is transmitted as fast (magnetic) mode in the $c_S < v_A$ region (visible in the transversal velocity). Its wavelength increases with height due to the rapid increase of the Alfvén speed. The left part of the wave front of the fast mode (closer to the axis) propagates faster than its right part (farther from the axis) which produces its reflection back to the photosphere at some height above the $c_S = v_A$ level. The same happens to the other fast mode propagating to the left, except that this fast wave refracts toward the sunspot axis. A

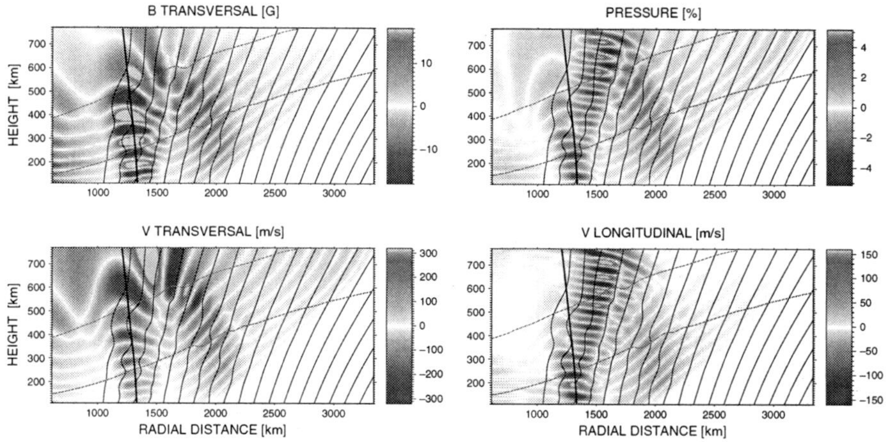

Fig. 1 Variations of the velocity, magnetic field and pressure at an elapsed time $t = 100$ sec after the beginning of the simulations. In each panel, the horizontal axis is the radial distance X from the sunspot axis and the vertical axis is height from the photospheric level. The black inclined lines are magnetic field lines. The two more horizontal lines are contours of constant c_S^2/v_A^2, the thick line corresponding to $v_A = c_S$ and the thin line to $c_S^2/v_A^2 = 0.1$. *Top left*: transversal variations of the magnetic field. *Top right*: relative pressure variations. *Bottom panels*: transversal and longitudinal variations of the velocity

part of the original fast (acoustic) mode energy is transformed to the slow (acoustic) mode in the $c_S < v_A$ region. The slow mode is visible in the longitudinal velocity snapshots. It propagates with a lower speed, close to the local speed of sound. This slow mode is channeled along the magnetic field lines higher up to the chromosphere increasing its amplitude. A similar behavior is observed in simulations with an initial condition in the form of an instantaneous pressure pulse rather than a monochromatic wave. From these simulations we conclude that in sunspots only a small fraction of the energy of the photospheric pulse can be transported to the upper layers, since an important part of the energy is returned back to the photosphere by the fast mode. Above a certain height in the low chromosphere, only slow (acoustic) modes propagating along the magnetic field lines can exist in the umbra defining the dominating frequency of waves according to the cut-off frequency of the sunspot atmosphere.

3 Waves in Small-Scale Flux Tubes

Acoustic waves propagating vertically in the quiet solar atmosphere change their dominating frequency with height from 3 mHz in the photosphere to 5 mHz in the chromosphere. An explanation for this effect was suggested by [11], who argue that this is a basic phenomenon due to resonant excitation at the atmospheric cut-off frequency. The low temperatures of the high photosphere give rise to a cut-off frequency around 5 mHz. However, numerous observations suggest that there is no such change for waves observed in chromospheric and coronal heights above solar facular and network regions [5, 31, 35, 44, 58]. What are the mechanisms that allow the 3 mHz waves (evanescent in the photosphere) to propagate up to chromospheric heights? Reference [45] suggest that the inclination of magnetic flux tubes in facular regions is essential for the leakage of p-modes strong enough to produce the dynamic jets observed in active region fibrils. However, this mechanism is hardly to be at work in the photosphere, where the plasma β is larger than 1 and the acoustic waves do not have a preferred direction of propagation defined by the magnetic field. In addition, it can not easily explain the observations of vertically propagating 3 mHz waves in facular and network regions at chromospheric heights. Alternatively, a decrease in the effective acoustic cut-off frequency can be produced by radiative energy losses in thin flux tubes [46].

We explored the latter mechanism and extended the theoretical analysis of [46] by means of non-adiabatic, non-linear 2D numerical simulations of magneto-acoustic waves in small-scale flux tubes with a realistic magnetic field configuration. The simulations are obtained by introducing a 3 mHz harmonic vertical perturbation at the axis of a magneto-static flux tube. Radiative losses were taken into account by means of Newton's law of cooling with a fixed value of the radiative relaxation time τ_{RR}. We compare two identical simulation runs that differ only by the value of τ_{RR}. The first run is in adiabatic regime ($\tau_{RR} \rightarrow \infty$) and the second run has $\tau_{RR} = 10$ s constant through the whole atmosphere. The magnetostatic flux tube model is constructed after the method of [43] with a maximum field strength of 740 G and radius of 100 km in the photosphere. The simulation box extends 2 Mm in the chromosphere. The details of these simulations are explained in [22, 23].

The vertical photospheric driver generates a fast magneto-acoustic wave. This wave propagates upwards through the $c_S = v_A$ layer, preserving its acoustic nature and being transformed into a slow magneto-acoustic wave higher up. An essential feature of these simulations is that the wave perturbation remains almost complete

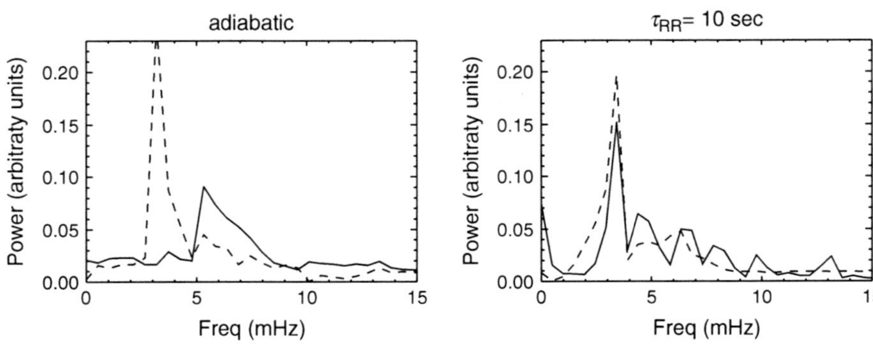

Fig. 2 *Left*: photospheric at 400 km (*dashed line*) and chromospheric at 1,500 km (*solid line*) power spectra for adiabatic simulations of waves in flux tubes. *Right*: same but for simulation with radiative relaxation time τ_{RR} equal to 10 s

within the same flux tube. Thus, it can deposit effectively the energy of the driver into the chromosphere. The power spectra of oscillations resulted from simulations at two heights in the photosphere and the chromosphere are displayed in Fig. 2. In the adiabatic case (left panel) there is a shift of the dominating frequency in the oscillation spectra from 3 mHz in the photosphere to 5 mHz in the chromosphere. In contrast, there is no such shift in the case of $\tau_{RR} = 10$ s (right panel). The latter power spectra look very similar to those obtained from spectropolarimetric observations of a facular region by [5]. It confirms that radiative losses play an important role in small-scale magnetic structures, such as those present in facular regions and are able to decrease the cut-off frequency and modify the transmission properties of the atmosphere.

4 Local Helioseismology in Magnetic Regions

Time-distance helioseismology is a branch of local helioseismology that makes use of wave travel times measured for wave packets traveling between various points on the solar surface through the interior. The inversion of these measurements is done under the assumption that the variations of the travel times are caused by mass flows and wave speeds below the surface [10,27,28,30,60]. The interpretation of results of time-distance seismology encountered major critics when applied to magnetic active regions of the Sun. Magnetic field in active regions modifies the wave propagation speeds and directions making difficult to separate magnetic and temperature effects (see e.g. [36]). To understand the influence of the magnetic field on travel time measurements, forward modeling of waves in magnetic regions is required. This has become the preferred approach in recent years.

With this aim, we performed 2D numerical simulations of magneto-acoustic wave propagation though a series of model sunspots with different field strength, from the deep interior to chromospheric layers [21]. The waves are excited by an external force localized in space just below the photosphere at −200 km, according to the models of wave excitation in the Sun [40, 53]. The spectral properties of the source resemble the spectrum of solar waves with the maximum power at 3.3 mHz. In the experiment described below the source is located at 12 Mm far from sunspot axis in the region where the acoustic speed is slightly larger than the Alfvén speed. The details of the numerical procedure are given in [24].

The magneto-static sunspot models are calculated using the method proposed in [21]. The sunspot models have a cool zone below the surface down to, about, −2 Mm depth. Below this depth the temperature gradient in horizontal direction is small and no hot zone is introduced in the present study. The photospheric field strength in three models used is of $B_{\rm phot}$=0.9, 1.5 and 2.4 kG.

Figure 3 shows a snapshot of the simulations. The fast magneto-acoustic modes (analog of p-modes in the quiet Sun) can be distinguished propagating below the surface in the pressure and velocity variations. In addition, there is a perturbation with much smaller vertical wavelength visible best in the magnetic field and

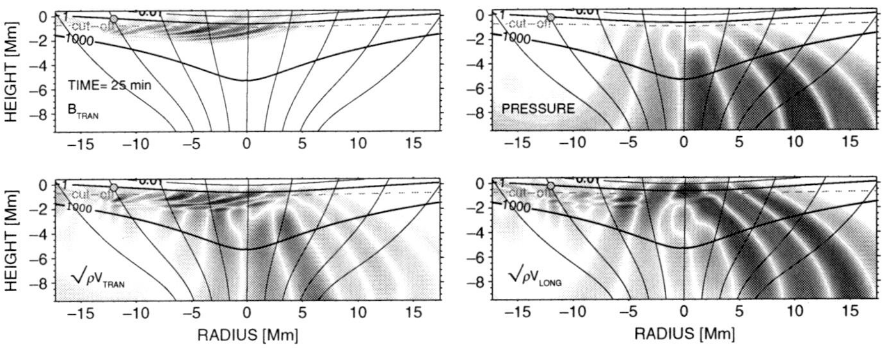

Fig. 3 *Top left*: transversal magnetic field variations. *Top right*: pressure variations. *Bottom panels*: transversal and longitudinal velocity variations. The snapshot is taken 25 min after the start of the simulations with $B_{\rm phot} = 1.5\,{\rm kG}$. The dots mark the location of the source. The velocities are scaled with a factor $\sqrt{\rho_0}$. Black contours of constant c_S^2/v_A^2 are marked with numbers. The dashed contour marks the position of the cut-off height for 3 mHz waves. Black inclined lines are magnetic field lines. Negative heights correspond to sub-photospheric layers

transversal velocity variations. A part of this perturbation is a slow MHD mode generated directly by the source at horizontal position $X = -12\,{\rm Mm}$. This mode propagates with a visibly low speed downwards along the sunspot magnetic field lines. In addition to the slow MHD waves generated directly by the source there is another wave type. The propagation speed of this small vertical wavelength disturbance is comparable to that of the fast modes. Unlike the slow MHD waves, these waves propagate horizontally across the sunspot. We conclude that these waves are magneto-gravity waves. The variations produced by these magneto-gravity waves decrease rapidly with depth and disappear almost completely below $-3\,{\rm Mm}$ (top left panel of Fig. 3).

What are the surface signatures of these modes and how do they affect the travel time measurements in solar active regions? As follows from the bottom right panel of Fig. 3, in the photospheric layers the dominant variations of the longitudinal (vertical) velocity are due to the fast magneto-acoustic mode propagating horizontally across the sunspot. Figure 4 gives the travel time differences between the phase travel times of the this mode measured in the sunspot photosphere relative to the non-magnetic quiet Sun. The latter is represented by standard solar model S [6]. The travel times are obtained from a Gabor's wavelet fit to the simulated time-distance diagrams. Negative values mean that waves in the magnetic simulations propagate faster. Here, it must be recalled that the model sunspots have a cool zone below the photosphere implying a lower acoustic speed. Thus, if the waves were purely acoustic in nature the travel time differences in Fig. 4 would have positive sign. Instead the propagation speed of the fast magneto-acoustic waves increases with the magnetic field, which has the natural consequences observed in Fig. 4: the waves in sunspot models with larger magnetic field propagate faster. The values of the travel times differences that we obtain from simulations agree rather well with those measured in solar active regions (see e.g. [7]). Thus, we conclude that the wave

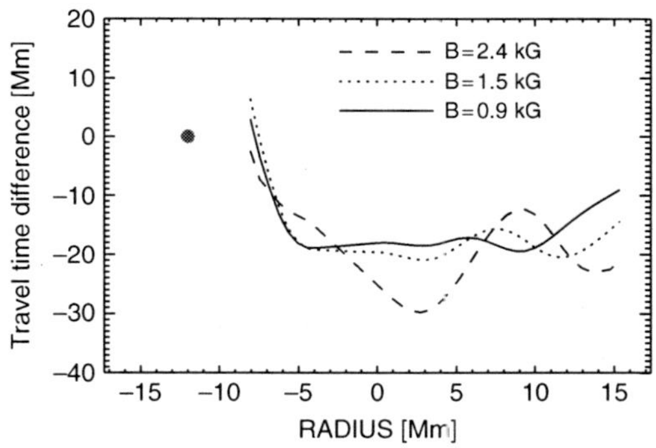

Fig. 4 Phase travel-time difference calculated between the travel times measured in simulations without magnetic field and simulations with different model sunspots (indicated in the figure) as a function of horizontal distance. The source location at $X_0 = 12$ Mm to the left of the sunspot axis is marked by a circle

propagation below solar active regions is governed by the magnetic field. Despite the new wave modes generated in the sunspot atmosphere do not affect directly the time-distance helioseismology measurements, the travel times of the fast MHD modes (analog of the p-modes) are affected by the magnetic field. A more complete analysis of these simulations is presented in [24].

5 Oscillations in Magnetic roAp Stars

Closely related to the problems of local helioseismology in magnetic regions is the problem of interpreting the oscillations in magnetic rapidly oscillating peculiar Ap stars. These stars posses a strong dipolar-like magnetic field of 1–25 kG and horizontal and vertical stratification of chemical composition. They pulsate with periods between 4 and 20 min. This offers a unique opportunity to study the interaction between convection, waves and strong magnetic field. Recent reviews on the properties of these stars and their pulsations can be found in [8, 25, 26, 33].

Several properties of the oscillations observed on these stars make them peculiar. The amplitudes of the pulsations vary with rotation period according to the magnetic field structure giving rise to oblique pulsator model [32]. The spectral lines of different chemical elements pulsate with order of magnitude different amplitudes and significant phase shifts between them, depending on their formation heights [49]. In addition, several stars pulsate with frequencies exceeding the acoustic cut-off predicted by stellar models [51]. Due to their strong magnetic fields, the atmospheres of roAp stars are regions where the magnetic pressure exceeds the gas pressure and the oscillations are magnetically dominated. Recent analytical modeling of high-frequency waves [51] suggest that the pulsations observed in the atmospheric layers can be a superposition of running acoustic waves (slow MHD) and nearly standing magnetic waves (fast MHD) that are nearly decoupled in the region $\beta \ll 1$.

In order to identify the wave modes observed in the atmospheres of roAp stars, we solved numerically the governing non-linear MHD equations in 2D geometry for a semi-empirical model atmosphere. The model has $T_{\rm eff} = 7750$ K and $\log g = 4.0$. We assumed that: (1) magnetic field varies on spatial scales much larger than the

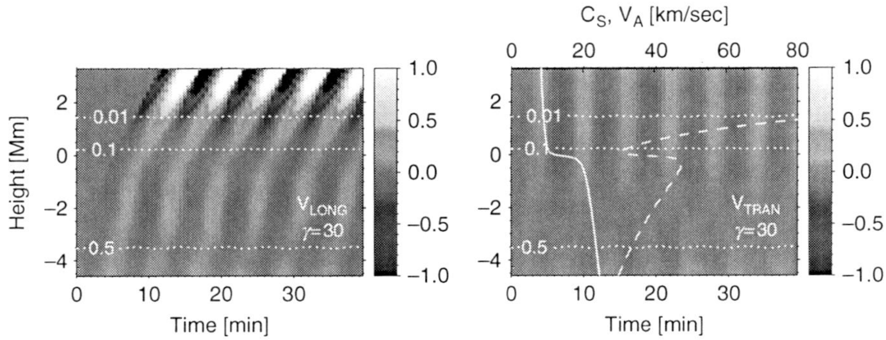

Fig. 5 Variations of the longitudinal (*left*) and transversal (*right*) velocities with height and time in a roAp star with $T_{\rm eff} = 7750$ K and $\log g = 4.0$. The units are km/sec. Results are for a dipolar magnetic field strength $B = 1000$ G, at a latitude where the inclination with respect to the local vertical is 30° and oscillation frequency 2.8 mHz. Zero height corresponds to the bottom photosphere. Contours of constant c_S^2/v_A^2 are plotted with dotted lines. Height dependences of the c_S (*solid line*) and v_A (dashed line) are plotted over the *right panel*, the scale is given by the upper axis

typical wavelength, allowing the problem to be solved locally for a plane–parallel atmosphere with a homogeneous inclined magnetic field; (2) waves in the atmosphere are excited by low-degree pulsation modes with radial velocities exceeding horizontal velocities. We studied a grid with the magnetic field strength B varying from 1 to 7 kG, its inclination varying between 0 and 60° and pulsation frequencies between 1.25 and 2.8 mHz (below and above the cut-off frequency of this simulated star).

Despite the simple magnetic field geometry, the simulations give rise to a complex picture of the superposition of several modes, varying significantly depending on the parameters of the simulations. An example of the velocity field developed in the simulation with $B = 1$ kG, $\gamma = 30°$ and pulsation period $T_0 = 360$ sec is given in Fig. 5. Longitudinal and transversal projections of the velocity with respect to the local magnetic field allow us to separate clearly the wave modes. The longitudinal velocity component shows the presence of the slow MHD (acoustic) wave. Under $\beta \ll 1$ conditions, this wave is propagating along the inclined magnetic field. The transversal velocity reveals the fast MHD (magnetic) wave. The rapid increase of the Alfvén speed with height makes the wavelength of this mode extremely large occupying the whole atmosphere. While at the bottom layers (below the photosphere) the amplitudes of the fast and slow modes are comparable, in the upper atmosphere the slow mode clearly dominates since its amplitude increases exponentially with height, much more than that of the fast mode. Two node heights can be observed in the case of the slow mode (at −3.5 and 0 Mm) and one node height in the case of the fast mode (at −2 Mm), all produced by wave reflection. The model atmosphere used in the simulations has strong density and temperature jumps at the photospheric level, producing efficient reflection. We can observe an evanescent pattern of the slow mode at heights between −3 and 0 Mm (left panel of Fig. 5). Below and above these heights the slow wave is propagating with a speed defined by the local speed

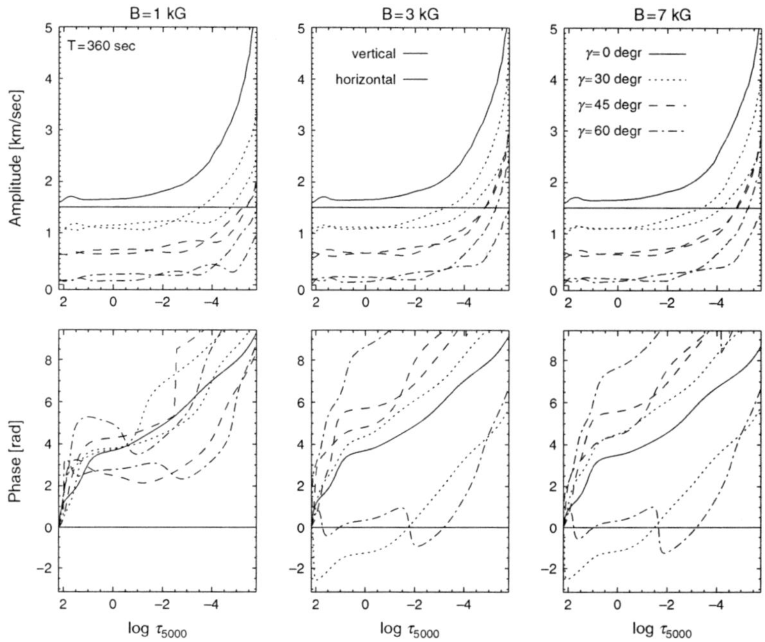

Fig. 6 Amplitudes (*top*) and phases (*bottom*) of the vertical (*thick lines*) and horizontal (*thin lines*) velocities as a function of optical depth in the atmosphere of a roAp star obtained from the simulations with dipolar magnetic field of different strength B and at different latitudes corresponding to inclinations γ (indicated in the figure). The results are for the oscillation frequency 2.8 mHz. The amplitude curves for each γ are separated by 0.5 km/s for better visualization

of sound. One can appreciate a considerably slower propagation in the upper layers due to the smaller sound speed. Strong slow wave shocks are formed above 2 Mm height with amplitudes up to 5 km/sec. The amplitudes of the velocities obtained in the simulations are similar to those observed [25].

Figure 6 gives the amplitudes and phases of the horizontal and vertical velocities as a function of the optical depth for different field strengths and inclinations obtained in the simulations with the pulsation period of 360 sec. The superposition of the fast and slow waves produces additional node-like structures at heights where these modes interfere destructively. Figure 6 shows that both amplitudes and phases of the velocity are complex functions of optical depth and of the parameters of the simulations. In general, the amplitude of the vertical velocity decreases with the inclination, while the amplitude of the horizontal velocity increases. In the case of the inclination $\gamma \neq 0$, the amplitudes of the waves at the top of the atmosphere are smaller for $B = 1$ kG compared to larger field strengths. This can be explained by the decrease of the cut-off frequency due to preferred wave propagation in the direction of the magnetic field [46]. The location of the node-like surfaces and waves propagating up and down at different heights can be appreciated from the phase curves at the bottom panels of Fig. 6. All these features are similar to observations.

The disc-integrated velocity signal produced by the atmospheric pulsations of such a star would depend in a complex way on the inclination of the magnetic axis with respect to the observational line of sight and needs a further study. However, we can conclude that the velocity signal observed in the upper atmospheric layers of roAp stars is mostly due to running slow mode acoustic waves. The node structures and the rapid phase variations at the lower atmospheric layers are due to multiple reflections and interference of the slow and fast MHD wave modes.

6 Conclusions

Magnetic field introduces an additional restoring force and makes the propagation of waves in stellar atmospheres more complex compared to the case of acoustic-gravity waves. We have developed a numerical code that allows the modeling of waves inside magnetic fields for a large variety of phenomena, from Sun to stars. We have applied this code to study the role of different solar magnetic structures concerning wave energy transport to the upper atmosphere; interpretation of local helioseismology measurements in solar active regions; pulsations of magnetic roAp stars. Puzzling physics of the interaction of waves with the magnetic field in a variety of magnetic field configurations in the Sun and stars is to be explored in the future.

References

1. Bi, S., Li, R., A&A 335, 673 (1998)
2. Birch, A.C., Kosovichev, A.G., Solar Phys. 192, 193 (2000)
3. Bloomfield, D.S., Lagg, A., Solanki, S.K., ApJ 671, 1005 (2007)
4. Bogdan, T.J., Carlsson, M., Hansteen, V., McMurry, A., Rosenthal, C.S., Johnson, M., Petty-Powell, S., Zita, E.J., Stein, R.F., McIntosh, S.W., Nordlund, A., ApJ 599, 626 (2003)
5. Centeno, R., Collados, M., Trujillo Bueno, J., in *Solar Polarization 4*, Casini, R., & Lites, B.W., eds., ASP Conference Series, Vol. 358, p. 465 (2006)
6. Christensen-Dalsgaard, J., Dappen, W., Ajukov, S.V., et al., Science 272, 1286 (1996)
7. Couvidat, S., Birch, A.C., Kosovichev, A.G., ApJ 640, 516 (2006)
8. Cunha, M.S., Communications in Astroseismology 150, 48 (2007)
9. Deubner, F.-L., Solar Phys. 39, 31 (1975)
10. Duvall, T.L.J., Jefferies, S.M., Harvey, J.W., Pomerantz, M.A., Nature 362, 430 (1993)
11. Fleck, B., Schmitz, F., A&A 250, 235 (1991)
12. Gizon, L., Birch, A.C., ApJ 571, 966 (2002)
13. Goldreich, P., Kumar, P., ApJ 363, 694 (1980)
14. Goldreich, P., Murray, N., Kumar, P., ApJ 424, 466 (1994)
15. Hasan, S.S., van Ballegooijen, A.A., A&A 680, 1542 (2008)
16. Hasan, S.S., van Ballegooijen, A.A., Kalkofen, W., Steiner, O., ApJ 631, 1270 (2005)
17. Hindman, B.W., Jain, R., ApJ 677, 769 (2008)
18. Jacoutot, L., Kosovichev, A.G., Wray, A., Mansour, N.N., ApJ 684, L51 (2008)
19. Kalkofen, W., ApJ 671, 2154 (2007)
20. Khomenko, E., Collados, M., ApJ 653, 739 (2006)
21. Khomenko, E., Collados, M., ApJ 689, 1379 (2008)

22. Khomenko, E., Centeno, R., Collados, M., Trujillo Bueno, J., ApJ 676, L85 (2008)
23. Khomenko, E., Collados, M., Feliipe, T., Solar Phys. 251, 589 (2008)
24. Khomenko, E., Kosovichev, A., Collados, M., Parchevsky, K., Olshevsky, V., ApJ 694, 411 (2009)
25. Kochukhov, O., Communications in Astroseismology 150, 39 (2007)
26. Kochukhov, O., Communications in Asteroseismology 157, 228 (2008)
27. Kosovichev, A.G., J. Comput. Appl. Math. 109, 1 (1999)
28. Kosovichev, A.G., Astron. Nachr. 323, 186 (2002)
29. Kosovichev, A.G., Duvall, T L.J., in *Solar Convection and Oscillations and their Relationship*, Pijpers, F., Christensen-Dalsgaard, J., & Rosenthal, C., eds., Astrophysics and Space Science Library, Kluwer, Vol. 225, p. 241 (1997)
30. Kosovichev, A.G., Duvall, T.L.J., Scherrer, P.H, Solar Phys. 192, 159 (2000)
31. Krijger, J.M., Rutten, R.J., Lites, B.W., Straus, T., Shine, R.A.,Tarbell, T.D., A&A 379, 1052 (2001)
32. Kurtz, D.W., MNRAS 200, 807 (1982)
33. Kurtz, D.W., Solar Phys. 25, 21 (2008)
34. Lighthill, M.J., Proc. R. Soc. London A. 211, 564 (1952)
35. De Moortel, I., Ireland, J., Hood, A.W., Walsh, R.W., A&A 387, L13 (2002)
36. Moradi, H., Cally, P., Solar Phys., in press (2008)
37. Moradi, H., Hanasoge, S.M., Cally, P.S., ApJ 690, 72 (2009)
38. Musielak, Z.E., Rosner, R., Stein, R.F., Ulmschneider, P., ApJ 423, 474 (1994)
39. Nagashima, K., Sekii, T., Kosovichev, A.G., Shibahashi, H., Tsuneta, S., Ichimoto, K., Katsukawa, Y., Lites, B., Nagata, S., Shimizu, T., Shine, R.A., Suematsu, Y., Tarbell, T.D., Title, A.M., PASP 59, S631 (2007)
40. Nordlund, Å., Stein, R.F., ApJ 546, 576 (2001)
41. Osterbrock, D.E., ApJ 134, 347 (1961)
42. Pizzo, V.J., ApJ 302, 785 (1986)
43. Pneuman, G.W., Solanki, S.K., Stenflo, J.O., A&A 154, 231(1986)
44. De Pontieu, B., Erdelyi, R., de Wijn, A.G., ApJ 595, L63 (2003)
45. De Pontieu, B., Erdelyi, R.J., Stewart, P.., Nature 430, 536 (2004)
46. Roberts, B., Solar Phys. 87, 77 (1983)
47. Rosenthal, C.S., Bogdan, T.J., Carlsson, M., Dorch, S. B.F., Hansteen, V., McIntosh, S.W., McMurry, A., Nordlund, A., Stein, R.F., ApJ 564, 508 (2002)
48. Rouppe van der Voort, L.H.M.R., Rutten, R.J., Sütterlin, P., Sloover, P.J., Krijger, J.M., A&A 403, 277 (2003)
49. Ryabchikova, T., Piskunov, N., Kochukhov, O., Tsymbal, V., Mittermayer, P., Weiss, W.W., A&A 384, 545 (2002)
50. Solanki, S.K., A&AR 11, 153 (2003)
51. Sousa, S.G., Cunha, M.S., MNRAS 386, 531 (2008)
52. Stein, R., Solar Phys. 2, 385 (1967)
53. Stein, R.F., Nordlund, Å, ApJ 546, 585 (2001)
54. Tziotziou, K., Tsiropoula, G., Mein, N., Mein, P., A&A 456, 689 (2006)
55. Tziotziou, K., Tsiropoula, G., Mein, N., Mein, P., A&A 463, 1153 (2007)
56. Ulmschneider, P., A&A 14, 275 (1971)
57. Ulrich, R.K., ApJ 162, 993 (1970)
58. Vecchio, A., Cauzzi, G., Reardon, K.P., Janssen, K., Rimmele, T., A&A 461, L1 (2007)
59. Wunnenberg, M., Kneer, F., Hirzberger, J., A&A 395, L51 (2002)
60. Zhao, J., Kosovichev, A.G., ApJ 591, 446 (2003)

The Search for Gravitational Waves: Opening a New Window into the Universe

Alicia M. Sintes

Abstract Several ground-based interferometric detectors are now in operation to detect gravitational waves. These include the Laser Interferometric Gravitational-wave Observatory (LIGO) at two sites in Livingston and Hanford, USA, and the VIRGO detector in Cascina, Italy. They have recently completed a first science run at or close to design sensitivity and are sensitive to gravitational waves from coalescing binaries at distances of tens to hundreds of megaparsecs depending on the total mass and the mass ratio of the system. This article briefly summarizes the status of operating gravitational wave facilities, plans for future detector upgrades and the status of the space-based gravitational wave detector LISA. It also describes some of the most promising sources of gravitational waves for ground-based detectors and LISA, and searches that are underway aimed at the first direct detection of gravitational radiation from astrophysical sources.

1 Introduction

Gravitation governs the large scale behavior of the Universe. Weak compared to the electromagnetic force, it is negligible at microscopic scales. The emission of gravitational waves from accelerated masses is one of the central predictions of the Theory of General Relativity [26, 27]. The confirmation of this conjecture would be fundamental on its own. Moreover, gravitational waves would provide us with information on strong field gravity through the study of the immediate environments of black holes, and they would provide an excellent cosmological probe, in particular to test the evolution of dark energy.

So far all our knowledge about astrophysics and cosmology is based on electromagnetic observations, and as such the observation of gravitational waves will

A. M. Sintes
Departament de Física, Universitat de les Illes Balears, Cra. Valldemossa Km. 7.5,
E-07122 Palma de Mallorca, Spain
e-mail: alicia.sintes@uib.es

J.M. Diego et al. (eds.), *Highlights of Spanish Astrophysics V*, Astrophysics and Space Science Proceedings, DOI 10.1007/978-3-642-11250-8_6,

open a new window into the universe and provide information about phenomena hitherto not accessible to direct observation. Astronomers have deepened dramatically their understanding of the Universe by correlating observations from various electromagnetic bands, an approach known as multi-wavelength astronomy. In the expected era of routine gravitational wave astronomy, gravitational wave observations of high-energy astrophysical systems will form a crucial component of a true multi-messenger toolset. Already now, gravitational wave astronomers are cooperating with radio astronomers. For instance, they use the observed radio timing data from pulsars to provide effective templates for the detection of gravitational waves from rotating neutron stars.

Any attempts to directly detect gravitational waves have not been successful yet and this has been subject of some controversy ever since their prediction by Einstein. But a growing network of gravitational-wave detectors such as LIGO, GEO600, and VIRGO is currently taking science data and we are heading into an era where this controversy will be resolved [33]. These laser interferometer detectors have arm lengths of up to several km and operate in a ultra high vacuum environment They incorporate high power stabilized laser sources, complicated optical configurations, suspended optical components and high performance seismic filters. A detector in space (LISA) is also planned jointly by ESA and NASA. In Spain, a scientific community oriented towards these detectors is currently developing, being the relativity group at the University of the Balearic Islands member of the LIGO Scientific Collaboration and of GEO.

Gravitational waves have the effect of stretching and contracting space-time. This effect is transverse to the direction of propagation and has two polarizations. Because gravity is very weak, the gravitational waves we expect to observe must be emitted by massive objects undergoing large accelerations. Gravitational waves are quite different from electromagnetic waves. Electromagnetic waves are easily scattered and absorbed by dust clouds between the object and the observer, whereas gravitational waves will pass through them almost unaffected. This gives rise to the expectation that the detection of gravitational waves will reveal new insights in strong field gravity by observing black hole signatures, large scale nuclear matter in neutron stars, inner processes in supernova explosions, and, of course, the possibility to discover new kinds of astrophysical phenomena, from our own Galaxy up to cosmological distances.

Gravitational waves come in many different frequencies. But unlike electromagnetic waves, it is not the microscopic processes deep inside the sources that determine the wavelength, but the global properties of the sources, leading to wavelengths of the order of the source sizes. These range from 10^{-17} Hz in the case of ripples in the cosmological background, through signals in the audio band from the formation of neutron stars in supernova explosions, to 10^{10} Hz from the cosmological background itself. The study of the full diversity of gravitational waves, in the different frequency bands, requires complementary approaches that include: cryogenic resonant detectors, laser interferometer detectors on earth and in space, Doppler tracking of spacecraft, timing of millisecond pulsars, etc. For an overview of frequency bands, detection methods and sources see [47] and references therein.

2 The Search for Gravitational Waves

Although the direct detection of gravitational waves have not been successful yet, their indirect influence has been measured in the binary neutron star system PSR 1913+16, discovered in 1974 by Russell Hulse and Joe Taylor [34–36]. This system consists of two neutron stars orbiting each other. One of these neutron stars is detected as a pulsar. The study of this binary has provided the strongest evidence to date for the existence of gravitational waves. General relativity predicts that such a system will radiate energy in the form of gravitational waves, causing the stars to slowly spiral towards each other. In 1982 Hulse and Taylor could report, after eight years of observation, that the system was losing energy and inspiraling at the rate predicted by Einstein's general relativity.

The direct observation of gravitational radiation is still a challenge for experimental physics, however after almost 40 years of experimental development, we now have the technology to hand. We need to emphasize the importance of a network analysis for the data provided by multiple instruments. In fact, the same astrophysical event could be seen by several detectors having an adapted sensitivity in the relevant frequency range. Combining the observations can provide key information such as the location of the source in the sky and the gravitational wave polarization, in addition to increasing the detection confidence, a critical issue at the time of first positive signals. In a broader context, gravitational wave observations would be combined with data from other information carriers–electromagnetic waves or neutrinos–and contribute to the multi-messenger approach discussed above.

2.1 Resonant Mass Detectors

The development of gravitational wave detectors has been a long process, dating to the work of Joseph Weber in the 1960s [50, 51]. Stimulated by predictions of the possibility of earth-incident gravitational waves with amplitude of order 10^{-17} at frequencies near 1 kHz, Weber set out to build a detector sufficiently sensitive to observe them. His idea was to use an aluminum bar of dimension two meter long, and one-half meter diameter, whose resonant mode of oscillation ($\sim$1.6 kHz) would overlap in frequency with the incoming waves. The bar, built at the University of Maryland, was fitted with piezoelectric transducers to convert the bar motion to an electrical signal, and provided a strain sensitivity of order 10^{-15} over millisecond time scales.

Over the ensuing decades, sensitivity has been improved by several orders of magnitude. Cryogenic technology has reduced mechanical thermal noise, and also allowed the use of ultra-quiet SQUID amplifiers. New vibration isolators and transducers have also played an important role in the improvement of the sensitivity. The AURIGA detector in Legnaro [18] is one of this latest generation of resonant-mass detectors. The Rome group operates two comparable detectors, NAUTILUS in Frascati and EXPLORER at CERN [17]. The Louisiana State group operated

ALLEGRO on the LSU campus in Baton Rouge [31] until the end of 2008. The current generation of resonant-mas detectors exhibits sensitivity of about $10^{-21}/\sqrt{\text{Hz}}$, in bands measuring tens of Hertz wide.

A way to improve the sensitivity is to increase the detector masses or to change their shape. The MINIGRAIL spherical detector [25] (Leiden, Netherlands) is exploring a new shape with a 1.4 ton sphere having a resonant frequency of 2.9 kHz. The proposal to build a 2-m diameter spherical detector of 33 tons called SFERA has been explored by Italian, Swiss and Dutch groups. With this kind of heavier sphere, the frequency band could be around 1 kHz. The advantage of the sphere is the measurement of all components of the gravitational wave tensor with the same detector. However, the expected sensitivity is no better than that of the upgraded interferometers.

Another possible detector is a dual-resonator detector (DUAL) [21]. At the quantum limit a DUAL detector of 16.4 tons, equipped with a wide area selective readout, would reach a sensitivity similar to that of the advanced versions of LIGO and VIRGO between 2 and 6 kHz, a frequency range where signals from merging or ring-down of compact objects are expected. The DUAL detector involves many new ideas and technologies. An R & D program carried out by the AURIGA group has started to investigate and demonstrate the feasibility of such innovative detectors.

2.2 Ground Based Interferometers

With the increasing theoretical confidence that gravitational wave strains were likely to be of the order of 10^{-21} or less and could encompass a wide range of frequencies, experimentalists sought a more sensitive and wider-band means of detection. Such a means became possible with the development of the laser interferometer, first proposed by [41] and [52]. This device used the configuration of the Michelson interferometer to achieve differential sensitivity to the instrument arm length changes caused by an incident gravitational wave. The first working laser interferometer [28] was 2 m in arm length and achieved 10^{-16} strain sensitivity in a 1 Hz bandwidth at 1 kHz. Subsequent advanced versions, using improved laser stability, optics, and isolation from background seismic noise, were built at Caltech [13], University of Glasgow [42], and Garching [44]. They were 40, 10 and 30 m in length and achieved strain sensitivities at several hundred Hz in a 1 Hz bandwidth of about 10^{-20}, 10^{-19} and 10^{-19}, respectively. After the above prototype demonstrations of high strain sensitivities, funding agencies in the US, Europe and Japan committed to the construction of large scale laser interferometers.

These detectors operate now with a sensitivity exceeding that of resonant bars, having a larger bandwidth (reaching from a low frequency cut-off at several tens of Hz up to several kHz) and extending the search to a much broader range of potential sources. The American LIGO observatories[1] [1] consist of a 4 km arm

[1] http://www.ligo.caltech.edu/

length interferometer in Livingston, Louisiana, a 4 km interferometer in Hanford, Washington, and a 2 km interferometer also at Hanford. The 2 km and 4 km interferometers at Hanford are in the same vacuum system. Construction of LIGO began in 1996 and progress has been outstanding with operation at instrumental design sensitivity having been achieved since the fall of 2005. The 3 km long French/Italian VIRGO detector[2] [14] near Pisa is also running at design sensitivity, and the 300 m long Japanese TAMA300 detector[3] [45] is operating at the Tokyo Astronomical Observatory. The German/British detector, GEO600[4] [53], through use of novel technologies has a sensitivity at frequencies above a few hundred Hz close to those of VIRGO and LIGO in their initial operation.

During the commissioning phase, work in the detectors was stopped several times to allow for data taking. These *Science Runs* involved the LIGO, GEO and TAMA detectors. Their data were analyzed for a variety of gravitational wave signals and new upper limits have been set on the strength of gravitational waves from a range of sources: coalescing compact binaries, pulsars, burst sources and a stochastic background of gravitational waves [2–12]. The fifth science run of the LIGO detectors started on 4th November 2005 with GEO having joined for data taking periods from January 2006 and VIRGO from May 2007. This run ended on 1st October 2007 when the LIGO system had one complete year of triple coincidence data from all three of its detectors and being this the longest stretch of data taking to date at initial design sensitivity.

Given our current understanding of the expected event rates, gravitational wave detection is not guaranteed with the initial interferometers. Thus a mature plan exists for planned upgrades to the existing detectors systems to create enhanced and advanced detector systems, such that the observation of gravitational waves within the first weeks or months of operating the advanced detectors at their design sensitivity is expected [37]. Thus plans for an upgraded LIGO, and VIRGO are already well formed. The upgrade is expected to commence in 2010, with full installation and initial operation of the upgraded system by 2014. On approximately the same timescale we can expect to see the rebuilding of GEO as a detector aiming at high sensitivity in the kHz frequency region [54] and the building of a long-baseline underground detector, LCGT[5] [38] in Japan. In Australia, the ACIGA consortium[6] operates an 80 m interferometric testbed and has plans for a future full-scale interferometer.

To go beyond this point, however, a number of challenges involving mechanical losses in coatings, thermal loading effects, and the use of non-classical light to bypass the standard quantum limit will have to be met. Cryogenic test mirrors and non-transmissive optics are likely to be adopted, using materials of high thermal conductivity such as silicon. Thus research groups in the field are already looking

[2] http://www.virgo.infn.it/

[3] http://tamago.mtk.nao.ac.jp/

[4] http://geo600.aei.mpg.de/

[5] http://www.icrr.u-tokyo.ac.jp/gr/gre.html

[6] http://www.anu.edu.au/Physics/ACIGA/

towards the third generation of ground-based detectors and a European design study for such a system, the Einstein gravitational wave Telescope ET[7], is already funded under the European Commission Framework Program 7. Third generation detectors would facilitate high precision tests of General Relativity that are not possible with solar-system or binary pulsar observations. By probing the highly curved structure of space-time near dense objects we would be able to answer fundamental questions about the final fate of gravitational collapsing (is it a rotating black hole or a naked singularity or some other exotic object?) and confirm if the emitted signals from such events are consistent with General Relativity to very high order in post-Newtonian perturbation theory.

2.3 *LISA: A Space-Based Interferometer*

The LISA (Laser Interferometer Space Antenna) Project[8] [24] is a planned space mission to deploy three satellites in solar orbit forming a large equilateral triangle with a base length of 5×10^6 km. The center of the triangle formation will be in the ecliptic plane 1 AU from the Sun and 20° behind the Earth. Each spacecraft will contain two free-floating proof masses forming the end points of three separate but not independent interferometers. Recently the NASA Beyond Einstein Program Advisory Committee (BEPAC) has recommended that LISA be the Flagship mission of the program, preceded by the Joint Dark Energy Mission (JDEM) and it is also competing in order to be selected as the first large mission of ESA Cosmic Vision program. Thus LISA is expected to be launched as a joint ESA/NASA mission in 2020 and to be producing data for up to ten years. Prior to LISA, the LISA Pathfinder mission[9], to be launched at the end of 2010 by ESA, will test some of the critical new technology required for the instrument.

The main objective of LISA mission is to observe low frequency (10^{-4}–10^{-1} Hz) gravitational waves from galactic and extragalactic binary systems, complementing those observations of the ground interferometers. LISA will be able to record the inspirals and mergers of binary black holes throughout the Universe, allowing a precise mathematical understanding of the most powerful transformation of energy in the cosmos. It will map isolated black holes with high precision, verifying that they can be completely specified by four numbers: mass and the three components of spin. With its enormous reach in space and time, LISA will observe how massive black holes form, grow, and interact over the entire history of galaxy formation. It will measure precise, gravitationally-calibrated, absolute distances up to very high red-shifts and such contribute in a unique way to measurements of the Hubble constant and of Dark Energy. It is also conceivable that LISA will discover new phenomena of nature, like phase transitions of new fields, extra dimensions or string networks produced in the relativistic early Universe.

---

[7] http://www.et-gw.eu/

[8] http://www.esa.int/science/lisa, http://lisa.nasa.gov/

[9] http://www.esa.int/science/lisapathfinder

2.4 *Big Bang Observatory: BBO*

BBO[10] is a follow-on mission to LISA to probe the frequency region of 0.01 – 10 Hz. Its primary goal would be the study of primordial gravitational waves from the era of the big bang, at a frequency range not limited by the confusion noise from compact binaries. BBO will also extend LISA's scientific program of measuring waves from the merging of intermediate mass black holes at any redshift, and will refine the mapping of space-time around super-massive black holes with inspiraling compact objects. The strain sensitivity of BBO at 0.1 Hz is planned to be $\sim 10^{-24}$. This will require a considerable investment in new technology, including kW power level stabilized lasers, picoradian pointing capability, multimeter sized mirrors with sub-angstrom polishing uniformity, and significant advances in thruster, discharging, and surface potential technology.

3 Astrophysical Sources

The gravitational waves we expect to observe must be emitted by massive astrophysical objects. The most predictable sources are binary star systems. However there are many sources of much greater astrophysical interest associated with black hole interactions and coalescences, neutron stars coalescences, low-mass X-ray binaries, such as Sco-X1, stellar collapses to neutron stars and black holes (supernova explosions), rotating asymmetric neutron stars such as pulsars, and the physics of the early universe. The signals from all these sources are at a level where detectors of very high strain sensitivity–of the order of 10^{-22} to 10^{-23} over relevant timescales–will be required to allow a full range of observations and such detectors may be on ground or in space. An extensive overview of promising sources can be found in [47]. Here we only give a brief summary.

3.1 *Compact Binaries*

Compact binaries are among the most promising sources of gravitational waves for ground-based detectors like LIGO and VIRGO, and the planned space-based detector LISA. Typical examples are the coalescences of binary neutron stars or black holes. Even more spectacular events could be observed from galaxy collisions and the subsequent mergers of super-massive black holes residing in the centers of the galaxies. Other known sources include double white dwarfs and ultra-compact X-ray binaries. A census of a significant portion of the visible Universe would allow us to study the evolution of the population of stars over cosmological time-scales and dynamical interactions in different stellar environments.

---

[10] http://universe.nasa.gov/new/program/bbo.html

The evolution of binary black holes is conventionally split into three stages: inspiral, merger and ring down. Gravitational waveforms from the inspiral and ringdown stages can be accurately computed by approximation/perturbation techniques in general relativity [20, 46]. The recent progress in numerical relativity has enabled to model also the merger phase of the coalescence of binary black holes [19, 22, 30] for some particular cases. Combining the results from analytical and numerical relativity will enable to coherently search for all the three stages of the binary black-hole coalescence [15, 16], which is significantly more sensitive than the current searches [9, 40], and improve the estimation of the parameters of the binary, which is particularly important for LISA [48, 49].

The observations of massive black hole coalescences will address several of LISA's science objectives. Firstly, detection of the signals from massive black hole binaries themselves will provide direct observations of the black holes. Measuring the spins and masses of the massive black holes will give us valuable information about the mechanism of their formation: rapid spins will imply that much of the black holes mass were built up by gas accretion from a disk, moderate spins imply building the massive black holes by a sequence of major mergers of comparable mass black holes and the low spins imply that massive black holes are mostly built by capturing smaller objects coming from random directions. Knowing the parameter values of the central objects will also enable more accurate studies of the dynamics of the stellar populations in the bulge.

Massive black holes will serve as laboratories for fundamental tests of gravitational theory. The measurement of their masses and spins will confirm (or disprove) some of the untested predictions of General Relativity. We should be able to probe predictions of General Relativity in the different stages of binary evolution starting with a moderately relativistic inspiral phase to a nonlinear strong-field regime. Detecting and characterizing the post-merger phase, where the resulting black hole sheds irregularities and deformations in a well-understood process resulting in ringdown radiation, will allow us to test the *no-hair* theorem for black holes. In addition, since many of the massive black hole mergers are likely to have electromagnetic counterparts, it is possible to constrain the values of cosmological parameters by combining the gravitational wave and electromagnetic observations [43]. In particular, using the distance-redshift relation from many binary black-hole standard sirens, LISA will be able to put interesting constraints on the equation of state of the dark energy [32]. The error bars on this depend on how accurately the red-shifted mass of the source and the luminosity distance are estimated, and how well the host galaxy of the electromagnetic counterpart is identified. This will have a tremendous impact on one of the outstanding issues of present-day cosmology.

3.2 Galactic Binaries

Low-mass binaries are binary systems containing white dwarfs, naked helium stars, neutron stars or black holes. LISA will be able to detect such binaries if they are located within our Galaxy. The most common sources are expected to be white-dwarf binaries emitting gravitational wave signals of nearly constant frequency and amplitude. Verification binaries are those systems known from previous astronomical observations, for which we know sky positions, approximate periods, distances and masses for many of these sources. Observation of these known binaries will provide a check of the operation of the instrument, as well as a strong test of General Relativity.

It is expected that LISA will detect and characterize around 10,000 individual compact binary systems, as well as establish an accurate estimate of the stochastic foreground produced by tens of millions of binaries in our Galaxy. The resolved systems will provide a unique map of ultra-compact binaries. The gravitational wave measurements of this population, largely inaccessible to electromagnetic detectors, will provide information about the formation and evolution of compact binaries in general and the physics of mass transfer and tides in white dwarfs in particular. In addition, up to 10% of the white dwarfs systems observed by LISA will be seen by optical follow-ups, allowing for interesting cross-comparisons and tests of fundamental physics, e.g. it will be possible to place improved bounds on the mass of the graviton [23].

3.3 *Massive Black Hole Captures*

Massive or super-massive black holes at the center of clusters or galaxies will capture stellar mass white dwarfs, neutron stars or black holes, leading to the so-called extreme mass-ratio inspirals (EMRI). EMRIs figure among the principal fundamental-physics goals of the LISA mission because their signals contain rich information about the geometry around the central black holes. The phase evolution of their signals, lasting for thousands or even hundreds of thousands of cycles, reflect in detail the near-geodesic orbits they follow around the black holes. From this phase information we expect to measure how closely the geometry matches the Kerr geometry predicted by general relativity [29], and thereby test the black-hole uniqueness theorems of Einstein's theory. Direct evidence of the existence of a horizon in the spacetime will come from seeing the signals cut off as they cross the horizon. If they do not cut off, then that will indicate that the central object is not massive black hole; explaining what it is will require exotic new physics.

3.4 *Stellar Collapse, Supernovae and Gamma-Ray Bursts*

The collapse of a massive star, after gravitation overwhelms the pressure sustained through nuclear burning, results in a supernovae explosion and the remnant in a neutron star or black hole. The core collapse, if it is sufficiently asymmetric, has sufficient mass dynamics to be a source of gravitational waves. However the physics of the process from collapse to compact object formation is not well understood and such events are rare (several per century per galaxy). This violent explosion will produce a burst signal, one that is short in time, but relatively large in amplitude.

Other possible sources may also produce intense gravity signals that are short in time. In addition to supernovae, there could be burst-like signals from the final merger of neutron star or black hole binary systems, instabilities in nascent rotating neutron stars, or kinks and cusps on cosmic strings. Unfortunately, the exact gravitational waveform for most of these types of events are not well-known, so general burst search templates need to be employed [6–8].

Coalescing compact binaries are currently believed to be sources of short gamma-ray bursts, and supernova explosions to be the sources of long gamma-ray bursts. The observation of the gravitational waves expected from such events will contribute to a better understanding of the processes leading to gamma-ray bursts. Particularly interesting would be coincident observations of neutrinos and gravitational waves from supernovae and gamma-ray bursts.

3.5 Spinning Neutron Stars

Rapidly spinning neutron stars, or pulsars, are the other key targets in the high-frequency band: they are cosmic laboratories of matter under extreme conditions of density, temperature and magnetic fields. The gravitational wave detectors will open a radically new window to explore such phenomena.

Although any departure from axial symmetry in a spinning neutron star will result in gravitational radiation, and despite having a fair idea of the neutron star population in the galaxy, we have little idea how smooth they are and therefore whether any will be visible with the current generation of detectors. We can however make a fair guess at how mountainous they are: the spin down rates of pulsars (radio or X-ray loud spinning neutron stars) are easily observed. The spin down must be at least partially due to magnetic dipole braking, as pulsars are highly magnetized objects, but if we allocate all we see to gravitational radiation we can, with some assumptions, place an upper limit on the likely mass quadrupole of each pulsar. Doing this, it seems unlikely that more than two or three known radio pulsars could be seen right now, though more extensive searches are ongoing [2]. The most likely candidate is the Crab pulsar, whose spin-down limit promises to be well and truly broken with current observational sensitivities [11].

Using the LIGO S5 data, limits on the strain signal strength as low as $h_0 \sim 4 \times 10^{-26}$ are achieved for some pulsars. This corresponds to limits on pulsar ellipticity as low as 10^{-7}. Searches for gravitational waves from radio quiet neutron stars are more difficult and indeed represent a massive computational challenge [10]. This has been met in part by the Einstein@Home project [12], a BOINC-based screensaver actively running on around 100,000 computers worldwide and currently delivering about 80 teraflops of processing power to the problem.

3.6 Stochastic Background

Finally, a variety of cosmological scenarios predict a cosmological background of gravitational waves, analogous to the electromagnetic cosmic microwave background. This would seem to be a background noise in each detector, but the signal could be extracted through a correlation of the outputs of two or more independent detectors (see for example [3–5]). Measuring the spectrum of the stochastic background would connect us to the Planck era and would be a good mean to discriminate the different cosmological models: inflation, excitations of scalar fields arising in string theories, QCD phase transitions, coherent excitations of our universe, regarded as a brane in a higher dimensional universe, etc. However, for most models the predicted amplitude of the stochastic background is well below the sensitivity of the current LIGO detectors. If detectable will open a new window to probe fundamental physics processes in regions and at energy scales hitherto not accessible.

Acknowledgements I would like to thank the organizing committee of the VIII Scientific Meeting of the Spanish Astronomical Society for the invitation to give this talk. I am also grateful to colleagues in GEO600 and the LIGO Scientific Collaboration for help, and the support by the Spanish Ministerio de Educación y Ciencia research projects FPA-2007-60220, HA2007-0042, CSD2007-00042, and the Conselleria d'Economia Hisenda i Innovació of the Government of the Balearic Islands.

References

1. Abbott, B., et al., Nucl. Instrum. Meth. A517, 154 (2004)
2. Abbott, B., et al., Phys. Rev. D 76, 042001 (2007)
3. Abbott, B., et al., ApJ 659, 918 (2007)
4. Abbott, B., et al., Phys. Rev. D 76, 022001 (2007)
5. Abbott, B., et al., Phys. Rev. D 76, 082003 (2007)
6. Abbott, B., et al., Phys. Rev. D 76, 062003 (2007)
7. Abbott, B., et al., Phys. Rev. Lett. 101, 211102 (2008)
8. Abbott, B., et al., ApJ 681, 1419 (2008)
9. Abbott, B., et al., Phys. Rev. D 77, 062002 (2008)
10. Abbott, B., et al., Phys. Rev. D 77, 022001(2008)
11. Abbott, B., et al., ApJ. 683, L45 (2008)
12. Abbott, B., et al., Phys. Rev. D 79, 022001 (2009)
13. Abramovici, A., et al., Phys. Lett. A 218, 157 (1996)
14. Acernese, F., et al., Class. Quantum Grav. 23, S63 (2006)
15. Ajith, P., et al., Class. Quant. Grav. 24, S689 (2007)
16. Ajith, P., et al., Phys. Rev. D 77, 104017 (2008)
17. Astone, P., et al., Class. Quantum Grav. 23, S57 (2006)
18. Baggio, L., et al., Phys. Rev. Lett. 94, 241101 (2005)
19. Baker, J.G., Centrella, J., Choi, D.I., Koppitz, M., van Meter, J., Phys. Rev. Lett. 96, 111102 (2006)
20. Blanchet, L., Damour, T., Esposito-Farese, G., Iyer, B.R., Phys. Rev. Lett. 93, 091101 (2004)
21. Bonaldi, M., et al., Phys. Rev. D 74, 022003 (2006)
22. Campanelli, M., Lousto, C.O., Zlochower, Y., Merritt, D., ApJ 659, L5 (2007)

23. Cutler, C., Hiscock, W., Larson, S., Phys. Rev. D 67, 024015 (2003)
24. Danzmann, K, Rüdiger, A., Class. Quantum Grav. 20, S1 (2003)
25. de Waard, A., et al., Class. Quantum Grav. 23, S79 (2006)
26. Einstein, A., Preuss. Akad. Wiss. Berlin, Sitzungsberichte der physikalish-mathematischen Klasse, 688 (1916)
27. Einstein, A., Preuss. Akad. Wiss. Berlin, Sitzungsberichte der physikalish-mathematischen Klasse, 154 (1918)
28. Forward, R.L., Phys. Rev. D 17, 379 (1978)
29. Glampedakis, K., Babak, S., Class. Quantum Grav. 23, 4167 (2006)
30. González, J.A., Sperhake, U., Brügmann, B., Hannam, M., Husa, S., Phys. Rev. Lett. 98, 091101 (2007)
31. Heng, I. S., et al., Class. Quantum Grav. 19, 1889 (2001)
32. Holz, D.E., Hughes, S.A., ApJ 629, 15 (2005)
33. Hough, J., Rowan, S., Sathyaprakash, B.S., J. Phys. B: At. Mol. Opt Phys. 38, S497 (2005)
34. Hulse, R.A., Taylor, J.H., ApJ 191, L59 (1974)
35. Hulse, R.A., Taylor, J.H., ApJ 195, L51 (1975)
36. Hulse, R.A., Taylor, J.H., ApJ 201, L55 (1975)
37. Kalogera, V., et al., ApJ 601, L179 (2004)
38. Kazuaki, K., and the LCGT Collaboration, Class. Quantum Grav. 23, S215 (2006)
39. Kazuaki, K., and the LCGT Collaboration, ApJ 447, L95 (2006)
40. Luna, M., Sintes, A.M., Class. Quantum Grav. 23, 3763 (2006)
41. Moss, G.E., Miller, L.R., Forward, R.L., Appl. Opt. 10, 2495 (1971)
42. Robertson, D.I., et al., Rev. Sci. Instr. 66, 4447 (1995)
43. Schutz, B.F., Nature 323, 310 (1986)
44. Shoemaker, D., et al., Phys. Rev. D 38, 423 (1988)
45. Takahashi, R., TAMA Collaboration, Class. Quantum Grav. 21, S403 (2004)
46. Teukolsky, S., Press, W., ApJ 193, 443 (1974)
47. Thorne, K.S., in *300 Years of Gravitation*, Hawking, S.W., & Israel, W., eds., Cambridge, UK: Cambridge University Press, (1987)
48. Trias, M., Sintes, A.M., Phys. Rev. D 77, 024030 (2008)
49. Trias, M., Sintes, A.M., Class. Quantum Grav. 25, 184032 (2008)
50. Weber, J., Phys. Rev. 117, 306 (1960)
51. Weber, J., Phys. Rev. Lett. 22, 1320 (1969)
52. Weiss, R., Quarterly Progress Report, MIT Research Lab of Electronics 105, 54 (1972)
53. Willke, B. , et al., Class. Quantum Grav. 21, S417 (2004)
54. Willke, B., et al., Class. Quantum Grav. 23, S207 (2006)

Part II
Sea Prize

Formation, Evolution and Multiplicity of Brown Dwarfs and Giant Exoplanets

J.A. Caballero

Abstract This proceeding summarizes the talk of the awardee of the Spanish Astronomical Society award to the best Spanish thesis in Astronomy and Astrophysics in the two-year period 2006–2007. The thesis required a tremendous observational effort and covered many different topics related to brown dwarfs and exoplanets, such as the study of the mass function in the substellar domain of the young σ Orionis cluster down to a few Jupiter masses, the relation between the cluster stellar and substellar populations, the accretion discs in cluster brown dwarfs, the frequency of very low-mass companions to nearby young stars at intermediate and wide separations, or the detectability of Earth-like planets in habitable zones around ultracool (L- and T-type) dwarfs in the solar neighborhood.

El que ama arde
y el que arde vuela a la velocidad de la luz
Lagartija Nick (Val del Omar)

1 "De Fuscis Pusillis Astris et Giganteis Exoplanetis" (Part I)

The recipient of the Spanish Astronomical Society (*Sociedad Española de Astronomía*) award to the best Spanish thesis in Astronomy and Astrophysics in the two-year period 2006–2007 was the thesis *Formation, evolution and multiplicity of brown dwarfs and giant exoplanets* (*"Formación, evolución y multiplicidad de enanas marrones y exoplanetas gigantes"*), by the author of this proceeding. It was supervised by R. Rebolo and V. J. S. Béjar and defended at the Universidad de La Laguna/Instituto de Astrofísica de Canarias in March 2006.

J.A. Caballero
Departamento de Astrofísica y Ciencias de la Atmósfera, Facultad de Física, Universidad Complutense de Madrid, E-28040 Madrid, Spain
email: caballero@astrax.fis.ucm.es

J.M. Diego et al. (eds.), *Highlights of Spanish Astrophysics V*, Astrophysics and Space Science Proceedings, DOI 10.1007/978-3-642-11250-8_7,

My thesis was an ambitious initiative to search for the answers to some key questions in Astrophysics: *How and where do the substellar objects form? What are their properties? How are they related to stars?* Such answers should be obtained through observations at 1–10 m-class telescopes, especially in the red optical and the near-infrared. Just to illustrate the amount and variety of data eventually collected or the difficulty in summarizing the thesis in a short talk or in this proceeding, during my PhD, I observed during 192 telescope nights with 18 different instruments in 11 different telescopes, not counting data acquired by other observers or with space missions (e.g. Hubble, Spitzer, XMM-Newton). I splitted the 459 pages of the thesis into five parts, 11 chapters and 3 appendices, which can be downloaded from a public ftp site[1]. Most of the chapters have been the basis of many refereed publications in main international journals. The used language was Spanish.

1.1 Brown Dwarfs and Objects Beyond the Deuterium-Burning Mass Limit (Chap. 1)

This was the necessary introductory chapter of the thesis. It dealt with the following subjects:

- Physical properties of substellar objects: basic definitions, hydrogen and deuterium-burning mass limits, lithium test; time evolution of physical parameters (luminosity, temperature, absolute magnitudes, colors); ultracool atmospheres, new spectral types L and T, meteorology.
- An historical view of the searches of substellar objects (with an interesting discussion on which was the first brown dwarf: Teide 1 [54, 55], GJ 229 B [51], PPL 15 [1, 64] , HD 114762 b [41], GD 165B [2, 39] or LP 944–20 [45, 65]).
- Theoretical scenarios of formation of substellar objects and planetary systems.
- Young star clusters, photometric searches and the substellar initial mass function.
- Ultracool companions to stars, multiplicity of L and T dwarfs, circumsubstellar discs (with compilations of late-type companions and very low-mass binaries).

The chapter ended with the main aims of the thesis, which were studying the mass function in the substellar domain of the σ Orionis cluster (τ ~3 Ma) down to a few $M_{\rm Jup}$, the relation between the cluster stellar and substellar populations, the accretion discs in young cluster brown dwarfs, the frequency of very low-mass companions to nearby young stars (τ ~100 Ma) at intermediate and wide separations, and the detectability of Earth-like planets in habitable zones around ultracool (L and T) dwarfs in the solar neighborhood.

[1] ftp://astrax.fis.ucm.es/pub/users/caballero/PhD.

2 The Substellar Population in σ Orionis and Its Relation with the Stellar Population (Part II)

2.1 The σ Orionis Cluster (Chap. 2)

The fourth brightest star in the Orion Belt, about 2 mag fainter than the three main stars, is σ Ori. The star, which is actually the hierarchical multiple Trapezium-like stellar system that illuminates the famous Horsehead Nebula, has taken a great importance in the last decade. Its significance lies in the very early spectral type of the hottest component (σ Ori A, O9.5 V) and in the homonymous star cluster that surrounds the system [31]. The σ Orionis star cluster (Fig. 1), re-discovered due to its large number of X-ray emitters [68], contains one of the best known brown dwarf and planetary-mass object populations [3, 4, 34, 69, 72], and is an excellent laboratory to study the evolution of X-ray emission, discs and angular momenta [30, 36, 53, 57, 59, 63]. Canonical, minimum and maximum values of main parameters of the cluster and *some* key references are provided in Table 1. In this chapter, I also described the work that the Canarias group had carried out in σ Orionis, with an emphasis on the discovery and characterization of S Ori 70, a mid-T-type object towards the cluster [10, 46, 60, 71, 73]. Finally, I presented a compilation of cluster members with spectroscopic confirmation that was the basis of two published catalogues of stars and brown dwarfs in the σ Orionis cluster [12, 16].

2.2 Multiobject Spectroscopy in σ Orionis: A Bridge Between the Stellar and Substellar Populations (Chap. 3)

We used the Wide Field Fibre Optical Spectrograph instrument and the robot positioner AutoFib2 (WYFFOS+AF2) at the 4.2 m William Herschel Telescope to acquire about 200 intermediate-resolution (R$\sim$8,000) spectra of sources in the direction of σ Orionis. We covered the wavelength range between 6,400 and 6,800 Å. We compiled a list of 80 cluster members with WYFFOS+AF2 spectroscopy, based on the presence of Li I λ6,707.8 Å in absorption and Hα λ6,562.8 Å in emission (late and mid-type stars) or spectral type determination (early-type stars). About one half of the objects were spectroscopically studied there for the first time. Using available data on the members, we investigated:

- the variation of the strength of the Li I line with spectral type (from 0.05 Å in late F stars to 0.70 Å in intermediate M stars), time and signal-to-noise ratio;
- the frequency of accretors according to the [67] criterion (46^{+16}_{-13} % of K and M stars – there might be a bias in the input sample towards Hα emitters) and the presence of asymmetries in the profiles of the Hα line;
- the existence of forbidden lines in emission ([N II] $\lambda\lambda$6548.0,6583.5 Å, [S II] $\lambda\lambda$6716.4,6730.8 Å);

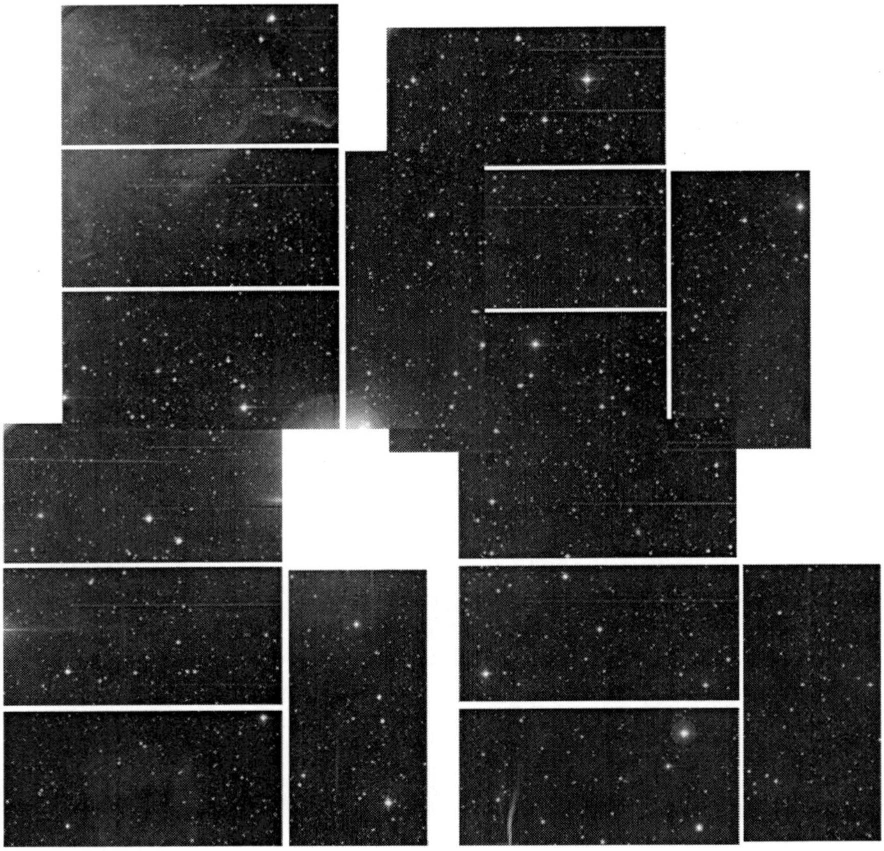

Fig. 1 False-color mosaic of a region surrounding the σ Orionis center (the bright star σ Ori falls in the central gap). The area corresponds to four pointings with the Wide Field Camera at the 2.5 m Isaac Newton Telescope. Colors red, green and blue are for passbands I, R and V, respectively. Note the bright R-band (i.e. Hα) emission to the northeast; the nebulosity is associated with the Horsehead Nebula

Table 1 Main parameters of the σ Orionis open cluster

Parameter	Canonical value	Min.:Max. values	Unit	Key references
Age, τ	3	1:8	Myr	[58, 70]
Distance, d	385	330:470	pc	[15, 48]
$E(B-V)$	0.07	0.00:0.10	mag	[6, 62]
[Fe/H]	-0.02 ± 0.13	-0.15:$+0.13$		[35]
Size, r_{max}	30	20:40	arcmin	[5, 14]
Total mass, ΣM	275[a]	150:225	$M_\odot$	[12, 61]
Disc frequency[b]	~33	5:>50	%	[24, 44]

[a] The value of $\Sigma M = 275\,M_\odot$ is from [18].

[b] The disc frequency in σ Orionis is mass-dependent and increases towards lower masses.

- the widening of photospheric lines (of up to 100 km s^{-1}) due to fast rotation;
- the relationship between the L'- and K_s-band flux excesses and the spectroscopic features associated with accretion from protoplanetary discs;
- the average of the radial velocity of the cluster members (+30.2 km s^{-1}) and the existence of radial velocity outliers (probably due to unresolved close companions and contaminants of overlapping young stellar populations in the Orion Belt);
- and the frequency of X-ray emitters catalogued by *ROSAT* and *ASCA* space observatories as a function of spectral type (the bulk of the K stars are X-ray emitters).

Our WYFFOS+AF2 data were used in the analysis of chemical abundances of late-type pre-main sequence stars in σ Orionis by [35], where we first determined the mean photospheric metallicity of the cluster. Besides, we presented a new Herbig-Haro object candidate (a few arcseconds to the southwest of the classical T Tauri star Mayrit 609206)[2] and preliminary results on topics that have been developed afterwards, such as radial distribution [14] and wide binarity [17].

2.3 A New Mini-Search in the Center of σ Orionis (Chap. 4)

Because of the intense brightness of the OB-type multiple star system σ Ori, the low-mass stellar and substellar populations close to the center of the very young σ Orionis cluster was poorly know. I presented an $IJHK_s$ survey in the cluster center, able to detect from the massive early-type stars down to cluster members below the deuterium burning mass limit. The near-infrared and optical data were complemented with X-ray imaging with the XMM-Newton and Chandra space missions. Ten objects were found for the first time to display high-energy emission. Previously known stars with clear spectroscopic youth indicators and/or X-ray emission defined a clear sequence in the I vs. $I - K_s$ diagram. I found six new candidate cluster members that followed this sequence. One of them, in the magnitude interval of the brown dwarfs in the cluster, displayed X-ray emission and a very red $J - K_s$ color, indicative of a disc[3]. Other three low-mass stars have excesses in the K_s band as well. The frequency of X-ray emitters in the area is 80 ± 20%. The spatial density of stars is very high, of up to 1.6 ± 0.1 arcmin^{-2}. There was no indication of lower abundance of substellar objects in the cluster center. Finally, I also

[2] Alternative names to Mayrit objects listed in this work, in order of appearance – Mayrit 609206: V505 Ori; Mayrit 11238: σ Ori C; Mayrit 13084: σ Ori D; Mayrit 530005: S Ori J053847.5–022711; Mayrit 528005 AB: [W96] 4771–899; Mayrit 3020 AB: σ Ori IRS1; Mayrit 306125 AB: HD 37525; Mayrit 208324: HD 29427; Mayrit 1359077: HD 37686; Mayrit 495216: S Ori J053825.4–024241.

[3] This object is actually an emission-line, Type 1, obscured quasar at $z = 0.2363 \pm 0.0005$ (UCM0536–0239; [20]).

reported two cluster stars with X-ray emission located at only 8,000–11,000 AU to σ Ori AFB, two sources with peculiar colors and an object with X-ray emission and near-infrared magnitudes similar to those of previously-known substellar objects in the cluster. (*A near-infrared/optical/X-ray survey in the center of σ Orionis*—[13]).

2.4 Multiplicity in σ Orionis: Adaptive Optics in the Near Infrared (Chap. 5)

Substellar objects, when companions to stars, are found in direct image at distances larger than ~40 AU to the primaries (e.g. [51, 56]). While many multiple stellar systems and isolated substellar objects are found in the σ Orionis cluster, no brown dwarf or planetary-mass object at projected physical separations from stellar members at less than about 10,000 AU has been *published* yet (but see the brown dwarf-exoplanet system candidate in [23]). Through a pilot programme of near-infrared adaptive optic imaging with Naomi+Ingrid at the William Herschel Telescope, we investigated the coronae between 150 and 7,000 AU from six stellar cluster members. The observed stars covered a wide range of spectral types, from O9.5 V to K7.0. Apart from the adaptive optic images, we used other near-infrared, optical and X-ray data to derive the real astrophysical nature of the detected visual companions. A total of 22 visual companions to the primary targets were detected in this pencil-beam survey. Six sources showed blue optical-near-infrared colors for their magnitudes, and they did not match in any color-magnitude diagram of the cluster. There is not enough information to derive the nature of other five sources (including a faint object 2 arcsec northeast of Mayrit 11238; see [9]). Eleven objects remained as cluster member candidates according to their magnitudes and colors: (a) three of them were previously known cluster members: Mayrit 11238, Mayrit 13084 (surrounding σ Ori AFB) and Mayrit 530005 (close to Mayrit 528005 AB); (b) one is the near-infrared counterpart of the mid-infrared and radio source Mayrit 3020 AB, a dust cloud next to σ Ori AFB discovered by [66]. The object was also detected in Chandra archive images taken with the HRC-I instrument (this result was advanced in [11]; see again [9]); (c) one of the Mayrit 306125 AB companions seemed to be a pre-main sequence photometric candidate star catalogued by [68]; (d) two bright objects were the previously unknown secondaries of the Mayrit 306125 AB and Mayrit 528005 AB close binary systems, at angular separations of 0.45 ± 0.04 and 0.40 ± 0.08 arcsec, respectively; and (e) the four remaining objects were visual companions to σ Ori AFB (1), Mayrit 208324 (2) and Mayrit 1359077 (1) at separations from 5.5 to 19.0 arcsec. A few of them display features of youth (e.g. discs). Even if their common spatial velocities are measured in the future, it is not known whether the systems will survive the gravitational field of the young cluster.

2.5 *The Mass Function Down to the Planetary Domain: The "Anaga" Survey (Chap. 6)*

We investigated the mass function in the substellar domain of the σ Orionis open cluster down to a few Jupiter masses. We performed a deep IJ-band search with Isaac at the 8.2 m Very Large Telescope UT1 and the Wide Field Camera at the Isaac Newton Telescope, covering an area of 790 arcmin^2 close to the cluster center. This survey was complemented with an infrared follow-up in the HK_s- and 3.6–8.0 μm-bands with IRAC at the Spitzer Space Telescope, CFHT-IR at the 3.6 m Canada-France-Hawai'i Telescope, Omega-2000 at the 3.5 m Calar Alto Teleskop and CAIN-II at the 1.5 m Telescopio Carlos Sánchez. Using color-magnitude diagrams, we selected 49 candidate cluster members in the magnitude interval 16.1 mag $< I <$23.0 mag. Accounting for flux excesses at 8.0 μm and previously known spectral features of youth, we identified 30 objects as bona fide cluster members. Four were first identified from our optical/near-infrared data. Eleven had most probable masses below the deuterium burning limit, which we therefore classified as candidate planetary-mass objects. The slope of the substellar mass spectrum ($\Delta N/\Delta M \approx aM^{-\alpha}$) in the mass interval 0.11 $M_\odot < M <$ 0.006 $M_\odot$ is $\alpha = +0.6 \pm 0.2$. Any mass limit to formation via opacity-limited fragmentation must lie below 0.006 $M_\odot$. The frequency of σ Orionis brown dwarfs with circumsubstellar discs is 47 $\pm$ 9%. The continuity in the mass function and in the frequency of discs suggests that very low-mass stars and substellar objects, even below the deuterium-burning mass limit, share the same formation mechanism. Besides, the technique used for calculating in detail the back and foreground contamination by field dwarfs of very late spectral types (intermediate and late M, L, and T) that we presented in this chapter was developed by [25] with the adoption of the latest models from the literature. (*The substellar mass function in σ Orionis. II. Optical, near-infrared and IRAC/*Spitzer *photometry of young cluster brown dwarfs and planetary-mass objects*—[24]).

3 Activity and Meteorology in Ultracool Objects: Discs and Atmospheres (Part III)

3.1 *Photometric Variability of Young Brown Dwarfs in σ Orionis (Chap. 7)*

We carried out multi-epoch, time-series differential I-band photometry of a large sample of objects in the south-east region of the σ Orionis open cluster. A field of ~1,000 arcmin^2 was monitored with the Wide Field Camera at the Isaac Newton Telescope during four nights over a period of two years. Using this dataset, we studied the photometric variability of twenty eight brown dwarf cluster candidates with masses ranging from the stellar-substellar boundary down to the planetary-mass domain. We found that about 50% of the sample showed photometric variability on timescales from less than one hour to several days and years. The amplitudes of the I-band light curves ranged from less than 0.01 up to ~0.4 mag. A correlation between the near-infrared excess in the K_s band, strong Hα emission and large-amplitude photometric variation was observed. We briefly discussed how these results may fit the different scenarios proposed to explain the variability of cool and ultracool dwarfs (i.e. magnetic spots, patchy obscuration by dust clouds, surrounding accretion discs and binarity). Additionally, we determined tentative rotational periods in the range 3 to 40 h for three objects with masses around 60 M_{Jup}, and the rotational velocity of 14 ± 4 km s^{-1} for one of them. The shortest periods can be explained by pulsational instability excited by central deuterium burning during the initial phases of evolution of young brown dwarfs. (*Photometric variability of young brown dwarfs in the σ Orionis open cluster*—[21]).

3.2 S Ori J053825.4–024241: A Classical T Tauri-Like Object at the Substellar Boundary (Chap. 8)

We presented a spectrophotometric analysis of Mayrit 495216 (S Ori J053825.4–024241), a candidate member close to the substellar boundary of the σ Orionis cluster. Our optical and near-infrared photometry and low-resolution spectroscopy indicated that Mayrit 495216 is a likely cluster member with a mass estimated from evolutionary models at $0.06^{+0.07}_{-0.02}$ $M_\odot$, which made the object a probable brown dwarf. The radial velocity of Mayrit 495216 was similar to the cluster systemic velocity. This target, which we classified as an M 6.0 ± 1.0 low-gravity object, showed excess emission in the near-infrared and anomalously strong photometric variability for its type (from the blue to the J band), suggesting the presence of a surrounding disc. The optical spectroscopic observations showed a continuum excess at short wavelengths and a persistent and resolved Hα emission (pseudo-equivalent width of about −250 Å) in addition to the presence of other forbidden and permitted emission lines, which we interpret as indicating accretion from the disc and possibly mass loss. We concluded that despite the low mass of Mayrit 495216, this object exhibits some of the properties typical of active classical T Tauri stars. (*Facilities*: LRIS at the 10.0 m Keck I Telescope, ALFOSC at the 2.6 m Nordic Optical Telescope, CAIN-II at the Telescopio Carlos Sánchez, ESACCD at the 1.0 m European Space Agency Optical Ground Station, CCD#1 at the 0.8 m Telescopio IAC-80). (*S Ori J053825.4–024241: a classical T Tauri-like object at the substellar boundary*–[22]).

4 Very Low-Mass Companions to Young Stars and Ultracool Dwarfs in the Solar Neighborhood (Part IV)

4.1 A Search for Very Low-Mass Objects Around Nearby Young Stars (Chap. 9)

There is a strong competition for searching for and characterizing resolved brown dwarfs and exoplanets in orbit to neighbor stars, being the ultimate goal of many astronomers the imaging of *exoearths* in solar-like systems. Given the overluminosity of very low mass objects during their contraction phase, most high spatial-resolution, photo(astro)metric searches have tended to explore nearby (d <100 pc), young ($\tau \sim$ 10–600 Myr) stars (e.g. [7, 27, 40, 42, 47, 49, 50, 52]). Following this idea, we imaged 51 stellar systems with features of youth (lithium, chromospheric activity, X-ray emission, membership in moving group) with the NICMOS instrument and the coronograph at the Hubble Space Telescope and with near infrared adaptive optics systems attached at 4 m-class telescopes: Alfa+Omega-Cass at the 3.5 m Calar Alto Teleskop, AdOpt@TNG+NICS at the 3.6 m Telescopio Nazionale Galileo and, especially, Naomi+Ingrid at the William Herschel Telescope. High resolution images were complemented with wide-field searches with CAIN-II at the Telescopio Carlos Sánchez and other near-infrared and optical instruments. Of the 51 investigated systems, 32 (44) are 100 Myr (600 Myr) old or younger. The survey was designed to detect all brown dwarfs at projected physical separations Δ > 50 AU and all exoplanets with M_2 > 0.008 $M_\odot$ at Δ > 100 AU. However, we did not detect any new substellar object. Complementing our results with those in the literature, the frequency of substellar objects with M_2 > 0.008 $M_\odot$ is less than 2% at any distance interval. Besides, we discovered three, possibly four, new stellar companions (M_2 ~0.35 – 0.80 $M_\odot$) and measured accurate astrometry (ρ, θ) of a dozen young, late-type, close binaries. From a personal point of view, the search was characterized by our *bad luck*: AB Dor C [29] was below the NICMOS coronographic mask; HN Peg B [43] was out of the Naomi+Ingrid field of view and had no optical images to complement with our wide-field near-infrared ones; the individual exposure time for the binary HD 160934 AC [37] was too long and we could not resolve it, etc.

4.2 Multiplicity of L Dwarfs: Binarity and Habitable Planets (Chap. 10)

On the one hand, stars do have planets. The least massive exoplanets found to date, with a few Earth masses ($M_\oplus$), orbit low-mass, M-type stars. On the other hand, the most massive moons in the Solar System, with up to 0.025 $M_\oplus$, orbit giant planets. Brown dwarfs, with masses in between the least massive stars and the most massive giant planets, also have planets in wide [28] and close orbits [38]. Besides, the frequency of (protoplanetary) discs in brown dwarfs is comparable, or even larger, than in stars (see, for example, [24]; Sect. 2.5). Therefore, it is natural to hypothesis the existence of terrestrial planets surrounding very low-mass stars and brown dwarfs with spectral types L and T. Because of their intrinsic dimmness, the habitable zones are very close to the Roche limit of the central objects. Such kind of systems can be detected with *current* technology. In this chapter, I showed preliminary results on photometric monitoring at medium-size telescopes to search for transits [8, 19], that resulted in a search for variability in brown dwarf atmospheres and for wide faint companions [20,32,33], and detailed a methodology for detecting *exoearths* in habitable zones around nearby L (and T) dwarfs with high-resolution near-infrared spectrographs (e.g. Nahual at the 10.4 m Gran Telescopio Canarias).

5 Conclusions, Appendices and Bibliography (Part V)

5.1 Summary (Chap. 11)

As a corollary of my thesis, the frequency of substellar companions is low, whether around nearby stars in the field or whether close to very young stars in the σ Orionis cluster. However, isolated brown dwarfs and planetary-mass objects in clusters represent a significative fraction of the total number of objects (but not of the total mass). The similarity in spatial distribution and the continuity in the rising mass spectrum and in the frequency of discs suggest that very low-mass stars and substellar objects, even below the deuterium-burning mass limit, share the same formation mechanism. If they formed in protoplanetary discs by gravitational instabilities, a very efficient ejection mechanism would be necessary during the first few million years. Thus, the isolated planetary-mass objects that we find free-floating in clusters likely formed from turbulent fragmentation in the primigenious gas cloud.

This thesis is a *full stop, new sentence* in the quotation *Smaller, Fainter, Cooler* (in humorous contraposition to *Bigger, Stronger, Faster*) of the brown dwarf and exoplanet searches.

Acknowledgements I thank R. Rebolo and V. J. S. Béjar for helpful comments an innumerable individuals and groups for their friendship and assistance during my PhD. Especial gratitude is for *uKi* and M4M4. Most of the thesis research was conducted during my residence at the Instituto de Astrofísica de Canarias. Partial financial support was provided by a number of projects of the Spanish Ministerio Educación y Ciencia, Ministerio de Ciencia y Tecnología, Comunidad Autónoma de Madrid, Universidad Complutense de Madrid, Spanish Virtual Observatory, and European Social Fund.

References

1. Basri, G., Marcy, G.W., Graham, J.R., ApJ 458, 600 (1996)
2. Becklin, E.E., Zuckerman, B., Nature 336, 656 (1988)
3. Béjar, V.J.S., Zapatero Osorio, M.R., Rebolo, R., ApJ 521, 671 (1999)

4. Béjar, V.J.S., Martín, E.L., Zapatero Osorio, M.R., et al., ApJ 556, 830 (2001)
5. Béjar, V.J.S., Caballero, J.A., Rebolo, R., Zapatero Osorio, M.R., Barrado y Navascués, D., Ap&SS 292, 339 (2004a)
6. Béjar,V.J.S., Zapatero Osorio, M.R., Rebolo, R., AN 325, 705 (2004b)
7. Biller, B.A., Close, L.M., Masciadri, E., et al., ApJS 173, 143 (2007)
8. Blake, C.H., Bloom, J.S., Latham, D.W., et al., PASP 120, 860 (2008)
9. Bouy, H., Huélamo, N., Martín, E.L., Marchis, F., Barrado Y Navascués, D., Kolb, J., Marchetti, E., Petr-Gotzens, M.G., Sterzik, M., Ivanov, V.D., Köhler, R., Nürnberger, D., A&A 493, 931 (2009)
10. Burgasser, A.J., Kirkpatrick, J.D., McGovern, M.R., et al., ApJ 604, 827 (2004)
11. Caballero, J.A., AN 326, 1007 (2005)
12. Caballero, J.A., A&A 466, 917 (2007a)
13. Caballero, J.A., AN 328, 917 (2007b)
14. Caballero, J.A., MNRAS 383, 375 (2008a)
15. Caballero, J.A., MNRAS 383, 750 (2008b)
16. Caballero, J.A., A&A 478, 667 (2008c)
17. Caballero, J.A., in *Highlights of Spanish Astrophysics V, Proceedings of the VIII Scientific Meeting of the Spanish Astronomical Society*, this volume, Poster #103 (2009)
18. Caballero, J.A., in preparation (2009)
19. Caballero, J.A., Rebolo, R., in *P*roceedings of the First Eddington Workshop on Stellar Structure and Habitable Planet Finding, Battrick, B., Favata, F., Roxburgh, I.W., Galadí, D., eds., ESA SP-485, Noordwijk: ESA Publications Division, p. 261 (2002)
20. Caballero, J.A., Béjar, V.J.S., Rebolo, R., in *Brown Dwarfs*, Proceedings of IAU Symposium #211, Martín, E.L., ed., San Francisco: Astronomical Society of the Pacific, p. 455 (2003)
21. Caballero, J.A., Béjar, V.J.S., Rebolo, R., Zapatero Osorio, M.R., A&A 424, 857 (2004)
22. Caballero, J.A., Martín, E.L., Zapatero Osorio, M.R., et al., A&A 445, 143 (2006a)
23. Caballero, J.A., Martín, E.L., Dobbie, P.L., Barrado y Navascués, D., A&A 460, 635 (2006b)
24. Caballero, J.A., Béjar,V.J.S., Rebolo, R., et al., A&A 470, 903 (2007)
25. Caballero, J.A., Burgasser, A.J., Klement, R., A&A 488, 181 (2008a)
26. Caballero, J.A., Valdivielso, L., Martín, E.L., Montes, D., Pascual, S., Pérez-González, P.G., A&A 491, 515 (2008)
27. Carson, J.C., Hiner, K.D., Villar, G.G., Blaschak, M.G., Rudolph, A.L., Stapelfeldt, K.R., AJ 137, 218 (2009)
28. Chauvin, G., Lagrange, A.-M., Dumas, C., et al., A&A 425, L29 (2004)
29. Close, L.M., Lenzen, R., Guirado, J.C., et al., Nature 433, 286 (2005)
30. Franciosini, E., Pallavicini, R., Sanz-Forcada, J., A&A 446, 501 (2006)
31. Garrison, R.F., PASP 79, 433 (1967)
32. Goldman, B., the CLOUDS Collaboration, in *Brown Dwarfs*, Proceedings of IAU Symposium #211, Martín, E.L., dd., (San Francisco: Astronomical Society of the Pacific), 461 (2003)
33. Goldman, B., Cushing, M.C., Marley, M.S., et al., A&A 487, 277 (2008)
34. González-García, B.M., Zapatero Osorio, M.R., Béjar, V.J.S., et al., A&A 460, 799 (2006)
35. González Hernández, J.I., Caballero, J.A., Rebolo, R., Béjar, V.J.S., Barrado Y Navascués, D., Martín, E.L., Zapatero Osorio, M.R., A&A 490, 1135 (2008)
36. Hernández, J., Hartmann, L., Megeath, T., et al., ApJ 662, 1067 (2007)
37. Hormuth, F., Brandner, W., Hippler, S., Janson, M., Henning, T., A&A 463, 707 (2007)
38. Joergens, V., Müller, A., ApJ 666, L113 (2007)
39. Kirkpatrick, J.D., Allard, F., Bida, T., et al., ApJ 519, 834 (1999)
40. Lafrenière, D., Doyon, R., Marois, C., et al., ApJ 670, 1367 (2007)
41. Latham, D.W., Stefanik, R.P., Mazeh, T., Mayor, M., Burki, G., Nature 339, 38 (1989)
42. Lowrance, P.J., Becklin, E.E., Schneider, G., et al., AJ 130, 1845 (2005)
43. Luhman, K.L., Patten, B.M., Marengo, M., et al., ApJ 654, 570 (2007)
44. Luhman, K.L., Hernández, J., Downes, J.J., Hartmann, L., Briceño C., ApJ 688, 362 (2008)
45. Luyten, W.J., Kowal, C.T., in *Proper motion survey with forty-eight inch Schmidt Telescope. XLIII.* One hundred and six faint stars with large proper motions. University of Minnesota, Minneapolis (1975)

46. Martín, E.L., Zapatero Osorio, M.R., ApJ 593, L113 (2003)
47. Masciadri, E., Mundt, R., Henning, T., Álvarez, C., Barrado y Navascués, D., ApJ 625, 1004 (2005)
48. Mayne, N.J., Naylor, T. MNRAS 386, 261 (2008)
49. McCarthy, C., Zuckerman, B., AJ 127, 2871 (2004)
50. Metchev, S.A., Hillenbrand, L.A., ApJ 617, 1330 (2005)
51. Nakajima, T., Oppenheimer, B.R., Kulkarni, S.R., et al., Nature 378, 463 (1995)
52. Neuhäuser, R., Guenther, E.W., Alves, J., et al., AN 324, 535 (2003)
53. Oliveira, J.M., Jeffries, R.D., van Loon, J.Th., Rushton, M.T., MNRAS 369, 272 (2006)
54. Rebolo, R., Zapatero Osorio, M.R., Martín, E.L., Nature 377, 129 (1995)
55. Rebolo, R., Martín, E.L., Basri, G., et al., ApJ 469, L53 (1996)
56. Rebolo, R., Zapatero Osorio, M.R., Madruga, S., et al., Science 282, 1309 (1998)
57. Reipurth, Bo, Bally, J., Fesen, R.A., Devine, D., Nature 396, 343 (1998)
58. Sacco, G.G., Franciosini, E., Randich, S., Pallavicini, R., A&A 488, 167 (2008)
59. Scholz, A., Eislöffel, J., A&A 419, 249 (2004)
60. Scholz, A., Jayawardhana, R., ApJ 672, L49 (2008)
61. Sherry, W.H., Walter, F.M., Wolk, S.J., AJ 128, 2316 (2004)
62. Sherry, W.H., Walter, F.M., Wolk, S.J., Adams, N.R., AJ 135, 1616 (2008)
63. Skinner, S.L., Sokal, K.R., Cohen, D.H., et al., ApJ 683, 796 (2008)
64. Stauffer, J.R., Hamilton, D., Probst, R.G., AJ 108, 155 (1994)
65. Tinney, C.G., MNRAS 296, L42 (1998)
66. van Loon, J.Th., Oliveira, J.M., A&A 405, L33 (2003)
67. White, R.J., Basri, G., ApJ 582, 1109 (2003)
68. Wolk, S.J., PhD thesis, State University New York at Stony Brook, NY, USA (1996)
69. Zapatero Osorio, M.R., Béjar, V.J.S., Martín, E.L., et al., Science 290, 103 (2000)
70. Zapatero Osorio, M.R., Béjar, V.J.S., Pavlenko, Ya., et al., A&A 384, 937 (2002a)
71. Zapatero Osorio, M.R., Béjar, V.J.S., Martín, E.L., et al., ApJ 578, 536 (2002b)
72. Zapatero Osorio, M.R., Caballero, J.A., Béjar, V.J.S., et al., A&A 472, L9 (2007)
73. Zapatero Osorio, M.R., Béjar, V.J.S., Bihain, G., et al., A&A 477, 895 (2008)

Part III
Galaxies and Cosmology

An Overview of the Current Status of CMB Observations

R.B. Barreiro

Abstract In this paper we briefly review the current status of the Cosmic Microwave Background (CMB) observations, summarizing the latest results obtained from CMB experiments, both in intensity and polarization, and the constraints imposed on the cosmological parameters. We also present a summary of current and future CMB experiments, with a special focus on the quest for the CMB B-mode polarization.

1 Introduction

In the last years, a series of high-quality cosmological data sets have provided a consistent picture of our universe, the so-called concordance model. This model presents a flat universe with an energy content of about 70% of dark energy, 25% of cold dark matter and only around 5% of baryonic matter. The data also indicate that the primordial density fluctuations are primarily adiabatic and close to Gaussian distributed with a nearly scale invariant power spectrum.

The Cosmic Microwave Background (CMB) observations are playing a key role in this era of precision cosmology (for a recent review see [9]). The data collected from a large number of experiments measuring the intensity and, more recently, the polarization of the CMB anisotropies are in very good agreement with the predictions of the inflationary paradigm. Most notably, the NASA WMAP (Wilkinson Microwave Anisotropy Probe) satellite, launched in June 2001, has constrained the cosmological parameters down to a few per cent [39]. The detection of the E-mode polarization of the CMB, first by DASI [41] and later by a handful of experiments, also provided strong support to the concordance model.

The major challenge in current CMB Astronomy is the detection of the primordial B-mode polarization, which would constitute a direct proof of the existence of

R.B. Barreiro
Instituto de Física de Cantabria (CSIC-UC), Avda. de los Castros s/n,
39005 Santander, Spain
e-mail: barreiro@ifca.unican.es

J.M. Diego et al. (eds.), *Highlights of Spanish Astrophysics V*, Astrophysics and Space Science Proceedings, DOI 10.1007/978-3-642-11250-8_8,

a primordial background of gravitational waves, as predicted by inflation. A large effort is currently being put within the CMB community in order to achieve this goal. Some experiments are already putting limits on the amplitude of the B-mode, while many others are in preparation. Complementary, a good number of CMB experiments are dedicated to the study of the CMB at very small scales, which will provide very valuable information about secondary anisotropies, such as those due to the Sunyaev–Zeldovich (SZ) effects or gravitational lensing. Moreover, the ESA Planck satellite [69], that has been launched in May 2009, will provide all-sky CMB observations, both in intensity and polarization, with unprecedented sensitivity, resolution and frequency coverage.

Another very active field of research is the study of the temperature distribution of the CMB. The standard inflationary scenario together with the cosmological principle predict that the CMB anisotropies should follow an isotropic Gaussian field. However, alternative theories predict the presence of non-Gaussian signatures in the cosmological signal. Interestingly, different works have found deviations of Gaussianity and/or isotropy in the WMAP data whose origin, at the moment, is uncertain (see [47] and references therein for a review). Future Planck data is expected to shed light on the origin of these anomalies.

The outline of the paper is as follows. Section 2 reviews some recent CMB observational results, both in intensity and polarization. Section 3 discusses current and future CMB experiments, including the Planck satellite.

2 Observational Results

In the last decade, there has been an explosion of data that has allowed a strong progress in the characterization of the CMB fluctuations. In particular, the unambiguous detection of the position of the first peak by different experiments (Boomerang [21], MAXIMA [30]) determined that the geometry of the universe is close to flat. In subsequent years, other experiments such as Archeops [3], VSA [27] and, most notably, the NASA WMAP satellite confirmed these results, imposing strong constraints on the cosmological parameters. Complementary, other cosmological data sets have also produced very valuable results, e.g. [25, 42, 54, 59, 63], which can be combined with the CMB to produce even tighter constraints [39]. In addition, a series of experiments are measuring the polarization power spectrum with increasing sensitivity, confirming further the current consistent picture of the universe.

WMAP consists of five instruments (with a total of 10 differencing assemblies) observing at frequencies ranging from 23 to 94 GHz, with a best resolution of 13 arcmin. The latest published results are based in 5-year of data, although the satellite continues in operation. The WMAP team found that the simple six-parameter ΛCDM model–a flat model dominated by dark energy and dark matter, seeded by nearly scale-invariant, adiabatic, Gaussian fluctuations–continues to provide a good fit to the data. In addition, the model is also consistent with other

Table 1 Cosmological parameters, with the corresponding 68% intervals, for the 6-parameter ΛCDM model derived using only WMAP 5-year data and combined WMAP, baryon acoustic oscillations and supernovae data (see [39] for details)

Parameter	WMAP	Combined
$100\Omega_b h^2$	2.273 ± 0.062	$2.267^{+0.058}_{-0.059}$
$\Omega_c h^2$	0.1099 ± 0.0062	0.1131 ± 0.0034
Ω_Λ	0.742 ± 0.030	0.726 ± 0.015
n_s	$0.963^{+0.014}_{-0.015}$	0.960 ± 0.013
τ	0.087 ± 0.017	0.084 ± 0.016
$\Delta^2_R(k_0)$[a]	$(2.41 \pm 0.11) \times 10^{-9}$	$(2.445 \pm 0.096) \times 10^{-9}$

[a] $k_0 = 0.002\,\mathrm{Mpc}^{-1}$.

cosmological data sets. Table 1 shows the cosmological parameters for the simple ΛCDM model as obtained by [39] using only WMAP and combining data from WMAP, baryon acoustic oscillations [54] and supernovae [42]. Moving beyond this simple model, the combined data set also constrains in additional parameters such as the tensor to scalar ratio $r < 0.22$ (95% confidence limit, hereafter CL) and put simultaneous limits on the spatial curvature of the universe $-0.0179 < \Omega_k < 0.0081$ and the dark energy equation of state $-0.14 < 1 + w < 0.12$ (both at the 95% CL). It is also interesting to point out that the best current limit on r from CMB data alone is $r < 0.33$ (95% CL) obtained using a combination of WMAP, QUAD and ACBAR data [7], while the tightest constraint obtained directly from the CMB B-mode of polarization has recently been provided by BICEP [12] and is $r < 0.73$ (95% CL).

Figure 1 shows the temperature power spectrum measured by different experiments. The solid line is the best-fit ΛCDM model to the WMAP 5-year data, which also agrees well with the additional CMB data sets up to $\ell \approx 2000$. However, some high resolution experiments have found an excess of power at multipoles $\ell \gtrsim 2000$, in particular CBI [66] and BIMA [20] (which observe at 30 GHz) and, at a lower level, ACBAR [58] (at 150 GHz). The spectrum of the reported excess could be consistent with Sunyaev–Zeldovich emission from cluster of galaxies but this would imply a value of σ_8 larger than the one favored by other measurements [39, 75]. Another possible origin of this excess is the presence of unsubtracted extragalactic sources [70]. Very recently, two experiments, QUAD and SZA, have reported new measurements of the CMB power spectrum at small scales, finding no excess. In particular, QUAD [26] reports that, after masking the brightest point sources, the results at 150 GHz are consistent with the primary fluctuations expected for the ΛCDM model. The SZA experiment [64], that observes at 30 GHz, finds that the level of SZ emission is in agreement with the expected value of $\sigma_8 \approx 0.8$. The latter work also suggests that the excess found by CBI and BIMA experiments could be due to an underestimation of the effect of extragalactic point sources. In any case, further observations will be needed to clarify the origin of this excess. Regarding polarization, several experiments have obtained very valuable data in recent years, providing a further test of the concordance model. In particular, the large angle anticorrelation seen by WMAP in the cross power spectrum between temperature and polarization

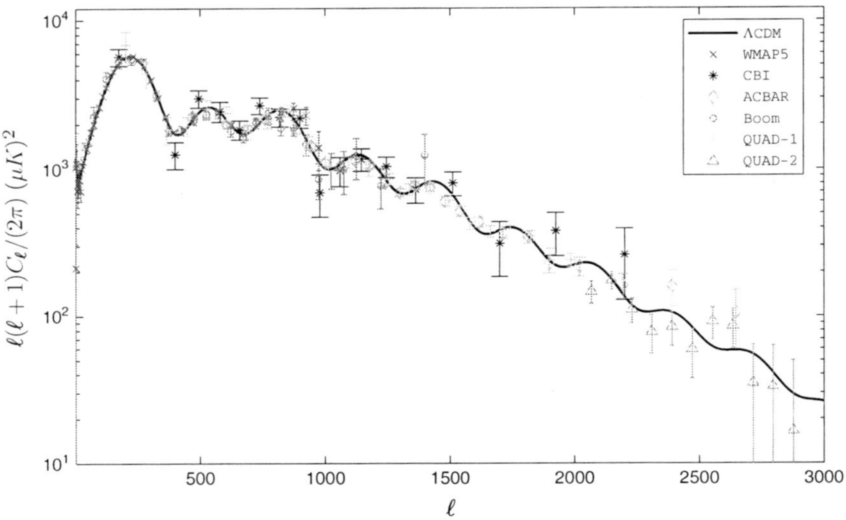

Fig. 1 CMB temperature power spectrum measured by different experiments: WMAP [51], CBI [66], ACBAR [58], Boomerang [37] and QUAD [7, 26]. The *solid line* corresponds to the best-fit model obtained using the WMAP 5-year data [39]

(TE) implies that the density fluctuations are primarily adiabatic, ruling out defect models and isocurvature models as the primary source of fluctuations [53]. In addition to WMAP [51], the TE cross power spectrum has also been measured by a number of experiments: DASI [46], CBI [65], BOOMERANG [55], QUAD [7] and BICEP [12]. A compilation of these measurements are shown in Fig. 2. Regarding the E-mode of polarization, after its first detection by DASI [41, 46], several experiments have delivered further measurements covering different ranges of angular scales: WMAP [51], CBI [65], CAPMAP [5], BOOMERANG [49], QUAD [7] and BICEP [12]. Figure 3 shows the E-mode power spectrum measured by these experiments, where acoustic oscillations are already seen. Conversely, no detection of the B-mode polarization has been found up to date, although several experiments have imposed upper limits, including the polarization experiments previously mentioned. In particular, BICEP [12] (at scales $\gtrsim 1°$) and QUAD [7] (at scales $\lesssim 1°$) have recently provided the tightest upper limits for the B-mode power spectrum (for a recent compilation of B-mode constraints see [12]).

Although most observational results show consistency with the concordance model, it is also interesting to point out that QUAD has recently found some tension between their polarization data and the simple ΛCDM model, which seems to be originated by the TE power spectrum [8]. Although this deviation is not highly significant, it will be interesting to see whether it is confirmed or not by future polarization experiments.

A number of works have also found deviations from Gaussianity and/or isotropy in the WMAP data, including, among others, a large cold spot in the southern hemisphere [15, 73], north–south asymmetries [19, 24, 33, 57], anomalies in the low

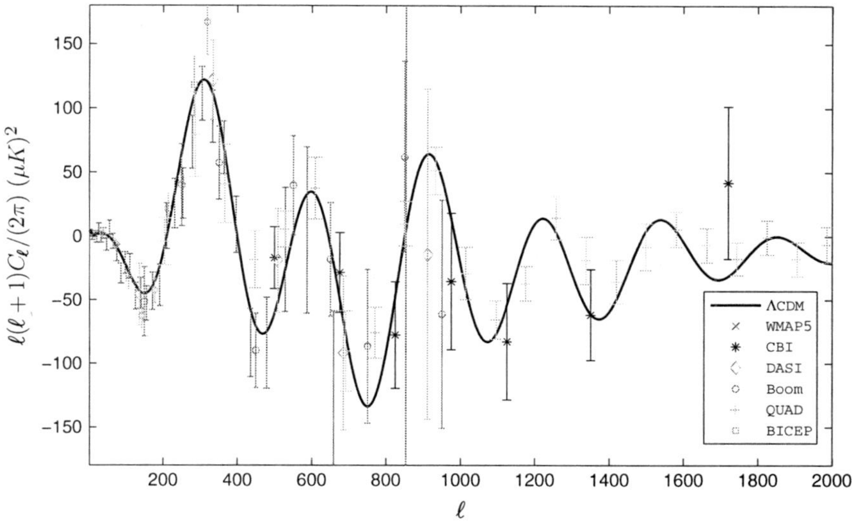

Fig. 2 TE cross power spectrum measured by different experiments: WMAP [51], CBI [65], DASI [46], Boomerang [55], QUAD [7] and BICEP [12]

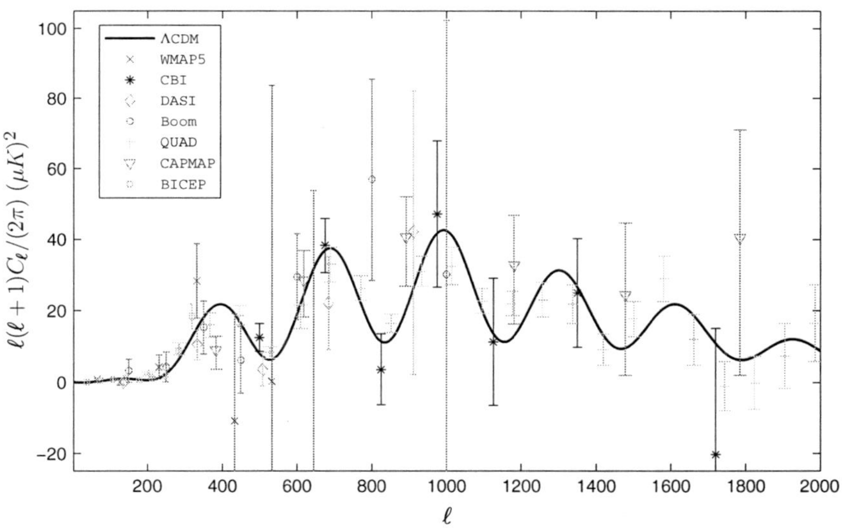

Fig. 3 CMB E-mode power spectrum measured by WMAP [51], CBI [65], DASI [46], Boomerang [49], QUAD [7], CAPMAP [5] and BICEP [12]

multipoles [4, 11, 13, 23, 43], anisotropies in the amplitude and orientation of CMB features [74, 76], an anomalously low CMB variance [48] or anomalous properties of CMB spots [1, 34, 44]. Although several possibilities have been considered to explain some of the anomalies, e.g. [16, 17, 29, 35, 36], their origin is still uncertain. The future Planck data, with a larger frequency coverage and better sensitivity than

WMAP, as well as a different scanning strategy, will allow one to carry out a more detailed study of the temperature distribution of the CMB, helping to shed light on these results.

Different groups have also placed constraints on some physically-motivated non-Gaussian models characterized by the $f_{\rm NL}$ parameter [2] finding, in general, consistency with Gaussianity, e.g. [18, 19, 32, 39, 56, 67, 72]. In particular, the best limits up to date are $-4 < f_{\rm NL}^{\rm local} < 80$ [67] and $151 < f_{\rm NL}^{\rm equil} < 253$ [39], for the local and equilateral models respectively, at the 95% CL. However, [79] have found a deviation from the Gaussian hypothesis at the 2.8σ for the local model, in disagreement with the previous mentioned results. Planck data, as well as future WMAP data with higher sensitivity, will help to confirm or discard the presence of such deviation.

It is also interesting to point out that the CMB polarization, and in particular the TB and EB cross-correlation spectra, can also be used to search for possible signatures of parity violation, e.g. [39, 78].

3 Summary of CMB Experiments

The most notable CMB experiment to operate in the near future is the ESA Planck satellite [69], that has been launched in May 2009. Planck will measure the CMB fluctuations over the whole sky, in intensity and polarization, with an unprecedented combination of sensitivity ($\Delta T/T \sim 2\times10^{-6}$), angular resolution (up to 5 arcmin), and frequency coverage (30–857 GHz). The main characteristics of Planck are summarized in Table 2. Planck will allow the fundamental cosmological parameters to be determined with a precision of $\sim$1% and will set constraints on fundamental physics at energies larger than 10^{15} GeV, which cannot be reached by any conceivable experiment on Earth. In addition, it will provide a catalogue of thousands of galaxy clusters through the SZ effect and very valuable information on the properties of radio and infrared extragalactic sources as well as on our own galaxy.

Complementary, a good number of ground-based and balloon-borne experiments are operating, or in preparation, in order to measure the intensity and polarization of the CMB with increasing sensitivity and resolution. Some of these experiments are devoted to the study of the CMB fluctuations at very small scales (a few arcminutes

Table 2 Summary of Planck instrument characteristics (taken from [69])

	LFI			HFI					
Detector technology	HEMT arrays			Bolometer arrays					
Center frequency (GHz)	30	44	70	100	143	217	353	545	857
Angular resolution (arcmin)	33	24	14	10	7.1	5.0	5.0	5.0	5.0
$\Delta T/T$ per pixel (Stokes I)[a]	2.0	2.7	4.7	2.5	2.2	4.8	14.7	147	6,700
$\Delta T/T$ per pixel (Stokes Q & U)[a]	2.8	3.9	6.7	4.0	4.2	9.8	29.8	–	–

[a] Goal (in μK/K) for 14 months integration, 1σ, for square pixels whose sides are given in the row angular resolution.

Table 3 Summary of the main characteristics of some B-mode polarization experiments

Experiment	Angular resolution (arcmin)	Frequency (GHz)	Goal (r)	Starting Year
		Ground Based		
ABS [68]	30	145	0.1	2010
BRAIN [10]	∼60	90, 150, 220	0.01	2010
C-BASS [52]	51	5	–	2009
Keck Array [50]	60–30	100, 150, 220	0.01	2010
MBI [71]	∼60	90	–	2008
QUIET [62]	28–12	40, 90	0.01	2008
QUIJOTE [60]	55–22	11, 13, 17, 19, 30	0.05	2009
PolarBear [45]	4–2.7	150, 220	0.025	2009
		Balloon Borne		
EBEX [28]	8	150, 250, 410	0.02	2009
PAPPA [38]	30	90, 210, 300	0.01	2010
PIPER	∼15	200, 270, 350, 600	0.007	2013
SPIDER [14]	58–21	100, 145, 225, 275	0.01	2010
		Satellite		
Planck [69]	33–5	30–353	0.05	2009

or below) and, in particular, to the study of the CMB secondary anisotropies, including those produced by the SZ effects and gravitational lensing. This will allow a further test of the concordance model as well as to clarify the possible excess of power found at small angular scales by previous CMB observations. Within this type of experiments we can mention AMI [80], SPT [40], ACT [61] or AMiBA [77].

However, the major challenge of current CMB astronomy is the detection of the primordial B-mode polarization, which will imply the existence of a primordial background of gravitational waves, as predicted by inflation. Table 3 summarizes some of the main on-going and future experiments targeted to study the CMB B-mode polarization. For comparison, we also include the Planck satellite in the table, as well as the C-Bass experiment which is devoted to the study of the synchrotron polarization and will provide complementary information to other experiments. The different experiments cover a wide range of frequencies, resolutions and technologies and will allow to detect (or to constrain) values of $r \approx 0.01$ in the next few years. In addition, design studies for the next generation of satellite missions are being conducted (BPol [22], EPIC [6]), which aim to achieve a sensitivity of $r \approx 0.001$, provided that foreground contamination can be properly removed.

4 Conclusions

During the last years, a consistent picture of our universe, the so-called concordance model, has emerged due to the availability of several high quality data sets. In particular, CMB observations have significantly contributed to improve our description of the universe. However, some fundamental questions still remain to be answered such as which is the nature of dark matter and dark energy, which parameters characterize the inflationary era or which is the origin of the WMAP anomalies. The future CMB data from the Planck satellite, as well as from other CMB experiments, will help to answer these open questions. In addition, the quest for the B-mode of polarization has already started and, if the scalar-to-tensor ratio is $r \approx 0.01$ or larger, the primordial background of gravitational waves–expected from inflation–could be detected in the next years. This would constitute a major breakthrough in our understanding of the early universe.

Acknowledgements The author thanks Patricio Vielva and Enrique Martínez-González for a careful reading of the manuscript. I acknowledge partial financial support from the Spanish Ministerio de Ciencia e Innovación project AYA2007-68058-C03-02.

References

1. Ayaita, Y., Weber, M., Wetterich, C., ArXiv e-prints (2009)
2. Bartolo, N., Komatsu, E., Matarrese, S., Riotto, A., Phys. Rept. 402, 103 (2004)
3. Benoît, A., et al., A&A 399, L19 (2003)
4. Bielewicz, P., Eriksen, H.K., Banday, A.J., Górski, K.M., Lilje, P.B., ApJ 635, 750 (2005)
5. Bischoff, C., et al., ApJ 684, 771 (2008)
6. Bock, J., et al., ArXiv e-prints (2008)
7. Brown, M.L., et al., for the QUaD collaboration, ArXiv e-prints (2009)
8. Castro, P.G., et al., for the QUaD collaboration, ArXiv e-prints (2009)
9. Challinor, A., Peiris, H., ArXiv e-prints (2009)
10. Charlassier, R., for the BRAIN Collaboration, ArXiv e-prints (2008)
11. Chiang, L.Y., Naselsky, P.D., Coles, P., ApJ 664, 8 (2007)
12. Chiang, H.C., et al., ArXiv e-prints (2009)
13. Copi, C.J., Huterer, D., Schwarz, D.J., Starkman, G.D., Phys. Rev. D 75(2), 023,507 (2007)
14. Crill, B.P., et al., in *Society of Photo-Optical Instrumentation Engineers (SPIE) Conference Series*, vol. 7010 (2008)
15. Cruz, M., Cayón, L., Martínez-González, E., Vielva, P., Jin, J., ApJ 655, 11 (2007)
16. Cruz, M., Turok, N., Vielva, P., Martínez-González, E., Hobson, M., Science 318, 1612 (2007)
17. Cruz, M., Martínez-González, E., Vielva, P., Diego, J.M., Hobson, M., Turok, N., MNRAS 390, 913 (2008)
18. Curto, A., Martinez-Gonzalez, E., Barreiro, R.B., ArXiv e-prints (2009)
19. Curto, A., Martínez-González, E., Mukherjee, P., Barreiro, R.B., Hansen, F.K., Liguori, M., Matarrese, S., MNRAS 393, 615 (2009)
20. Dawson, K.S., Holzapfel, W.L., Carlstrom, J.E., Joy, M., LaRoque, S.J., ApJ 647, 13 (2006)
21. de Bernardis, P., et al., Nature 404, 955 (2000)
22. De Bernardis, P., Bucher, M., Burigana, C., Piccirillo, L., Exp. Astron. 23, 5 (2009)
23. de Oliveira-Costa, A., Tegmark, M., Zaldarriaga, M., Hamilton, A. Phys. Rev. D 69(6), 063,516 (2004)
24. Eriksen, H.K., Banday, A.J., Górski, K.M., Hansen, F.K., Lilje, P.B., ApJ 660, L81 (2007)
25. Freedman, W.L., et al., ApJ 553, 47 (2001)
26. Friedman, R.B., et al., for the QUaD collaboration, ArXiv e-prints (2009)
27. Grainge, K., et al., MNRAS 341, L23 (2003)
28. Grainger, W., et al., in *Society of Photo-Optical Instrumentation Engineers (SPIE) Conference Series*, vol. 7020 (2008)
29. Groeneboom, N.E., Eriksen, H.K., ApJ 690, 1807 (2009)

30. Hanany, S., et al., ApJ 545, L5 (2000)
19. Hansen, F.K., Banday, A.J., Gorski, K.M., Eriksen, H.K., Lilje, P.B., ArXiv e-prints (2008)
32. Hikage, C., Matsubara, T., Coles, P., Liguori, M., Hansen, F.K., Matarrese, S., MNRAS 389, 1439 (2008)
33. Hoftuft, J., Eriksen, H.K., Banday, A.J., Gorski, K.M., Hansen, F.K., Lilje, P.B., ArXiv e-prints (2009)
34. Hou, Z., Banday, A.J., Gorski, K.M., ArXiv e-prints (2009)
35. Inoue, K.T., Silk, J., ApJ 664, 650 (2007)
36. Jaffe, T.R., Hervik, S., Banday, A.J., Górski, K.M., ApJ 644, 701 (2006)
37. Jones, W.C., et al., ApJ 647, 823 (2006)
38. Kogut, A., et al., New Astron. Rev. 50, 1009 (2006)
39. Komatsu, E., et al., ApJS 180, 330 (2009)
40. Kosowsky, A., New Astron. Rev. 50, 969 (2006)
41. Kovac, J.M., Leitch, E.M., Pryke, C., Carlstrom, J.E., Halverson, N.W., Holzapfel, W.L., Nature 420, 772 (2002)
42. Kowalski, M., et al., ApJ 686, 749 (2008)
43. Land, K., Magueijo, J., MNRAS 378, 153 (2007)
44. Larson, D.L., Wandelt, B.D., ApJ 613, L85 (2004)
45. Lee, A.T., et al., in *American Institute of Physics Conference Series*, vol. 1040, p. 66 (2008)
46. Leitch, E.M., Kovac, J.M., Halverson, N.W., Carlstrom, J.E., Pryke, C., Smith, M.W.E., ApJ 624, 10 (2005)
47. Martínez-González, E., ArXiv e-prints (2008)
48. Monteserín, C., Barreiro, R.B., Vielva, P., Martínez-González, E., Hobson, M.P., Lasenby, A.N., MNRAS 387, 209 (2008)
49. Montroy, T.E., et al., ApJ 647, 813 (2006)
50. Nguyen, H.T., et al., in *Society of Photo-Optical Instrumentation Engineers (SPIE) Conference*, vol. 7020 (2008)
51. Nolta, M.R., et al., ApJS 180, 296 (2009)
52. Pearson, T.J., & the C-BASS collaboration, Bull. Am. Astron. Soc., vol. 38, p. 883 (2007)
53. Peiris, H.V., et al., ApJS 148, 213 (2003)
54. Percival, W.J., Cole, S., Eisenstein, D.J., Nichol, R.C., Peacock, J.A., Pope, A.C., Szalay, A.S., MNRAS 381, 1053 (2007)
55. Piacentini, F., et al., ApJ 647, 833 (2006)
56. Pietrobon, D., Cabella, P., Balbi, A., de Gasperis, G., Vittorio, N., ArXiv e-prints (2008)
57. Räth, C., Schuecker, P., Banday, A.J., MNRAS 380, 466 (2007)
58. Reichardt, C.L., et al., ApJ 694, 1200 (2009)
59. Riess, A.G., et al., ApJ 699, 539 (2009)
60. Rubino-Martin, J.A., et al., ArXiv e-prints (2008)
61. Ruhl, J., et al., in *Society of Photo-Optical Instrumentation Engineers (SPIE) Conference Series*, Bradford, C.M., Ade, P.A.R., Aguirre, J.E., Bock, J.J., Dragovan, M., Duband, L., Earle, L., Glenn, J., Matsuhara, H., Naylor, B.J., Nguyen, H.T., Yun, M., Zmuidzinas, J., eds., Vol. 5498, p. 11 (2004)
62. Samtleben, D., for the QUIET collaboration, ArXiv e-prints (2008)
63. Seljak, U., Slosar, A., McDonald, P., J. Cosmology Astro-Particle Physics 10, 14 (2006)
64. Sharp, M.K., et al., ArXiv e-prints (2009)
65. Sievers, J.L., et al. ApJ 660, 976 (2007)
66. Sievers, J.L., et al., ArXiv e-prints (2009)
67. Smith, K.M., Senatore, L., Zaldarriaga, M., ArXiv e-prints (2009)
68. Staggs, S.T., et al., In: *Proceedings of Mitigating Systematic Errors in Space-based CMB Polarization Measurements* (2008)
69. The Planck Collaboration, ArXiv Astrophys. e-prints (2006)
70. Toffolatti, L., Negrello, M., González-Nuevo, J., de Zotti, G., Silva, L., Granato, G.L., Argüeso, F., A&A 438, 475 (2005)
71. Tucker, G.S., et al., in *Society of Photo-Optical Instrumentation Engineers (SPIE) Conference Series*, vol. 7020 (2008)

72. Vielva, P., Sanz, J.L., ArXiv e-prints (2008)
73. Vielva, P., Martínez-González, E., Barreiro, R.B., Sanz, J.L., Cayón, L., ApJ 609, 22 (2004)
74. Vielva, P., Wiaux, Y., Martínez-González, E., Vandergheynst, P., MNRAS 381, 932 (2007)
75. Vikhlinin, A., et al., ApJ 692, 1060 (2009)
76. Wiaux, Y., Vielva, P., Barreiro, R.B., Martínez-González, E., Vandergheynst, P., MNRAS 385, 939 (2008)
77. Wu, J.H.P., et al., ApJ 694, 1619 (2009)
78. Xia, J.Q., Li, H., Zhao, G.B., Zhang, X., ApJ 679, L61 (2008)
79. Yadav, A.P.S., Wandelt, B.D., Phys. Rev. Lett. 100(18), 181,301 (2008)
80. Zwart, J.T.L., et al. MNRAS 391, 1545 (2008)

The Anisotropic Redshift Space Galaxy Correlation Function: Detection on the BAO Ring

Enrique Gaztañaga and Anna Cabre

Abstract In a series of papers we have recently studied the clustering of Luminous Red Galaxies (LRG's) in the latest spectroscopic Sloan Digital Sky Survey (SDSS) data release, which has 75,000 LRG's sampling 1.1 Gpc^3/h^3 to $z = 0.47$. Here we focus on the evidence in detecting a local maxima shape as a circular ring in the 2-point correlation function $\xi(\sigma, \pi)$, separated in perpendicular σ and line-of-sight π directions. We find a significant detection of such a peak at $r \simeq 110\,\text{Mpc/h}$. The overall shape and location of the peak is consistent with it originating from the recombination-epoch baryon acoustic oscillations (BAO). This agreement provides support for the current understanding of how large scale structure forms in the universe. We study the significance of such feature using large mock galaxy simulations to provide accurate errorbars.

1 Gravitational Instability

Is the large scale structure that we see in the galaxy distribution produced by gravitational growth from some small initial fluctuations? We will explore two ways of addressing this question with measurements of the 2-point galaxy correlation:

$$\xi(r,t) = \langle \delta(r_1,t)\delta(r_2,t) \rangle \,, \tag{1}$$

where $r = |r_2 - r_1|$ and $\delta(r) = \rho(r)/\bar{\rho} - 1$ is the local density fluctuation about the mean $\bar{\rho} = \langle \rho \rangle$, and the expectation values are taken over different realizations of the model or physical process. In practice, the expectation value is over different spatial regions in our Universe, which are assumed to be a fair sample of possible realizations. The measured redshift distance of a galaxy differs from the true radial distance by its peculiar velocity along the line-of-sight. We can split the distance $\boldsymbol{r}$

E. Gaztañaga and A. Cabre
Institut de Ciencies de l'Espai (IEEC/CSIC), Barcelona, www.ice.cat,
e-mail: gazta@ice.cat,cabre@ice.cat

J.M. Diego et al. (eds.), *Highlights of Spanish Astrophysics V*, Astrophysics and Space Science Proceedings, DOI 10.1007/978-3-642-11250-8_9,

into its component along the line-of-sight (LOS) π and perpendicular to the LOS σ, where $r^2 = \pi^2 + \sigma^2$. Azimuthal symmetry implies ξ is in general a function of π and σ alone: $\xi(\sigma, \pi)$.

Consider the fully non-linear fluid equations that determine the gravitational evolution of density fluctuations, δ, and the divergence of the velocity field, θ, in an expanding universe for a pressureless irrotational fluid. In Fourier space [see (37)–(38) in [2]]:

$$\begin{aligned} \dot{\delta} + \theta &= -\textstyle\int dk_1 dk_2\, \alpha(k_1, k_2)\theta(k_1)\delta(k_2)\,, \\ \dot{\theta} + \mathcal{H}\theta + \tfrac{3}{2}\Omega_{\rm m} H^2 \delta &= -\textstyle\int dk_1 dk_2\, \beta(k_1, k_2)\theta(k_1)\theta(k_2)\,, \end{aligned} \tag{2}$$

where derivatives are over conformal time $d\tau \equiv dt/a$ and $\mathcal{H}(\tau) \equiv d\ln a/d\tau = aH$ is given by the expansion rate $H = \dot{a}/a$ of the cosmological scale factor a. On the left hand side $\delta = \delta(k, \tau)$ and $\theta = \theta(k, \tau)$ are functions of the Fourier wave vector k. The integrals are over vectors k_1 and k_2 constrained to $k = k_{12} \equiv k_2 - k_1$. The right hand side of the equation include the non-linear terms which are quadratic in the field and contain the mode coupling functions α and β. When fluctuations are small we can neglect the quadratic terms in the equations and we then obtain the linear solution $\delta_{\rm L}$. The first equation yields $\dot{\delta}_{\rm L} = -\theta_{\rm L}$, which combined with the second equation yields the well known harmonic oscillator equation for the linear growth:

$$\ddot{\delta}_{\rm L} + \mathcal{H}\dot{\delta}_{\rm L} - \frac{3}{2}\Omega_{\rm m}\mathcal{H}^2\delta_{\rm L} = 0\,. \tag{3}$$

Because the Fourier transformation is linear, this equation is valid in Fourier or in configuration space. In linear theory each Fourier mode and each local fluctuation evolves independently of the others, moreover, they all grow linearly out of the initial fields with the same growth function, i.e. $\delta_{\rm L}(t) = D(t)\delta_0$, where δ_0 is the value of the field at a point (or a given Fourier mode) at some initial time and $D(t)$ is the linear growth function, which is a solution to the above harmonic equation. In a flat universe dominated by cold dark matter (CDM) we have that $H \propto a^{-3/2}$ and $D(t)$ goes as the scale factor $D(t) \propto a$. In an accelerated phase, such as ΛCDM, the growth halts or grows less rapidly with a. Thus measurements of $D(t)$ can be used as an independent diagnostics for accelerated expansion.

1.1 BAO Signature

Consider now our observable, the 2-point function in (1). In linear theory $\xi(r,t) = D(t)^2\xi(r,0)$. This means that on large scales, i.e. $r > 10$ Mpc/h, where $\delta < 1$, we then expect the shape of $\xi(r)$ today to be the same as the shape it took in the early universe. The prediction, ignoring redshift space distortions, is shown in Fig. 1. The BAO ring corresponds to the local maxima at a radius of about 110 Mpc/h. This

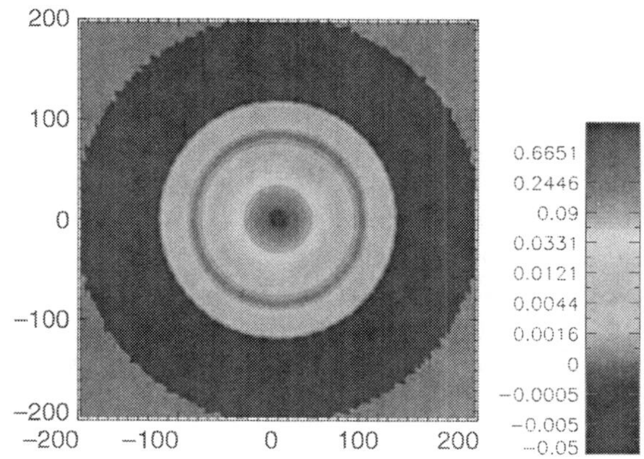

Fig. 1 The prediction of $\xi(\sigma, \pi)$ without redshift space distortions. The vertical axis shows the radial direction, π, while the horizontal panel shows the transverse direction σ. There is a prominent local maximum corresponding to the BAO ring at a radius of about 110 Mpc/h. (*green circle* between *blue rings*)

will be tested here by comparing the linear theory prediction for the baryon acoustic oscillation (BAO) with the measurements in the large scale galaxy distribution.

The mean BAO signature in the 2-point correlation has been detected in LRG's of the SDSS galaxy survey [4] using the monopole, i.e. the average signal of $\xi(\pi, \sigma)$ in circles of constant $\pi^2 + \sigma^2$. In [3, 5] and below we will show that the galaxy distribution has a BAO ring very similar to that predicted in the initial conditions of the CDM model.

1.2 Growth Factor

On the other hand, if we knew the shape and amplitude of the initial conditions $\xi(r, 0)$, we could then estimate $D(t)^2 \simeq \xi(r,t)/\xi(r,0)$ from measurements of $\xi(r,t)$ and compare it to the linear solution of (3) to test gravitational instability. But this approach is difficult in practice because there is a bias in the amplitude of galaxy clustering compare to the one in dark matter fluctuations. We could instead test gravitational growth with independence of time or initial conditions by using the linear relation between density and velocities, $\dot{\delta}_L = -\theta_L = f\delta_L$, where $f \equiv \dot{D}/D = d\ln D/d\ln a$ is the velocity growth factor. For a flat dark matter dominated universe $D = a$ and $f = 1$, while for a flat accelerated universe $f = \Omega_m(z)^\gamma < 1$, where γ is the gravitational growth index [8] and $\Omega_m(z)$ is the matter density at a redshift z where $a = 1/(1+z)$. The growth index separates out two physical effects on the growth of structure: $\Omega_m(z)$ depends on the expansion history while γ depends on the underlaying theory of gravity. The value $\gamma = 0.55$ corresponds to standard gravity, while γ is different for modified gravity, for example $\gamma = 0.68$ in the braneworld cosmology. We will show next how f can be measured using redshift space distortions in the galaxy correlation function $\xi(r)$.

2 Analysis of Data

In this work we use the most recent spectroscopic SDSS data release, DR6 [1]. We use the same samples and methodology here as presented in [3] of this series. LRG's are targeted in the photometric catalog, via cuts in the $(g-r, r-i, r)$ color–color–magnitude cube. We need to k-correct the magnitudes in order to obtain the absolute magnitudes and eliminate the brightest and dimmest galaxies. We have seen that the previous cuts limit the intrinsic luminosity to a range $-23.2 < M_r < -21.2$, and we only eliminate from the catalog some few galaxies that lay out of the limits. Once we have eliminated these extreme galaxies, we still do not have a volume limited sample at high redshift. For the 2-point function analysis we account for this using a random catalog with identical selection function but 20 times denser (to avoid shot-noise). The same is done in simulations.

There are about 75,000 LRG's with spectroscopic redshifts in the range $z = 0.15 - 0.47$ over 13% of the sky. We break the full sample into 3 independent subsamples with similar number of galaxies: low $z = 0.15 - 0.30$, middle $z = 0.30 - 0.40$ and high $z = 0.40 - 0.47$. In this analysis we will just show results for the $z = 0.15 - 0.30$ sample.

To estimate the correlation $\xi(\sigma, \pi)$, we use the estimator of [7],

$$\xi(\sigma, \pi) = \frac{DD - 2DR + RR}{RR}, \tag{4}$$

with a random catalog $N_R = 20$ times denser than the SDSS catalog. The random catalog has the same redshift (radial) distribution as the data, but smoothed with a bin $dz = 0.01$ to avoid the elimination of intrinsic correlations in the data. The random catalog also has the same mask. We count the pairs in bins of separation along the line-of-sight (LOS), π, and across the sky, σ. The LOS distance π is just the difference between the radial comoving distances in the pair. The transverse distance σ is given by $\sqrt{s^2 - \pi^2}$, where s is the net distance between the pair. We use the small-angle approximation, as if we had the catalog at an infinite distance, which is accurate until the angle that separates the galaxy pair in the sky is larger than about 10° for $\xi(\sigma, \pi)$ (see [10] and [9]). This condition corresponds to transverse scales larger than $\sigma = 80$ Mpc/h for our mean catalog.

The right panel in Fig. 2 shows the measurements of $\xi(\pi, \sigma)$ in the $z = 0.15 - 0.30$ sample. As we will show below there is a remarkable agreement with the predictions (left panel) and there is good evidence for a BAO ring.

2.1 Errors and Simulations

There are two sources of error or variance in the estimation of the two-point correlation: (a) *shot-noise* which scales as one over the square root of the number of pairs in each separation bin, and (b) *sampling variance* which scales with the amplitude of the correlation. It is easy to check that for the size and number density of our

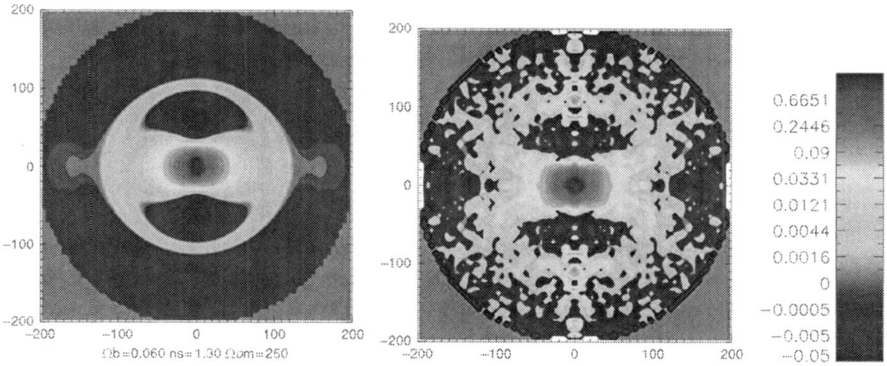

Fig. 2 A comparison of $\xi(\sigma, \pi)$ in data and models for the $z = 0.15 - 0.30$ slice. The vertical axis shows the radial direction, π, while the horizontal panel shows the transverse direction σ. The left panel includes a model of redshift space distortions which gives the best fit to the monopole in the galaxy data (see [3]). Right panel shows the LRG SDSS measurements using the same color scheme. The data shows a prominent BAO ring at a radius of about 110 Mpc/h, in good agreement with the model

sample, the shot-noise term dominates over the sampling variance error. This has been confirmed in detail by using numerical simulations (see [3] for more details). The simulation contains 2048^3 dark matter particles, in a cube of side 7680 Mpc/h (which we call MICE7680), $\Omega_M = 0.25$, $\Omega_b = 0.044$, $\sigma_8 = 0.8$, $n_s = 0.95$ and $h = 0.7$. We have divided this big cube in 3^3 cubes of side 2×1275 Mpc/h, and taking the center of these secondary cubes as the observation point (as if we were at $z = 0$), we apply the selection function of LRG, which arrives until $z = 0.47$ ($r = 1275$ Mpc/h). We can obtain 8 octants from the secondary sphere included in the cube, so at the end we have 8 mock LRG catalogs from each secondary cube, which have the same density per pixel as LRG in order to have the same level of shot noise, and the area is slightly smaller (LRG occupies 1/7 of the sky with a different shape). The final number of independent mock catalogs is $M = 216$ (27×8). We also apply redshift distortions in the line-of-sight direction $s = r + v_r/H(z)/a(z)$, using the peculiar velocities v_r from the simulations. The error covariance is found from the dispersion of M realizations:

$$C_{ij} = \frac{1}{M} \sum_{k=1}^{M} (\xi(i)^k - \widehat{\xi}(i))(\xi(j)^k - \widehat{\xi}(j)) , \quad (5)$$

where $\xi(i)^k$ is the measure in the k-th simulation ($k = 1, \ldots, M$) and $\widehat{\xi}(i)$ is the mean over M realizations (which we have checked that agrees with overall mean, indicating that volume effects are small). The case $i = j$ gives the diagonal error (variance). In our analysis we model the errors using dark matter groups. These groups are chosen to have the same number density and amplitude of clustering as the observed LRG's. The resulting errorbars from simulations are typically in good agreement with Jack–knife errors from the actual data (see [3] for details). We have also checked in [3] that we can recover the theory predictions for $\xi(\sigma, \pi)$ within the errors by using mock simulations with similar size as the real data.

2.2 Redshift Space Distortions

Radial displacements caused by peculiar velocities lead to redshift distortions, with two important contributions. The first, on large scale fluctuations, caused by coherent bulk motion. We see walls denser and voids bigger and emptier, with a squashing effect in the 2-point correlation function along the line-of-sight: known as the Kaiser [6] effect. At small scales, random velocities inside clusters and groups of galaxies produce a radial stretching pointed at the observer, known as *fingers of God* (FOG). Although such distortions complicate the interpretation of redshift maps as positional maps, they have the advantage of bearing unique information about the dynamics of galaxies. In particular, the amplitude of distortions on large scales yields a measure of the linear redshift distortion parameter f.

In the large-scale linear regime, and in the plane-parallel approximation, the distortion caused by coherent infall velocities takes a particularly simple form. On average, large scale fluctuations in redshift space δ_s are enhanced with respect to real space δ because of the radial velocity infall, so that $\delta_s \simeq \delta - \theta/3 = (1 + f/3)\delta$ so that they are larger by a factor $(1 + f/3)$. This enhancement is anisotropic. In Fourier space:

$$P_s(\boldsymbol{k}) = (1 + f\mu_k^2)^2 P(k) , \tag{6}$$

where $P(k)$ is the power spectrum of density fluctuations δ, μ is the cosine of the angle between $\boldsymbol{k}$ and the line-of-sight, the subscript s indicates redshift space, and f is the velocity growth rate in linear theory.

The correlation $\xi(\sigma, \pi)$ is related to the power spectrum by a Fourier transform:

$$\xi(\sigma, \pi) = \int P_s(\boldsymbol{k}) e^{-ikr} \frac{d^3k}{(2\pi)^3} . \tag{7}$$

After integration in (7), these linear distortions in $P_s(\boldsymbol{k})$ produce a distinctively anisotropic $\xi(\sigma, \pi)$. At scales smaller than about 50 Mpc/h there is a clear squashing in the correlation function caused by the peculiar velocity divergence θ field, this effect can be used to estimate f, for example by fitting the normalized quadrupole to the data (e.g. see [3]). For the sample SDSS $z = 0.15 - 0.30$ sample considered here, we find $f = 0.48 - 0.83$, which corresponds to $\Omega_m = 0.24 - 0.32$ when we assume standard gravity ($\gamma = 0.55$).

Redshift distortions in the linear regime produce a lower amplitude and sharper baryon acoustic peak in the LOS than in the perpendicular direction because of the coherent infall into large scale overdensities. This is illustrated in the left panel of Fig. 2. A characteristic feature of this effect is a valley of negative correlations (in blue) on scales between $\pi = 50 - 90$ Mpc/h, which as we will show is in good

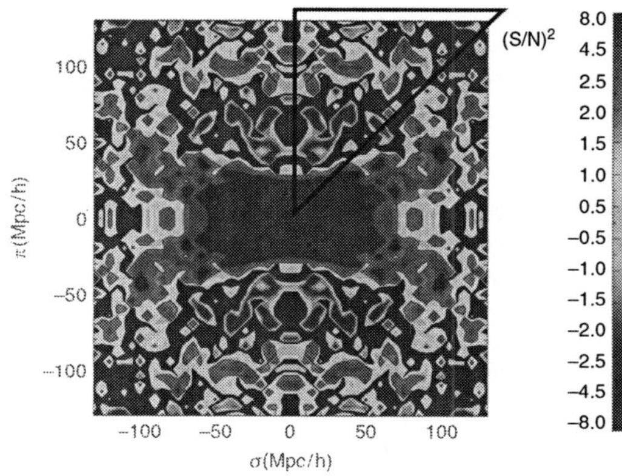

Fig. 3 Signal-to-noise ratio in $\xi(\sigma, \pi)$ for the $z = 0.15 - 0.30$ slice. The color scheme denotes $(S/N)^2$ multiplied by the sign of the signal i.e. negative values correspond to a negative signal. The triangle highlights the region $\pi > \sigma$, which receives little weight in the monopole. The mean $(S/N)^2$ in radial bins in this region is shown in Fig. 4

agreement with our measurements from real data. Such a valley is absent without redshift distortions.

In Fig. 3, we show the signal-to-noise of ξ in the $\sigma - \pi$ plane for the redshift slice $z = 0.15 - 0.3$. This complements the $\xi(\sigma, \pi)$ signal plot in the right panel of Fig. 2. The signal-to-noise shown in Fig. 3 is for each pixel of size 5 Mpc/h by 5 Mpc/h (the same pixel size is used in Fig. 2). Note that there is covariance between pixels, and so this figure should be interpreted with some care (see [3]). Nonetheless, it demonstrates the high quality detection of a BAO ring in the $\sigma - \pi$ plane. The triangle highlights the region $\pi > \sigma$, which receives not much weight in the monopole, but where the BAO ring still shows up nicely. Note that the $(S/N)^2$ shown is modulated by the sign of the signal: the (blue) valley of negative correlations at $\pi \sim 50 - 90$ Mpc/h–in accord with the predictions of the Kaiser effect–are detected with significance as well. The overall coherent structure of a negative valley before a positive BAO peak (at just the right expected scales) is quite striking, and cannot be easily explained away by noise or systematic effects.

The evidence for a BAO peak in the monopole is quite convincing (see [3–5]). The data follows the model prediction and produces a clear Ω_b detection which otherwise (without the BAO peak) is degenerate with other cosmological parameters. But the monopole signal is dominated by pairs in the perpendicular direction $\sigma > \pi$ and here we would like to assess if the BAO peak is also significant in the radial direction. We do this by studying the signal-to-noise ratio in $\xi(\sigma, \pi)$ for $\pi > \sigma$. In Fig. 3 this corresponds to the region inside the over-plotted triangle. We do the mean signal-to-noise inside the region $\pi > \sigma$ as a function of the radius $s^2 = \pi^2 + \sigma^2$, in radial shells $d \pm ds$ of width $ds = 2.5$ Mpc/h:

$$mean(S/N)^2 = \sum_{s \pm ds} (Sign)\,(S/N[\pi, \sigma])^2\,, \tag{8}$$

where (*Sign*) is the sign of the signal. When the signal is negative this gives a negative contribution to the mean signal-to-noise square. If the signal is dominated by noise, positive and negative fluctuations will tend to cancel and reduced the mean

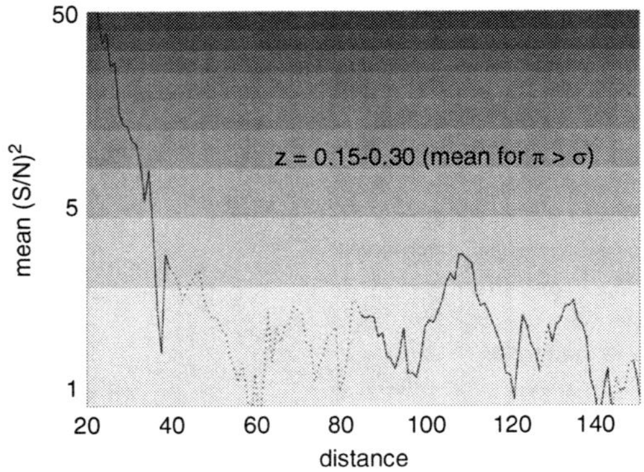

Fig. 4 Mean square signal-to-noise ratio averaged in radial bins with $\pi > \sigma$, shown as a function of the radial distance π. Note that we include the sign of the signal in doing the mean, see (8), which cancels the noise contribution (mean negative values are shown as a *dotted line*)

$(S/N)^2$. Results are shown in Fig. 4. The mean signal-to-noise is always larger than unity both in the negative valley between 50 – 90 Mpc/h and also around the BAO peak, where the mean $(S/N)^2$ approaches 2. This clearly indicates that the BAO peak is also significant in the radial direction and it also has the shape that is predicted by the models, with a negative valley and a positive peak that extend in a coherent way over the expected lengths. It is unlikely that noise or systematic errors could reproduce these correlations. Similar results are found for the other redshift slices, with less significant detection for the middle slice.

Acknowledgements We acknowledge the use of MICE simulations (www.ice.cat/mice) developed at the MareNostrum supercomputer (www.bsc.es) and with support from PIC (www.pic.es), the Spanish Ministerio de Ciencia y Tecnologia (MEC), project AYA2006-06341 with EC-FEDER funding, Consolider-Ingenio CSD2007-00060 and research project 2005SGR00728 from Generalitat de Catalunya. AC acknowledges support from the DURSI department of the Generalitat de Catalunya and the European Social Fund.

References

1. Adelman-McCarthy, J.K., Agüeros, M.A., Allam, S.S., Allende Prieto, C., Anderson, K.S.J., Anderson, S.F., Annis, J., Bahcall, N.A., Bailer-Jones, C.A.L., Baldry, I.K., et al., ApJS 175, 297 (2008)
2. Bernardeau, F., Colombi, S., Gaztañaga, E., Scoccimarro, R., Phys. Rev. 367, 1 (2002)
3. Cabré, A., E. Gaztañaga, E., ArXiv e-prints 2460, 0807.2460 (2008)
4. Eisenstein, D.J., Zehavi, I., Hogg, D.W., Scoccimarro, R., Blanton, M.R., Nichol, R.C., Scranton, R., Seo, H.-J., Tegmark, M., Zheng, Z., et al., ApJ 633, 560 (2005)
5. Gaztanaga, E., Cabré, A., Hui, L., ArXiv e-prints 807, 0807.3551 (2008)
6. Kaiser, N., MNRAS 227, 1 (1987)
7. Landy, S.D., Szalay, A.S., ApJ 412, 64 (1993)
8. Linder, E.V., Phys. Rev. D 72, 043529 (2005)
9. Matsubara, T., ApJ 535, 1 (2000)
10. Szapudi, I., ApJ 614, 51 (2004)

UKIDSS: Surveying the Sky in the Near-IR

E.A. González-Solares, B.P. Venemans, R.G. McMahon, S.J. Warren, D.J. Mortlock, M. Patel, P.C. Hewett, S. Dye, R.G. Sharp, and the UKIDSS Collaboration

Abstract The UKIRT Infrared Deep Sky Survey (UKIDSS) is observing about 7,000 square degrees of the northern sky in the near-IR. Summed together it is 12 times larger in effective volume than the 2MASS survey. The scientific aims of UKIDSS include the detection of the nearest and faintest substellar objects and brown dwarfs, probe the substellar initial mass function, detect clusters of galaxies at $z \sim 2$ and detect high redshift quasars at $z \sim 6-7$. We give here a short introduction to UKIDSS focusing mainly on the search for high-z quasars.

1 High-z QSOs and the Epoch of Reionization

After the recombination epoch at $z \sim 1,000$ the universe becomes mostly neutral until the first generation of stars and quasars reionize the interstellar galactic medium (IGM) and ended the cosmic dark ages [14]. Cosmological models predict reionization at redshifts between 6 and 20. When and how this reionization occurs and what are the objects responsible for it are fundamental questions for our understanding of the evolution of the Universe.

E.A. González-Solares, B.P. Venemans, R.G. McMahon, and P.C. Hewett
Institute of Astronomy, University of Cambridge, Madingley Road, Cambridge CB3 0HA, UK
e-mail: eglez@ast.cam.ac.uk

S.J. Warren, D.J. Mortlock, and M. Patel
Astrophysics Group, Imperial College London, Blackett Laboratory, Prince Consort Road, London, SW7 2AZ, UK

S. Dye
Cardiff University, School of Physics and Astronomy, Queens Buildings, The Parade, Cardiff CF24 3AA, UK

R.G. Sharp
Anglo-Australian Observatory, P.O. Box 296, Epping, NSW 1710, Australia

J.M. Diego et al. (eds.), *Highlights of Spanish Astrophysics V*, Astrophysics and Space Science Proceedings, DOI 10.1007/978-3-642-11250-8_10,

In the last few years we have seen the first direct observational constrains on the epoch of reionization. The WMAP polarization results indicate a largely ionized IGM by $z \sim 10\pm3$ [11, 15]. The lack of complete Gunn–Peterson absorption in quasars at $z < 6$ (characterized by the suppression of flux at wavelengths shorter than the Ly_α emission line [8]) indicates that the IGM is highly ionized at that epoch [1,5]. However the high-z quasar observations suggest that the Universe could have been mostly neutral as late as $z = 6$–8 [7].

Quasars are indeed useful probes of reionization because they can be detected at very high redshifts and have a strong instrinsic UV radiation that is absorbed at the Lyman lines by neutral hydrogen. The absorption spectra of these quasars reveal the state of the intergalactic medium close to reionization epoch. Observations of high-z quasars are also important to study the formation of supermassive black holes and their host galaxies. The high luminosities and broad line widths of the most distant quasars require black holes masses greater than 10^9 solar masses. Forming such massive black holes within the first billion years of the Universe provide a challenge to models of galaxy formation, black hole formation and black hole growth.

The Sloan Digital Sky Survey (SDSS [19]) has been very successful in detecting high redshift quasars. A total of 27 quasars at $z > 5.7$ have been found in 7,700 deg^2 at magnitudes $z < 20$ (AB), the highest of which is at $z = 6.42$ [6]. The Canada France High-z Quasar Survey (CFHQS) has detected 4 quasars at $z > 6$ in 400 deg^2 at magnitudes $z < 22.5$ (AB) [17].

The detection of deep Gunn–Peterson troughs in the spectra of high redshift quasars indicate an accelerated rate of evolution at $z > 5.7$ consistent with the IGM transition at the end of the overlapping stage of reionization [7]. However these results do not indicate that the IGM has achieved a high level of neutrality at $z \sim 6$. The large dispersion of the IGM along different lines of sight strongly suggests that the reionization is a complex process and not likely a uniform transition over a very narrow redshift range [7].

Increasing the sample of quasars at $z \sim 6$ and reaching the $z \sim 7$ barrier is necessary if we want to gain understanding on the process of reionization. However quasars at $z > 6.5$ are difficult to find in the optical because Ly_α moves out of the z band which is the reddest one available in CCD based surveys and because they are very rare [18]. Near-IR surveys are thus crucial to continue finding higher redshift quasars and will complement constrains from CMB polarization measurements (e.g. Planck) and 21 cm experiments (e.g. LOFAR, SKA).

2 The UKIRT Infrared Deep Sky Survey

UKIDSS [12] is a survey programme combining a set of five surveys which is using the Wide Field Camera (WFCAM [2]) in the 3.8 m United Kingdom Infrared Telescope (UKIRT) at Mauna Kea, Hawaii. WFCAM is composed by four 2048×2048 pixels Rockwell Hawaii-II arrays with a pixel size of 0.4 arcsec. In order to cover a contiguous area of sky four pointings are needed which together cover an area of 0.77 deg^2. See [3] for more information on WFCAM and UKIDSS.

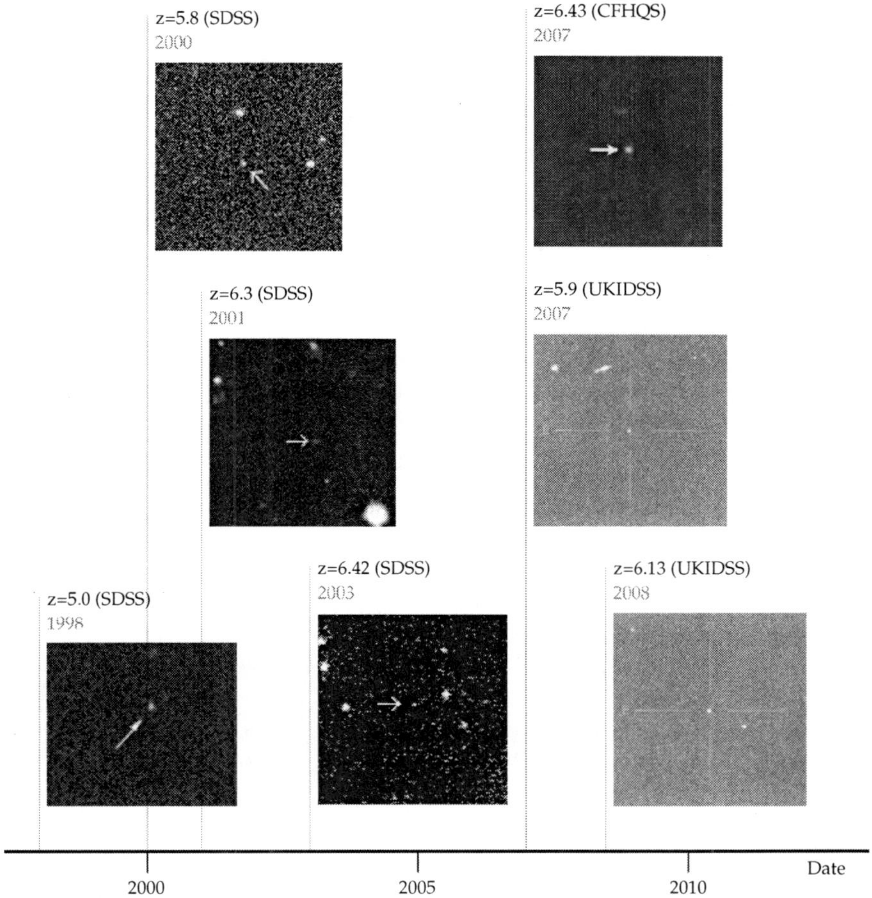

Fig. 1 Highest redshift quasars discovered versus date since the start of the SDSS survey. Also added the first two high redshift quasars detected from UKIDSS

Table 1 UKIDSS Surveys

Survey Name	Area (deg^2)	Bands	Depth[a]
Large Area (LAS)	4,000	YJHK	20.1
Galactic Plane (GPS)	1,800	JHK	20.7
Galactic Clusters (GCS)	1,400	ZYJHK	20.4
Deep Extragalactic (DXS)	35	JK	22.7
Ultra Deep (UDS)	0.77	JHK	24.7

[a] K band, AB, 5σ for a point source in an aperture of 2 arcsec diameter.

About 250 GB of data are obtained per night at UKIRT which are transferred to the Cambridge Astronomy Survey Unit (CASU) for processing. Final data products are photometrically and astrometrically calibrated multi-extension FITS files together with catalogues containing image derived parameters (i.e. positions,

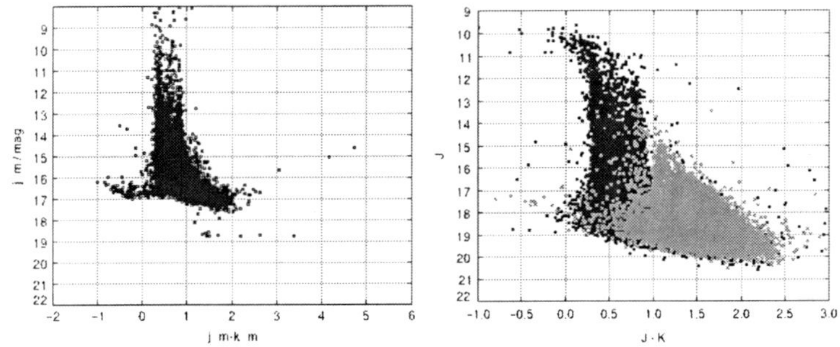

Fig. 2 J vs. $J-K$ (Vega) color-magnitude diagram corresponding to the same 1 deg radius of sky from the 2MASS Point Source Catalogue (*left*; ~5,200 objects) and from the UKIDSS LAS survey (*right*; ~70,000 objects). Light grey points correspond to objects classified as galaxies while black ones are classified as point-like

fluxes measured in different apertures, classification, morphological parameters, etc.) [10]. Images and catalogues are ingested into a queryable relational database at the WFCAM Science Archive (WSA [9]). Data products are made available to all ESO countries and after a 18 month period to the rest of the world. At the moment of writing this contribution DR4 is available to ESO countries and DR2 to the world.

Each UKIDSS survey is using a different set of filters and reaches different depths. The Large Area Survey (LAS), the Galactic Clusters Survey (GCS), and the Galactic Plane Survey (GPS) cover approximately 7000 deg^2 to a depth of $K \sim 20$; the Deep Extragalactic Survey (DXS) covers 35 deg^2 to $K \sim 22.7$, and the Ultra Deep Survey (UDS) covers 0.77 deg^2 to $K \sim 24.7$ (Table 1).

The LAS survey in particular aims to cover 4000 deg^2 within the SDSS footprint with 40 s exposures in Y, J, H and K bands and is the ideal survey to search for high-z quasars. Figure 4 shows the area observed by LAS versus date. It is however worth noting that quasars at $z \sim 6$ are very rare in the sky. For example SDSS detected 4 dropout quasars in 1,550 deg^2 which contained also 15 million objects and 6.5 million cosmic rays in the z band [4]. Moreover, quasars are faint and have low signal-to-noise ratio. Contamination from L, T and mainly M stars is also very important. With about 20,000 objects per square degree in the LAS surveys it is clear that an efficient mechanism of selecting high redshifts quasars is needed.

Candidate high redshift quasars are selected on the basis of their blue $Y-J$ UKIDSS colors and their red $i-Y$ colors (see [16] for detailed information and color cuts). List driven photometry is performed on the SDSS images (coadded when several single epoch images exist) to confirm or remove candidates. Deeper observations in i and/or z are then obtained using the INT Telescope at The Observatorio del Roque de los Muchachos and the NTT telescope at La Silla Paranal Observatory. The first luminous high redshift quasar in UKIDSS was confirmed using FORS2 in the VLT and shows the Ly$_\alpha$ emission line and the continuum break resulting from Lyman forest absorption at a redshift of $z \sim 5.86$ [16]. Following this discovery another $z = 6.13$ quasar has been found in the LAS survey [13].

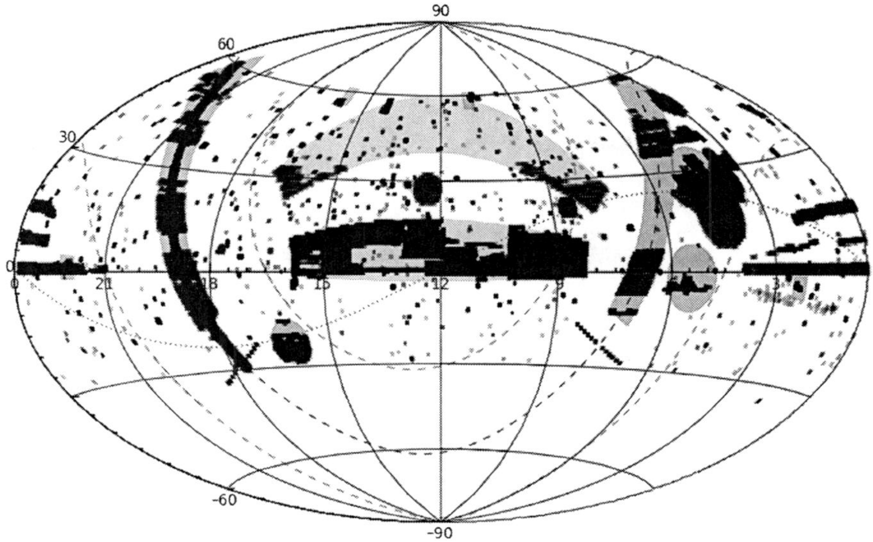

Fig. 3 Areas observed with WFCAM up to October 2008. Light grey areas show the planned UKIDSS survey footprints and the dark areas what it has been actually observed. Dark small areas generally correspond to observations carried out as part of non-survey programmes. A color figure is available from http://casu.ast.cam.ac.uk

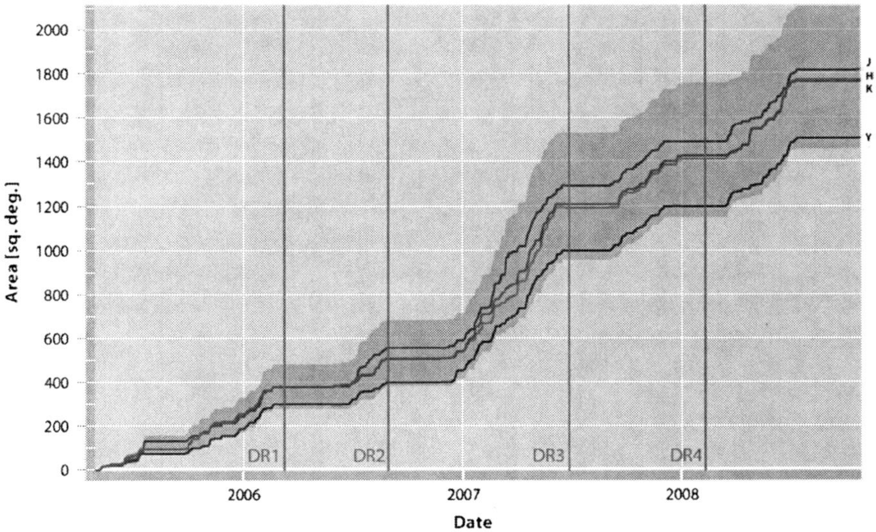

Fig. 4 Area observed by the Large Area Survey in different bands. The lower limit of the *grey shade* indicates the area observed in all bands. Also shown the dates of each of the four UKIDSS data releases to date

Table 2 VISTA Surveys

Survey Name	Area (deg^2)	Bands	Depth[a]
Ultra-VISTA	0.75	YJHK	25.6
VIKING	1,500	ZYJHK	21.2
Magellanic Clouds (VMC)	184	YJK	20.3
Variables Via Lactea (VVV)	520	ZYJHK	18.1
Vista Hemisphere (VHS)	20,000	YJHK	20.0
Deep Extragalactic (VIDEO)	15	ZYJHK	23.5

[a]K band, AB, 5σ for a point source in an aperture of 2 arcsec diameter.

3 The VISTA Hemisphere Survey (VHS)

The Visible and Infrared Survey Telescope for Astronomy (VISTA) is located in Paranal, near the VLT site, and is a wide-field survey telescope with a primary mirror of 4.1 m. Designed for both optical and infrared observations it is now equipped with a near-IR camera, VIRCAM composed by 16 detectors $2{,}048 \times 2{,}048$ pixels in size. In this case, 6 pointings are needed to cover a contiguous area of sky, or tile, of about 1.5 deg^2. The first years of observations are dedicated mainly ($\sim$75% of total tile) to large scale public surveys.

The VISTA Hemisphere Survey (VHS; P.I. R. McMahon) will result, when combined with other large VISTA surveys, in coverage in the whole southern celestial hemisphere ($\sim$20,000 deg^2) to a depth $\sim$4 mag fainter than 2MASS in at least two wavebands, J and K. In the South Galactic Cap, $\sim$5,000 deg^2 will be imaged deeper, including H band, and will have supplemental deep multi-band grizY imaging data provided by the Dark Energy Survey (DES). The remainder of the high galactic latitude sky will be imaged in YJHK combined with the ugriz wavebands from VST ATLAS survey (P.I. T. Shanks). The survey when completed will have observed 100 times to volume observed by 2MASS and 10 times the volume observed by UKIDSS. One of the main goals is then to study the physics of the epoch of reionization and the discovery of the firsts quasars at $z > 7$. Other scientific aims include the detection of the nearest and lowest mass stars and the study of the evolution of the large scale structure in the Universe.

References

1. Becker, R.H., Fan, X., White, R.L., Strauss, M.A., Narayanan, V.K., Lupton, R.H., et al., AJ 122, 2850 (2001)
2. Casali, M., Adamson, A., Alves de Oliveira, C., Almaini, O., Burch, K., Chuter, T., et al., A&A 467, 777 (2007)
3. Dye, S., Warren, S.J., Hambly, N.C., Cross, N.J.G., Hodgkin, S.T., Irwin, M.J., et al., MNRAS 372, 1227 (2006)
4. Fan, X., Narayanan, V.K., Lupton, R.H., Strauss, M.A., Knapp, G.R., Becker, R.H., et al., AJ 122, 2833 (2001)
5. Fan, X., Narayanan, V.K., Strauss, M.A., White, R.L., Becker, R.H., Pentericci, L., Rix, H.W., AJ 123, 1247 (2002)

6. Fan, X., Strauss, M.A., Schneider, D.P., Becker, R.H., White, R.L., Haiman, Z., et al., AJ 125, 1649 (2003)
7. Fan, X., Strauss, M.A., Becker, R.H., White, R.L., Gunn, J.E., Knapp, G.R., et al., AJ 132, 117 (2006)
8. Gunn, J.E., Peterson, B.A., ApJ 142, 1633 (1965)
9. Hambly, N.C., Collins, R.S., Cross, N.J.G., Mann, R.G., Read, M.A., Sutorius, E.T.W., et al., MNRAS 384, 637 (2008)
10. Irwin, M.J., MNRAS 214, 575 (1985)
11. Kogut, A., Spergel, D.N., Barnes, C., Bennett, C.L., Halpern, M., Hinshaw, G., et al., ApJS 148, 161 (2003)
12. Lawrence, A., Warren, S.J., Almaini, O., Edge, A.C., Hambly, N.C., Jameson, R.F., et al., MNRAS 379, 1599 (2007)
13. Mortlock, D.J., et al., MNRAS submitted (2008)
14. Rees, M.J., in *After the Dark Ages: When Galaxies were Young (the Universe at $2<z<5$)*, American Institute of Physics Conference Series, Holt, S., & Smith, E., eds., Vol. 470, p. 13 (1999)
15. Spergel, D.N., Bean, R., Doré, O., Nolta, M.R., Bennett, C.L., Dunkley, J., et al., ApJS 170, 377 (2007)
16. Venemans, B.P., McMahon, R.G., Warren, S.J., González-Solares, E.A., Hewett, P.C., Mortlock D.J., et al., MNRAS 376, L76 (2007)
17. Willott, C.J., Delorme, P., Omont, A., Bergeron, J., Delfosse, X., Forveille, T., et al., AJ 134, 2435 (2007)
18. Wyithe, S., Warszawski, L., Geil, P.M., Oh, S.P., ArXiv e-prints (2008)
19. York, D.G., Adelman, J., Anderson Jr., J.E., Anderson, S.F., Annis, J., Bahcall, N.A., et al., AJ 120, 1579 (2000)

Galaxies Hosting AGN Activity and Their Environments

Isabel Márquez and Josefa Masegosa

Abstract Nuclear activity in galaxies (AGN) has recently started attracting a broader attention, in one hand, because of the relationship found between the mass of the central spheroidal regions of galaxies and the black holes they host. On the other hand, the output energy provided by AGN (feedback) seems to be crucial for current cosmological simulations to reproduce the observations. One of the main issues concerning AGN studies is to understand the triggering mechanisms for the onset of their non-thermal emission. Both the origin of the material accreted onto the black hole and the physical mechanisms for the loose of angular momentum required for this funneling to be effective, have to be elucidated. Among the many aspects of the investigation that are still a matter of debate, we review those related to the role played by gravitational interactions and the relevance of the host galaxy. The different relationships between AGN activity, the morphological type of the host galaxy and its environment at different scales are discussed, to understand whether the AGN activity is more related to interacting effects or otherwise it can be due to the secular evolution in the hosting galaxies.

1 Introduction

The properties of galaxies have long been known to depend on the environment in which they are located, with ellipticals mostly residing in rich clusters, and spirals mainly found in their outskirts (i.e. [4, 10]). Galaxy properties can change with time both by secular processes and by the effects of interactions with neighbors or with the surroundings. Secular evolution has been suggested as a possible mechanism to make galaxies evolve from later to earlier types; this could be the case specially within the group of spiral galaxies. The response of the disk to an initial small external perturbation would produce gravitational instabilities in the disk

I. Márquez and J. Masegosa
IAA (CSIC), Apdo 3004, 18080 Granada, Spain
e-mail: isabel@iaa.es, pepa@iaa.es

J.M. Diego et al. (eds.), *Highlights of Spanish Astrophysics V*, Astrophysics and Space Science Proceedings, DOI 10.1007/978-3-642-11250-8_11,

giving rise to the formation of a bar with the subsequent transfer of material to the center; eventually, the bar itself will be destroyed due to the huge accumulation of gas. Strong gravitational interaction can produce similar effects, but at more violent levels and, depending on the relative masses of the systems involved, giving rise to the formation of elliptical galaxies [5, 59].

The importance of the presence of a bar in a spiral galaxy has been established already in the pioneering numerical simulations of barred galaxies [26, 70]. A bar is described as a component that rigidly rotates over a differentially rotating disk, what gives rise to different resonances usually associated to ring features. Bars are easily formed in minor mergers [5, 23], and some of their effects on the host galaxy are to dynamically heat the disk and to produce net inflows to the central regions. Such inflows directly explain the observed flatness in metallicity gradient in barred galaxies. The infalling processes also relate to the various star forming (SF) features observed in barred galaxies, since they are expected to be different for old/young and strong/faint bars. The final stages of the infalling process, those of material accumulated onto the central region and bar destruction, have been explored by means of numerical simulations: starting from a late type spiral that respond to a small perturbation of the disk, a bar is generated, which provokes the transport of some disk material to the center; this enhances the bulge component and destroys the bar itself, so that the galaxy finally ends up as an earlier unbarred galaxy.

Therefore, special care is needed when trying to disentangle internal from external drivers. The use of control samples matching both in morphology and environmental status will be crucial at this respect.

2 AGN in Galaxies

tHE processes sustaining AGN activity relate to angular momentum looses that have to operate in order to make the material to be funneled to the center. The large scale processes driving material to the central regions have been explored as eventually related to those required much closer to the nuclear black hole. In this contribution, we analyze how AGN are related to nature, i.e. the properties of the galaxy hosting an AGN, and how AGN relate to nurture, i.e. with externally triggered modifications. All in all, we will keep in mind that the power of the AGN is another parameter to consider, as clearly illustrates the case of the strongest AGN (quasars) corresponding to the most massive galaxies or the strongest interactions.

3 The Role of the Local/Large Scale Environment

The studies of local (few kpc) and large (some hundred Mpc) environments of AGN have pointed to differing results concerning whether they are similar to those of inactive galaxies. Several studies have concluded that no difference is found [68, 69, 76].

Other investigations suggest that companions are more frequent for Seyfert galaxies [8, 9] with clear differences between Seyfert 1 and Seyfert 2 [11, 34–36] and even depending on the power of the AGN [20, 21]. Their main limitations relate to the sample selection, completeness and matching procedure for control galaxies. The wavelength used for selecting the AGN samples could also play a role, as optically selected AGN could be biased against obscured objects.

The recent dramatic increase in the number of AGN (from several hundreds to several thousands) has resulted in the possibility of approaching this study with statistical significance. The main result with the first data release is that of a constant fraction of AGN with projected galaxy density [60]. In contrast, for higher densities the fraction of passive galaxies is enhanced and that for SF galaxies decreases.

Reference [31] found that type 2 AGN hosts are almost exclusively massive galaxies, with the AGN fraction strongly declining for $M_{\mathrm{star}} < 10^{10}\ \mathrm{M}_{\odot}$; they have similar sizes and stellar masses than normal early-types but show slightly different stellar ages than the parent general population. They also found that high power AGN (those with $\log(L[\mathrm{OIII}]) > 7.0$) show somewhat younger stellar populations. Reference [32] obtain a larger fraction of high power AGN in lower density environments, whereas at a fixed AGN power, the hosts are similar for any density.

Reference [73] conclude that Sy1 and Sy2 galaxies have similar large scale environments with a higher percentage of Sy2 appearing in close pairs. They suggest that some of the conflicting previous results could be a consequence of not taking into account the necessity to distinguish between close and large scale environments when studying the eventual effects of the gravitational interactions as related to AGN activity. A result that seems to be confirmed in more recent works [12, 37, 74].

4 AGN and Bars

Previous to studies of huge samples in the last few years, the connection between the presence of AGN activity and the morphology of the host galaxy had been stressed from earlier studies [22]. AGN hosts are more frequently found in early types, with a peak of the morphology distribution in Sb spirals [22, 25, 32, 33, 44, 46, 61, 77]. Already in these works the need of explaining the internal mechanisms to be related to the onset of nuclear activity in galaxies was one of their first aims, and the presence of an asymmetric component of the gravitational potential was invoked as a main driver. Whereas in the case of interacting galaxies the departure from the symmetry is immediately provided by the tidal forces, isolated galaxies deserve closer inspection, since the asymmetry should come from the host itself. But in spirals the presence of such internal asymmetric component is very frequent, since at least two thirds of spiral galaxies are barred [14, 19, 56, 62].

Still the question was whether there was a clearcut relation between the presence of a bar and the AGN activity. Whereas a consensus has not been reached at this respect (see [33, 38, 40]), most works conclude that there is not an excess of bars among Seyfert galaxies [25,56,61,62]. Since bars evolve with time, some complications are expected in such a simple expectation, what could be related to the result of a higher percentage of outer rings in Seyfert 1 galaxies [27]. The main concerns of all these works are related to the sample numbers, the sample selection procedures, and the way the control samples are defined. But in addition to all these eventual biases, the mechanism that is expected to drive the feeding material to the nuclear source has to operate down to the scales close enough to the nucleus. Large scale bars seem to be limited to produce such transport only till the region of the innermost resonance, at scales of about 1 kpc, where the material is trapped and no more inflow occurs. Several mechanisms have been proposed that could help in getting rid of the angular momentum at this point and drive the material closer to the center [29,43]; one of the first one proposed was that of nuclear bars, nested with respect to the large scale, primary bar [16,30,63,65,72,79]. The results provided by the observations nevertheless agree that no more nuclear bars are found in Seyfert galaxies [47, 48, 67], but on the contrary there seems to exist an excess of nuclear spirals [53, 54, 71] or nuclear disks [66] that can be stellar and decoupled from the main galactic disk [13]. Only a slight excess (at 2σ level) is found: the central region of galaxies hosting HII, Seyfert 2, Seyfert 1 and LINER nuclei are progressively less asymmetric [28].

4.1 Isolated Galaxies and the DEGAS Project

In an attempt to clarify the role played by the internal structure of the host galaxies, we started a project devoted to the characterization of ISOLATED Seyfert galaxies and its comparison with a matched control sample of ISOLATED spirals. The project DEGAS (Dynamics and nuclear Engine of Galaxies of Spiral type) was aimed at constructing a sample of nearby Seyfert galaxies from the Véron-Cetty & Véron catalogue [75], with intermediate inclinations, with no reported belonging to any group or pair, with no companion at a projected distance of 600 kpc and with redshift difference smaller than 500 km/s, and with no projected companion in the DSS plates. The control sample was selected from the RC3 catalogue, imposing the same conditions for the absence of companions, and matching in redshift distribution, inclination and morphological type (including the percentage of barred galaxies). The final sample amounted to 18 (15) Seyfert (control) galaxies.

The main results from the analysis of their NIR J and Ks images were that Seyfert and non-Seyfert hosts shared similar bulges and disk properties (sizes, luminosities and surface brightnesses), with primary bars also equivalent in both samples; secondary bars were found both in Seyferts (9 out of 12) and control (6 out of 10) galaxies. The gas kinematics were characterized through long slit spectroscopic data along several position angles; the resulting kinematical properties of Seyfert and control galaxies appeared to be indistinguishable from those

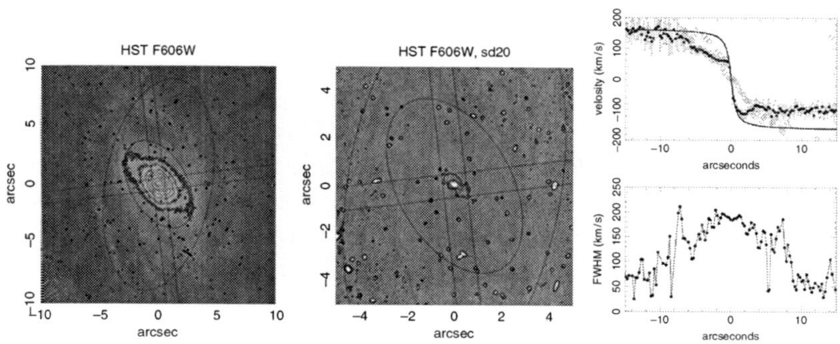

Fig. 1 *Left*: IC184 image (HST/F606W) band with the two slits superimposed, the ellipses correspond (PA and ellipticities) to the two bars detected in the NIR. *Middle*: Sharp divided image of the center. *Top Right*: Velocity curve of the gas (*open, green circles*) and stars (*black circles*) along $PA = 7$. *Bottom Right*: FWHM of the stellar component along $PA = 7$. Taken from [50]

of early spiral types: same rotation curve shape, same position in the Tully–Fisher diagram, same disk metallicities, same kind of kinematical peculiarities in the central regions [49, 51].

A subsample was selected to study their morphological properties, stellar and gaseous kinematics. Optical and/or NIR images from HST were used to characterize the main properties of the host. Additional long slit spectra along several position angles in the region of CaT were used to obtain stellar rotation curves, velocity dispersions and a stellar population tracer through the equivalent width of the CaT absorption lines, EW(CaT). Such study allowed us to detect the presence of stellar velocity dispersion drops in the central 1–3 arcsec in 5 galaxies (see an example in Fig. 1), spatially coincident with an increase in EW(CaT), that could hint the presence of young stars (red supergiants). Nine other galaxies in our sample had previously found to show such a drop. The analysis of the HST imaging of the total 14 (9+5) galaxies resulted in most of them hosting such a nuclear disk-like structure, spatially coincident with the region where the velocity dispersion drop occurs [50]. This result was interpreted in terms of models that predict the formation of velocity dispersion drops once the gas coming from the outer, cooler disk, is driven to the center by the large-scale bar effects, giving rise to a decoupled nuclear disk [80]. The gas velocity dispersion is hence smaller, so it is that of the stars formed from it. An update of this modeling, provides time scales (less than 500 Myr for the formation of the drop, and lifetime depending on the availability of fuel)[78]. But some controversy still remains on whether the stellar populations in drops are younger or not [42,64]. More nuclear rings and a higher percentage of Seyfert galaxies have been found in a morphological approach to galaxies with drops [7]. But the kinematical information is an unavoidable requirement to understand the reported connections (see for instance [17]).

5 A General Picture for Low-Luminosity AGN

A relatively recent consensus has been achieved on the presence of a black hole (BH) in the center of any massive galaxy [15, 45, 57], irrespective of whether it hosts an AGN or not. In addition, the properties of such BHs seem to be shared by both active and non active galaxies. Both types show the same relationship between the mass of the BH and the mass of the large scale spheroid hosting it [58]. From the preceding discussion the properties of the host galaxies seem to be equivalent, both morphology and kinematics, at least at scales of the order of the disk, bulge and/or bar components. The differences, if any, have hence to occur at scales much closer to the center, not still resolved by present day observations of the analyzed samples. If even at much smaller scales no differences are found, an alternative explanation would be that the AGN activity can be switched on and off (as it is also suggested to explain recurrent radio activity [39]). We note that this may be the case explored by numerical simulations that reproduce the evolution of galaxies, moving along and across the Hubble diagram, with any galaxy being able to become barred, or active, or both, and spend some time as an early type or late type [6].

A number of complications appear when trying to analyze in depth the different mechanisms related to the fueling of low-luminosity AGN. Four reasons are proposed [52] to explain why surveys have been unsuccessful up to now in resolving this question: (a) the current classifications for fueling mechanisms are too broad, and additional refining is required, in particular for describing bars, since strong or faint bars are expected to produce different effects, (b) there are correlations between the fueling mechanism and the fueling rate, that are easier to identify for higher accretion rates (related to mergers), but deserves much closer inspection of the central parsec region at the lowest accretion rates (where other processes like dynamical friction on molecular clouds, stellar disruptions, many forms of turbulence, mass loss, etc, have an increasing relative importance), (c) multiple fueling mechanisms may be operating, as it is the case for bars and interactions (but even when only isolated galaxies are considered, as it was the case for the DEGAS project, the results are not conclusive), (d) the two main time scales operating are the AGN lifetime and the fueling time, so the time dependence is important and has to be taken into account. The analysis therefore requires a broader description of the physical situation of the central regions, and the dynamical information is crucial at this respect. The NUGA (Nuclei of Galaxies) project, based on high spatial resolution CO data, derive an scenario for self-regulated activity in low-luminosity AGN [18].

6 Final Considerations

As cited in Ho's review, "BH growth, its observational manifestation as nuclear activity, and the consequences of feedback from AGN, are now widely accepted as unavoidable pieces of the overall puzzle of cosmological structure formation" [24]. The detailed mechanisms connecting them are nevertheless not completely understood. Both the detailed study of nearby galaxies and the statistical approach of massive surveys will be complementary in the study of the relationship between the structure of the galaxy, the power of the AGN it hosts and the environment. On one hand, detailed, high resolution studies of individual galaxies providing dynamical clues for the different phases, are still needed. For nearby galaxies, projects like NUGA are expected to produce fruitful results in the near future. High resolution imaging of a large number of distant galaxies has recently started the exploration of how frequent single or even double bars are at $z = 0.1$ (see [41] within the GOODS survey), what opens the possibility of studying evolutionary effects. On the other hand, precious informations remain to be extracted from existing massive surveys like SDSS, in terms of a much more detailed characterization of the morphological types, but also on the presence and strengths of bars (see [2, 3] and EFIGI project, [1]) and the characterization of the interaction state for AGN hosts and for comparable control samples. Finally, focused numerical simulations with all the required ingredients, as those provided by Wozniak et al. for reproducing velocity dispersion drops, both for small (host) to large (environment) scales will help to understand the physical processes that give rise to the presence and onset of AGN activity in galaxies. All these paths have been somewhat explored along this meeting.

Acknowledgements IM gratefully acknowledges the kind invitation from the SOC of this meeting to make this review. Financial support also provided by the Spanish Ministry of Science and Education (AYA2006-01325 and AYA2007-62190) and the Junta de Andalucía (TIC-114).

References

1. Baillard, A., et al., in ASP Conf. Series, Vol. 35, p. 236 (2006)
2. Ball, L., et al., MNRAS 348, 1038 (2004)
3. Ball, L., et al., MNRAS, 373, 845 (2006)
4. Balogh, M., et al., ApJ 504, 75 (1998)
5. Barnes, J.E., Henquist, L.E., ApJ 471, 115 (1996)
6. Bournaud, F., Combes, F., A&A 392, 83 (2002)
7. Comeron, S., et al., A&A 485, 695 (2008)
8. Dahari, O., AJ 90, 1772 (1985)
9. Dahari, O., ApJS 57, 643 (1985)
10. Dressler, A., ApJ 236, 351 (1980)
11. Dultzin-Hacyan, D., et al., ApJ 513, 111 (1999)
12. Dultzin, D., et al., in preparation (2009)
13. Emsellem, E., et al., A&A 368, 52 (2001)
14. Eskridge, P.B., Frogel, J.A., AAS 193, 812 (1998)
15. Ferrarese, L., Merrit, D., ApJ 539, L9 (2000)
16. Friedli, D., et al., A&AS 118, 461 (1996)
17. Ganda, K., et al., MNRAS 367, 46 (2006)
18. García-Burillo, S., et al., A&A 441, 1001 (2005)
19. Grosbol, P., et al., A&A 423, 849 (2004)
20. Hao, H., et al., AJ 129, 1783 (2005)
21. Hao, H., et al., AJ 129, 1795 (2005)
22. Heckman, T.M., A&A 87, 152 (1980)
23. Hernquist, L., Mihos, J.C., ApJ 448, 41 (1995)

24. Ho, L.C., ARA&A 46, 475 (2008)
25. Ho, L.C., Filippenko, A.V., Sargent, W.L.W., ApJ 487, 568 (1997)
26. Hohl, F., ApJ 168, 343 (1971)
27. Hunt, L.K., Malkan, M.A., ApJ 516, 660 (1999)
28. Hunt, L.K., Malkan, M.A., ApJ 616, 707 (2004)
29. Jogee, S., in *Physics of AGN at All Scales*, Lecture Notes in Physics, Vol. 693, p. 143 (2004)
30. Jungwiert, B., et al., A&AS 125, 479 (1997)
31. Kauffmann, G., et al., MNRAS 346, 1055 (2003)
32. Kauffmann, G., et al., MNRAS 353, 713 (2004)
33. Knapen, J.H., Shlosman, I., Peletier, R.F., ApJ 529, 93 (2000)
34. Koulouridis, E., et al., ApJ 639, 37 (2006)
35. Koulouridis, E., et al., ApJ 651, 93 (2006)
36. Krongold, Y., et al., Rev. Mex. A&A 39, 225 (2003)
37. Kuo, C.-Y., et al., ApJ 679, 1047 (2008)
38. Laine, S., et al., ApJ 567, 97 (2002)
39. Lara, L., et al., NewAR 46, 89 (2002)
40. Laurikainen, E., et al., ApJ 607, 103 (2004)
41. Lisker, T., et al., MNRAS 370, 477 (2006)
42. Lyubenova, V.N., et al., A&A 485, 425 (2008)
43. Maciejewski, W., in *The Central kpc of SB and AGN'*, ASP Conf. Ser., Vol. 249, p. 153 (2001)
44. Maia, M.A.G., et al., AJ 126, 1750 (2003)
45. Magorrian, J., et al., AJ 115, 2285 (1998)
46. Márquez, I., Moles, M., AJ 108, 90 (1994)
47. Márquez, I., et al., A&AS 140, 1 (1999)
48. Márquez, I., Durret, F., Masegosa, J., et al., A&A 360, 431 (2000)
49. Márquez, I., Masegosa, J., Moles, M., et al., A&A 393, 389 (2002)
50. Márquez, I., Masegosa, J., Durret, F., et al., A&A 409, 459 (2003)
51. Márquez, I., et al., A&A 416, 475 (2004)
52. Martini, P., in *IAU Symp.*, Cambridge, Cambridge University Press, Vol. 222, p. 235 (2004)
53. Martini, P., Pogge, R.W., AJ 118, 2646 (1999)
54. Martini, P., et al., ApJ 562, 139 (2001)
55. Martini, P., et al., ApJ 644, 116 (2006)
56. McLeod, B.A., Rieke, M.J., ApJ 454, 77 (1995)
57. McLure, R.J., Dunlop, J.S., MNRAS 331, 795 (2002)
58. McLure, R.J., Dunlop, J.S., MNRAS 352, 1390 (2004)
59. Mihos, J.C., Richstone, D.O., Bothun, G.D., ApJ 400, 153 (1992)
60. Miller, C.J., et al., ApJ 597, 142 (2003)
61. Moles, M., Márquez, I., Pérez, E., ApJ 438, 604 (1995)
62. Mulchaey, J., Regan, M., ApJL 482, 135 (1997)
63. Norman, C., Silk, J., ApJ 266, 502 (1983)
64. Peletier, R.F., et al., MNRAS 379, 445 (2007)
65. Pfenniger, D., ASPC 197, 413 (2001)
66. Pogge, R.W., Martini, P., ApJ 569, 624 (2002)
67. Regan, M.W., Mulchaey, J.S., AJ 117, 2676 (1999)
68. Schmitt, H.R., AJ 122, 2243 (2001)
69. Schmitt, H.R., et al., ApJ 555, 663 (2001)
70. Sellwood, j.A., A&A 99, 362 (1981)
71. Shlosman, I., Heller, C., ApJ 565, 921 (2002)
72. Shlosman, I., et al., Nature 338, 45 (1989)
73. Sorrentino, G., et al., A&A 451, 809 (2006)
74. Tang, Y.-W., et al., ApJ 679, 1094 (2008)
75. Véron-Cetty, M.-P., Véron, P., *A Catalogue of Quasars and active nuclei*, ESO Scientific Report, 6th edition (1993)
76. Virani, S.N., et al., AJ 120, 1739 (2000)
77. Wake, D.A., et al., ApJ 610, 85 (2004)
78. Wozniak, H., Champavert, N., MNRAS 369, 853 (2006)
79. Wozniak, H., et al., A&AS 111, 115 (1995)
80. Wozniak, H., et al., A&A 409, 469 (2003)

The QUIJOTE CMB Experiment

J.A. Rubiño-Martín, R. Rebolo, M. Tucci, R. Génova-Santos, S.R. Hildebrandt, R. Hoyland, J.M. Herreros, F. Gómez-Reñasco, C. López Caraballo, E. Martínez-González, P. Vielva, D. Herranz, F.J. Casas, E. Artal, B. Aja, L. dela Fuente, J.L. Cano, E. Villa, A. Mediavilla, J.P. Pascual, L. Piccirillo, B. Maffei, G. Pisano, R.A. Watson, R. Davis, R. Davies, R. Battye, R. Saunders, K. Grainge, P. Scott, M. Hobson, A. Lasenby, G. Murga, C. Gómez, A. Gómez, J. Ariño, R. Sanquirce, J. Pan, A. Vizcargüenaga, and B. Etxeita

Abstract We present the current status of the QUIJOTE (Q-U-I JOint TEnerife) CMB Experiment, a new instrument which will start operations early in 2009 at Teide Observatory with the aim of characterizing the polarization of the CMB and other processes of galactic and extragalactic emission in the frequency range 10–30 GHz and at large angular scales. QUIJOTE will be a valuable complement at low frequencies for the PLANCK mission, and will have the required sensitivity to detect a primordial gravitational-wave component if the tensor-to-scalar ratio is larger than $r = 0.05$.

J.A. Rubiño-Martín, R. Rebolo, M. Tucci, R. Génova-Santos, S.R. Hildebrandt, R. Hoyland, J.M. Herreros, F. Gómez-Reñasco, and C. López Caraballo
Instituto de Astrofisica de Canarias (IAC), C/Via Lactea, s/n, E-38200, La Laguna, Tenerife, Spain

E. Martínez-González, P. Vielva, D. Herranz, and F.J. Casas
Instituto de Fisica de Cantabria (IFCA), CSIC-Univ. de Cantabria, Avda. los Castros, s/n, E-39005 Santander, Spain

E. Artal, B. Aja, L. dela Fuente, J.L. Cano, E. Villa, A. Mediavilla, and J.P. Pascual
Departamento de Ingenieria de COMunicaciones (DICOM), Laboratorios de I+D de Telecomunicaciones, Plaza de la Ciencia s/n, E-39005 Santander, Spain

L. Piccirillo, B. Maffei, G. Pisano, R.A. Watson, R. Davis, R. Davies, and R. Battye
Jodrell Bank Centre for Astrophysics, School of Physics and Astronomy, University of Manchester, Oxford Road, Manchester M13 9PL, UK

R. Saunders, K. Grainge, P. Scott, M. Hobson, and A. Lasenby
Astrophysics Group, Cavendish Laboratory, University of Cambridge, Madingley Road, Cambridge CB3 0HE, UK

G. Murga, C. Gómez, A. Gómez, J. Ariño, R. Sanquirce, J. Pan, A. Vizcargüenaga, and B. Etxeita
IDOM, Avda. Lehendakari Aguirre, 3, E-48014 Bilbao, Spain
e-mail: jalberto@iac.es

J.M. Diego et al. (eds.), *Highlights of Spanish Astrophysics V*, Astrophysics and Space Science Proceedings, DOI 10.1007/978-3-642-11250-8_12,

1 Introduction

The study of the Cosmic Microwave Background (CMB) anisotropies is one of the main pillars of the Big Bang model. With the latest results from WMAP satellite [5], and the information provided by ground-based experiments such as VSA [3], ACBAR [13] or CBI [12], it has been possible to determine cosmological parameters with accuracies better than 5% (see e.g. [4]).

However, the CMB contains far more information encoded in its polarization signal. Since the first detection of polarization by the DASI experiment [8], other experiments have started to measure the angular power spectrum of the polarization. Although those measurements are still having a relatively poor signal-to-noise, they show excellent agreement with the predictions of the standard ΛCDM model.

The standard theory predicts that the CMB is linearly polarized, the physical mechanism responsible for its polarization being Thomson scattering during the recombination or reionization epochs. Generally speaking, the polarization tensor can be decomposed in terms of a E-field (gradient) and a B-field (rotational) components [6, 17]. Due to parity conservation, this implies that we have three angular power spectra to describe the polarization field: the TE (cross-correlation of temperature and E mode), the EE and BB power spectra. All the other combinations (TB and EB) should be zero.

If the fluctuations in CMB intensity are seeded by scalar perturbations (i.e. fluctuations in the density alone), one would only expect primordial E modes in the CMB polarization. However, vector and tensor perturbations, like those due to gravitational waves in the primordial Universe (e.g. [11]), are mechanisms that could generate primordial B-modes in the polarization on large angular scales. Therefore if we can measure these modes we may have a unique way to carry out a detailed study of the inflationary epoch. In particular, the energy scale V at which inflation occurred can be expressed in terms of r, the ratio of tensor to scalar contributions to the power spectrum, as [9]

$$r = 0.001 \left(\frac{V}{10^{16}\,\mathrm{GeV}} \right)^4 . \tag{1}$$

The current upper limit of $r \lesssim 0.3$ from WMAP data [7] translates into $V \lesssim 4 \times 10^{16}$ GeV.

Because of the importance of detecting primordial gravitational waves [2, 10], there is a huge interest to develop ground-based experiments to measure (or constrain) the amplitude of B-modes power spectrum of the CMB polarization. Here we present one of these efforts.

The QUIJOTE (Q-U-I JOint TEnerife) CMB Experiment is a scientific collaboration between the Instituto de Astrofísica de Canarias, the Instituto de Física de Cantabria, the IDOM company, and the universities of Cantabria, Manchester and Cambridge, with the aim of characterizing the polarization of the CMB, and other galactic and extragalactic physical processes in the frequency range 10–30 GHz and at angular scales larger than 1°. Updated information of the project can be found at the following web page: http://www.iac.es/project/cmb/quijote.

2 Science Goals

The QUIJOTE-CMB experiment has two primary scientific goals:

- To detect the imprint of gravitational B-modes if they have an amplitude $r \geq 0.05$.
- To provide essential information of the polarization of the synchrotron and the anomalous microwave emissions from our Galaxy at low frequencies (10–20 GHz).

In order to achieve these scientific objectives, we need to cover a total sky area of the order of 3,000–10,000 square degrees, and to reach sensitivities of $\sim$3–4 μK per one degree beam after one year of operation with the low frequency instrument (11–19 GHz), and $\lesssim 1\,\mu$K per beam with the second instrument at 30 GHz. Although the final observing strategy is still under discussion, a possible solution is presented in Fig. 1, where we show the scientific goal for the angular power spectrum of the E and B modes after 3 years of operation, assuming a sky coverage of 5,000 square degrees. In this particular case, the final noise figure for the 30 GHz map is $\sim$0.5 μK/beam.

According to those nominal sensitivities, QUIJOTE will provide one of the most sensitive 11–19 GHz measurements of the polarization of the synchrotron and anomalous emissions on degree angular scales. This information is extremely

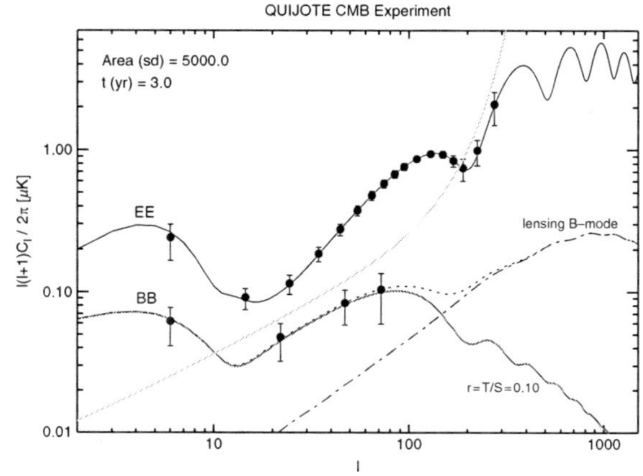

Fig. 1 QUIJOTE scientific goal for the angular power spectrum of the CMB E and B mode signals. It is shown the case for 3 years operation time, and a sky coverage of $\sim$5,000 square degrees. The red line corresponds to the primordial B-mode contribution in the case of $r = 0.1$. Yellow line indicates the associated QUIJOTE noise power spectrum. Dots with error bars correspond to averaged measurements over a certain multipole band

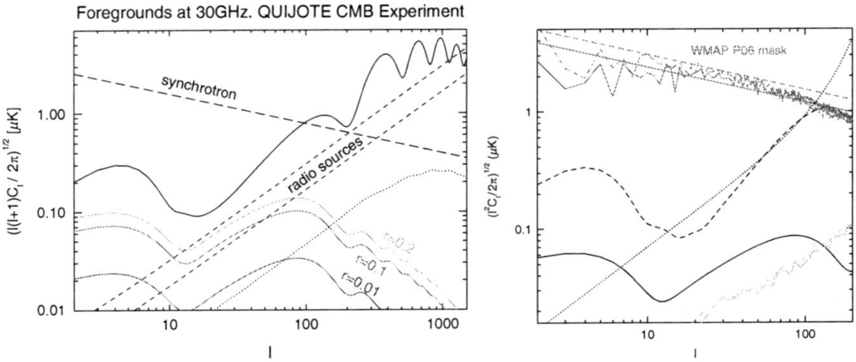

Fig. 2 *Left*: Expected foreground contamination in the 30 GHz QUIJOTE frequency band. It is shown the contribution of polarized synchrotron emission and radio-sources for the case of subtracting sources down to 1 Jy in total intensity (*upper dashed-line* for radio-sources) and 300 mJy (lower dashed-line). *Right*: Correction of the synchrotron emission at 30 GHz. The magenta line shows the predicted synchrotron level from a simulation of the expected signal. Assuming a pure power-law dependence for the synchrotron emission in the QUIJOTE frequency range, the blue line shows the synchrotron residual after correction using the low frequency channels (11–19 GHz)

important given that B-modes are known to be sub-dominant in amplitude as compared to the Galactic emission (see e.g. [14]). This is illustrated in the left panel of Fig. 2, where we present the amplitude of the expected synchrotron and radio-source contribution at 30 GHz, computed according to the models described in [14].

The QUIJOTE low frequency maps will complement the measurements of the Planck satellite[1], helping in the characterization of the Galactic emission. In particular, QUIJOTE will provide a key contribution to assess the level of a possible contribution of polarized microwave anomalous emission [1, 16].

Using the low frequency maps, we plan to correct the high frequency QUIJOTE channel (30 GHz) to search for primordial B-modes. To illustrate this, Fig. 2 shows the residual synchrotron contribution after a pixel-by-pixel correction of the high frequency map assuming a pure power-law behavior for the synchrotron emission law. The issue of radio-sources is discussed below, in the context of the source subtraction facility.

3 Experimental Details

3.1 Project Baseline

QUIJOTE-CMB will observe at five frequencies, namely 11, 13, 17, 19 and 30 GHz, with an angular resolution of $\sim 1°$. It will operate from the Observatorio del Teide (2,400 m) in Tenerife (Spain), which has been shown to be an excellent place for

[1] PLANCK: http://www.rssd.esa.int/index.php?project=Planck.

Table 1 QUIJOTE-CMB Experiment. Instruments characteristics

	Instrument I					Instrument II
Frequency (GHz)	11.0	13.0	17.0	19.0	30.0	30.0
Bandwidth (GHz)	2.0	2.0	2.0	2.0	8.0	8.0
Number of channels	8	8	8	8	2	38
Beam FWHM (deg)	0.92	0.92	0.60	0.60	0.37	0.37
T_{sys} (K)	20.0	20.0	20.0	20.0	30.0	20.0
Sensitivity (Jy $s^{1/2}$)	0.24	0.34	0.24	0.30	0.43	0.07
Sens per beam (mK $s^{1/2}$)	0.22	0.22	0.22	0.22	0.34	0.05

CMB observations (same site as Tenerife, COSMOSOMAS, VSA and JBO-IAC Interferometer experiments).

The project has two phases. *Phase I*, which is completely funded, consists in the construction of a first telescope and two instruments which can be exchanged in the focal plane. The first instrument will be a multichannel instrument providing the frequency coverage between 11 and 19 GHz, plus a single pixel at 30 GHz, and it is expected to start observations at the beginning of 2009. The second instrument will consist of 19 polarimeters working at 30 GHz, and it is expected to start operations at the end of 2009. Table 1 summarizes the basic parameters describing these two instruments[2]. The temperature sensitivity per beam is computed as

$$\Delta Q = \Delta U = \sqrt{2}\frac{T_{\rm sys}}{\sqrt{\Delta\nu\ t\ N_{\rm chan}}}\,, \tag{2}$$

being $N_{\rm chan}$ the number of channels, $\Delta\nu$ the bandwidth and $T_{\rm sys}$ the system temperature (i.e. including the sky). *Phase I* also includes a source subtractor facility to monitor and correct the contribution of polarized radio-sources in the final maps. The overall time baseline for the project is to achieve the main science goal ($r = 0.1$) by end of 2011, and $r = 0.05$ by 2015.

Finally, *Phase II* (which is still not funded) considers the construction of a second telescope identical to the first one, and a third instrument with 30 polarimeters at 40 GHz.

3.2 Telescope and Enclosure

The QUIJOTE-CMB telescope uses a crossed-Dragonian design, where the primary has a 3 m projected aperture, and the secondary 2.6 m. The system is under-illuminated to minimize sidelobes and ground spillover. In addition, a cylindrical absorbing screen surrounding the optics (see Fig. 3) minimizes the spillover signal.

[2] Note that our definition for Stokes parameters is such that $Q = T_x - T_y$.

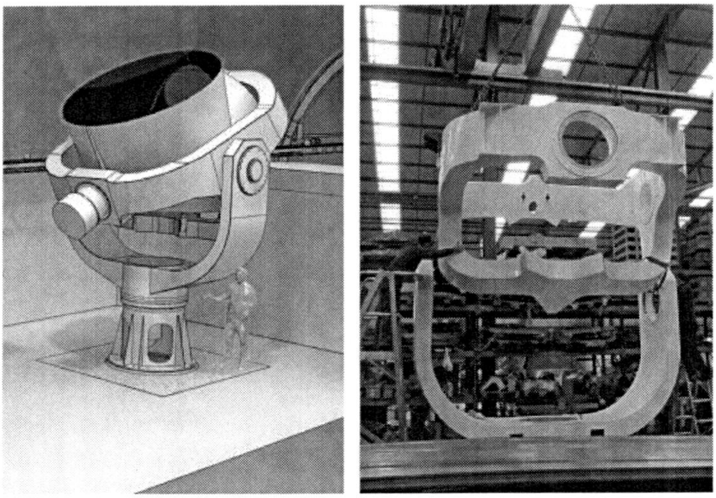

Fig. 3 *Left*: Design of one of the QUIJOTE telescopes inside the enclosure. *Right*: Assembling the first QUIJOTE telescope (October 7th, 2008)

Both mirrors have been designed to operate up to 90 GHz i.e. rms $\leq$ 20μm and maximum deviation of $d = 100\,\mu$m.

The whole system is mounted on a platform that can rotate around the vertical axis at a frequency of 0.25 Hz. The supporting structure has been designed using an alto-azimuthal concept which enables the telescope to point to any position in the sky with elevation above the horizon higher than 30°.

3.3 First Instrument

This is a multi-channel instrument with five separate polarimeters (providing 5 independent sky pixels): two which operate at 10–14 GHz, two which operate at 16–20 GHz, and a central polarimeter at 30 GHz. The science driver for this first instrument is the characterization of the galactic emission. The optical arrangement includes 5 conical corrugated feedhorns (designed by the University of Manchester) staring into a dual reflector crossed-dragonian system, which provides optimal cross-polarization properties (designed to be ≤ -35 dB) and symmetric beams. Each horn feeds a novel cryogenic on-axis rotating polar modulator which can rotate at speeds of up to 40 Hz (see Fig. 4). This rotational rate is fast enough to switch out 1/f noise in the lower frequency LNAs (since polar modulation occurs at four times the rotational rate, i.e. 160 Hz). The 30 GHz Front-End module (FEM) has additional phase switching to provide stability. The orthogonal linear polar signals are separated through a wide-band cryogenic Ortho-Mode-Transducer (OMT) before being amplified through two similar LNAs (a Faraday type module in the case of 30 GHz). These two orthogonal signals are fed into a room-temperature Back-End

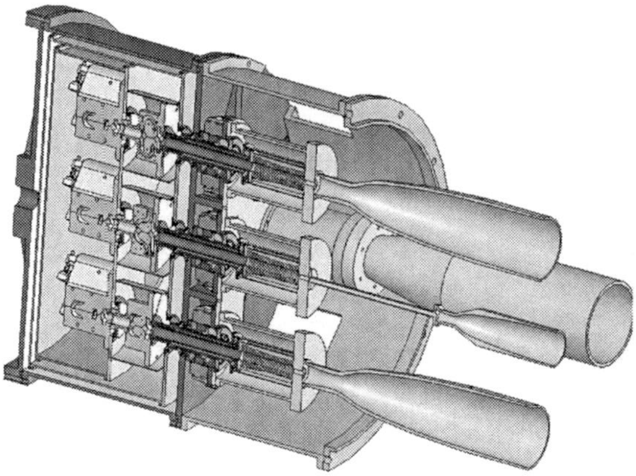

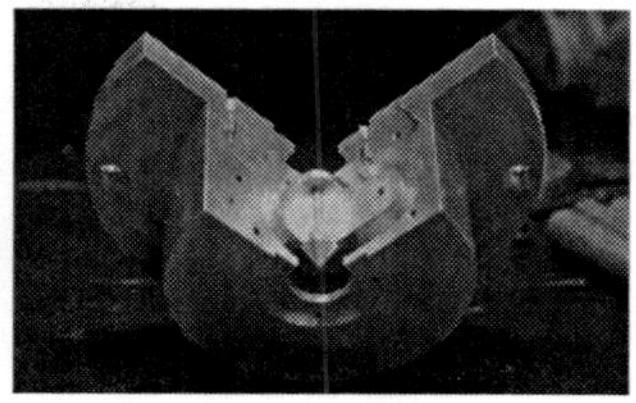

Fig. 4 *Left*: Detailed design of the first instrument. It is shown four horns (the 30 GHz is the central one, two horns at 17–19 GHz, and one of the 11–13 GHz horns in the back). After each antenna, we see the waveguide, the polar modulator and the OMT. *Right*: A prototype of the final polar modulator

module (BEM) where they are further amplified and spectrally filtered before being detected by square-law detectors. All the polarimeters except the 30 GHz receiver have simultaneous Q and U detection i.e. the 2 orthogonal linear polar signals are also correlated through a 180° hybrid and passed through two additional detectors. The band passes of these lower frequency polarimeters have also been split into an upper and lower band which gives a total of 8 channels per polarimeter (see Table 1).

The FEM for the low frequency channels is being built by IAC. The receivers for these channels use MMIC 6–20 GHz LNAs (designed by S. Weinreb and built in Caltech). The gain for these amplifiers is approximately 30 dB, and the noise temperature is less than 9 K across the band. The 30 GHz FEM is being built at the University of Manchester, and the design uses an existing Faraday module (same as the one used for OCRA-F[3]). The BEM for the 30 GHz instrument is being built by DICOM, with collaboration of IFCA at the simulation level. The cryogenics and the mechanical systems are provided by CMS[4] (Jeff Julian), IDOM and IAC.

3.4 *Second Instrument*

This instrument will be devoted to primordial B-modes science. It will consist in 19 polarimeters operating at 30 GHz. The conceptual design is a re-scaled version of the first instrument.

---

[3] OCRA-F: http://www.jodrellbank.manchester.ac.uk/research/ocra/ocraf.html.

[4] CMS: http://www.cryo-mechanicalsystems.com/.

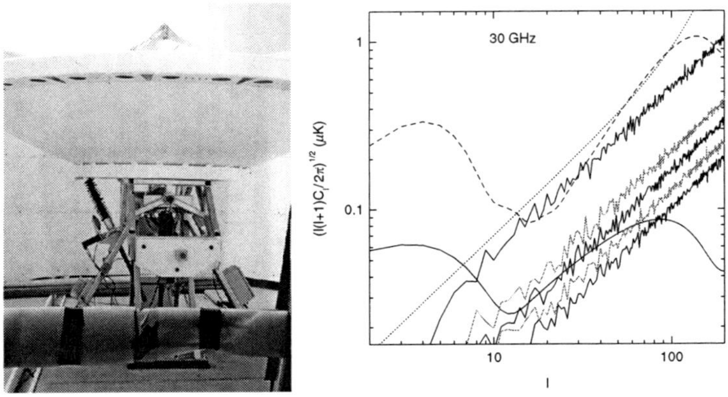

Fig. 5 *Left*: One antenna of the VSA source subtractor. This facility will be re-used to measure polarization of radio sources to correct the QUIJOTE 30 GHz maps. *Right*: Radio-source contribution at 30 GHz (*top solid oscillating line*), and the residuals after removing/masking sources with fluxes (in total intensity) higher than 300 or 100 mJy (*black lines*). *Solid red lines* are obtained removing sources with polarized intensity higher than 50 or 10 mJy

3.5 Source Subtractor Facility

An upgraded version of the VSA source subtractor (VSA-SS) facility [15], which is being carried out by the Cavendish Laboratory and the University of Manchester, will be used to monitor the contribution of radio-sources in the QUIJOTE maps. The VSA-SS is a two element interferometer, operating at 30 GHz, with 3.7 m dishes and a separation of 9 m (see Fig. 5). We have estimated that at 30 GHz it is enough to correct the emission of all sources with fluxes in total intensity higher than 300 mJy in order to make the residual source contribution equal or smaller than the expected B-mode signal for the case of $r = 0.1$ (see Fig. 5). In that case, the total number of sources to be monitored in the whole QUIJOTE surveyed area is around 500. The expected flux sensitivity per source is $\sim$2–3 mJy.

4 Conclusions

QUIJOTE-CMB will provide unique information about the polarization emission (synchrotron and anomalous) from our Galaxy at low frequencies. This information will be valuable for future B-mode experiments. In particular, QUIJOTE will complement at low frequencies the information obtained by Planck. Using the information from the low frequencies, QUIJOTE will be able to detect the B-mode signal due to primordial gravity waves in the 30 GHz map if $r \geq 0.05$.

References

1. Battistelli, E.S., Rebolo, R., Rubiño-Martín, J.A., Hildebrandt, S.R., Watson, R.A., Gutiérrez, C., Hoyland, R.J., ApJ 645, L141 (2006)
2. Bock, J., et al., *Task Force on Cosmic Microwave Background Research*, arXiv:astro-ph/0604101 (2006)
3. Dickinson, C., et al., MNRAS 353, 732 (2004)
4. Dunkley, J., Komatsu, E., Nolta, M.R., Spergel, D., Larson, D., Hinshaw, G., Page, L., Bennett, C.L., Gold, B., Jarosik, N., Weiland, J.L., Halpern, M., Hill, R.S., Kogut, A., Limon, M., Meyer, S.S., Tucker, G.S., Wollack, E., Wright, E.L., ApJS 180, 306 (2009)
5. Hinshaw, G., Weiland, J.L., Hill, R.S., Odegard, N., Larson, D., Bennett, C.L., Dunkley, J., Gold, B., Greason, M.R., Jarosik, N., Komatsu, E., Nolta, M.R., Page, L., Spergel, D.N., Wollack, E., Halpern, M., Kogut, A., Limon, M., Meyer, S.S., Tucker, G.S., Wright, E.L., ApJS 180, 225 (2009)
6. Kamionkowski, M., Kosowsky, A., Stebbins, A., Phys. Rev. D 55, 7368 (1997)
7. Komatsu, E. Dunkley, J., Nolta, M.R., Bennett, C.L., Gold, B., Hinshaw, G., Jarosik, N., Larson, D., Limon, M., Page, L., Spergel, D.N., Halpern, M., Hill, R.S., Kogut, A., Meyer, S.S., Tucker, G.S., Weiland, J.L., Wollack, E., Wright, E.L., ApJS 180, 330 (2009)
8. Kovac, J.M., Leitch, E.M., Pryke, C., Carlstrom, J.E., Halverson, N.W., Holzapfel, W.L., Nature 420, 772 (2002)
9. Partridge, B., in *The Cosmic Microwave Background: from quantum fluctuations to the present Universe*, Rubiño-Martin, J.A., Rebolo, R. & Mediavilla, E., eds., Cambridge University Press, p. 1 (2009)
10. Peacock, J.A., Schneider, P., Efstathiou, G., Ellis, J.R., Leibundgut, B., Lilly, S.J., Mellier, Y., in *ESA-ESO Working Group on Fundamental Cosmology*, Peacock, J.A., et al., ed., ESA (2006)
11. Polnarev, A.G., Soviet Astronomy 29, 607 (1985)
12. Readhead, A.C.S., et al., ApJ 609, 498 (2004)
13. Reichardt, C.L., Ade, P.A.R., Bock, J.J., Bond, J.R., Brevik, J.A., Contaldi, C.R., Daub, M.D., Dempsey, J.T., Goldstein, J.H., Holzapfel, W.L., Kuo, C.L., Lange, A.E., Lueker, M., Newcomb, M., Peterson, J.B., Ruhl, J., Runyan, M.C., Staniszewski, Z., ApJ 694, 1200 (2009)
14. Tucci, M., Martínez-González, E., Vielva, P., Delabrouille, J., MNRAS 360, 935 (2005)
15. Watson, R.A., et al., MNRAS 341, 1057 (2003)
16. Watson, R.A., Rebolo, R., Rubiño-Martín, J.A., Hildebrandt, S., Gutiérrez, C.M., Fernández-Cerezo, S., Hoyland, R.J., Battistelli, E.S., ApJ 624, L89 (2005)
17. Zaldarriaga, M., Seljak, U., Phys. Rev. D 55, 1830 (1997)

Part IV
The Galaxy and Its Components

The AB Doradus System Revisited: The Dynamical Mass of AB Dor A

J.C. Guirado, I. Martí-Vidal, J.M. Marcaide, L.M. Close, J.-F. Lestrade, D.L. Jauncey, S. Jiménez-Monferrer, D.L. Jones, R.A. Preston, and J.E. Reynolds

Abstract We report new radio interferometric observations of the quadruple pre-main-sequence (PMS) system AB Doradus. From these observations, combined with existing VLT near-infrared relative astrometry, we have refined the estimates of the dynamical masses of the system. In particular, we find component masses of $0.86 \pm 0.09\,M_{\odot}$ and $0.090 \pm 0.003\,M_{\odot}$ for AB Dor A and AB Dor C, respectively. These dynamical masses, coupled with temperatures and luminosities, allow for comparison with theoretical stellar models. The case of AB Dor C, in terms of calibration of evolutionary models of low-mass young stars has been widely reported

---

J.C. Guirado, I. Martí-Vidal, J.M. Marcaide, and S. Jiménez-Monferrer
Universidad de Valencia, Dr. Moliner 50, 46100 Burjassot, Valencia, Spain
e-mail: Jose.C.Guirado@uv.es, J.M.Marcaide@uv.es, sergio.jimenez@uv.es

I. Martí-Vidal
Max-Planck-Institut für Radioastronomie, Auf dem Hügel 69, D-53121 Bonn, Germany
e-mail: imartiv@mpifr-bonn.mpg.de

L.M. Close
Steward Observatory, University of Arizona, Tucson, Arizona 85721, USA
e-mail: lclose@as.arizona.edu

J.-F. Lestrade
Observatoire de Paris/LERMA, Rue de l'Observatoire 61, 75014 Paris, France
e-mail: jean-francois.lestrade@obspm.fr

D.L. Jauncey and J.E. Reynolds
Australian Telescope National Facility, P.O. Box 76, Epping, NSW 2121, Australia
e-mail: David.Jauncey@atnf.csiro.au, John.Reynolds@atnf.csiro.au

D.L. Jones and R.A. Preston
Jet Propulsion Laboratory, California Institute of Technology, 4800 Oak Grove Drive, Pasadena, CA 91109, USA
e-mail: dayton.jones@jpl.nasa.gov, rap@sgra.jpl.nasa.gov

J.M. Diego et al. (eds.), *Highlights of Spanish Astrophysics V*, Astrophysics and Space Science Proceedings, DOI 10.1007/978-3-642-11250-8_13,

in previous studies. In this contribution, we compare the measured properties of AB Dor A with several solar-composition models for PMS stars. The models used in this comparison predict the dynamical mass to within the quoted uncertainties.

1 Introduction

Studies of the fundamental parameters of pre-main-sequence (PMS) stars are relevant, since they provide tests of stellar evolution models that are used to derive theoretical masses of young, low mass objects. However, finding PMS stars suitable to measure, independently, mass and luminosity is difficult (only 27 system for masses $<1.2\,M_\odot$ [9]), and many of them are affected by some extra uncertainties in parameters such as distance and/or age.

The case of the PMS star AB Doradus is of particular interest. Hipparcos/VLBI astrometry of the main star AB Dor A [7] revealed the presence of a low-mass companion, AB Dor C, which was later imaged by VLT infrared observations [3]. AB Dor A has also a wide separation companion 9" away, AB Dor B, which is actually a close binary (AB Dor Ba/AB Dor Bb, at 0.070" separation [10]). The combination of all astrometric observations of AB Doradus provides precise limits to the dynamical masses of the components of the system. In this contribution we report about new VLBI observations of the different pairs in the AB Doradus system, with special focus on the new constraint of the dynamical mass of the main component AB Dor A.

2 Observations and Data Reduction

We observed AB Doradus in November 2007 at 8.4 GHz using antennas from the Australia Long Baseline Array (LBA). These antennas were the Australian Compact Array (ATCA), Parkes, Mopra, Hobart, Ceduna, and Hartebeesthoek. With these observations, we resumed the astrometric monitoring carried out in this systems during the period 1992–1996 (see [7]). The data were disk-based recorded over 8 (4) 16 MHz bands at a bit rate of 512 (256) Mbps at ATCA, Parkes, Mopra, and Hartebeesthoek (Ceduna and Hobart). We interleaved observations of the strong background radio source PKS 0516-621 using, typically, a duty cycle consisting of 170 s on AB Doradus, and 70 s on PKS 0516-621. The data were processed at the software correlator in Swinburne University [5]. The post-correlation data reduction was performed using the NRAO Astronomical Image Processing System (*AIPS*). Visibility amplitude calibration was done using the antenna gains and the system temperatures measured at each station and, afterwards, it was refined by interpolation of the antenna gains obtained from the hybrid mapping of the calibrator. The phase calibration was performed following standard phase-reference calibration techniques (AB Dor A with respect to the calibrator). The data were exported

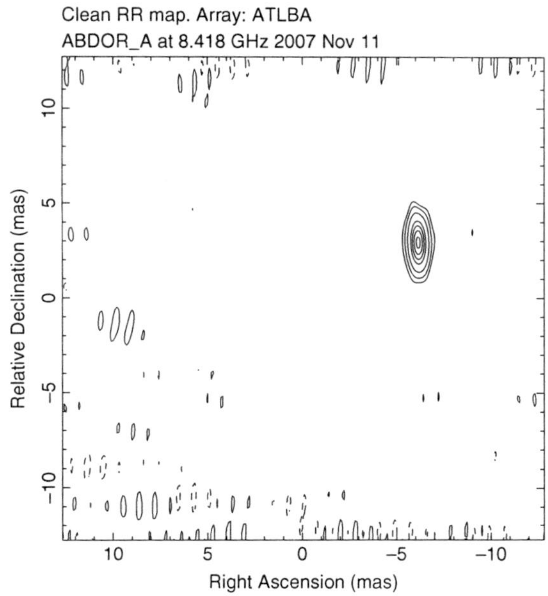

Fig. 1 LBA map of AB Dor A phase-referenced to the position of PKS 0521-621 at epoch 2007.863. Contours are −6,6,12,24,48,64,80,95% of the peak flux density in the image (8 mJy/beam). The FWHM beam size is 2.24 × 0.66 mas at a position angle of 2.4°. The shift of the source with respect to the origin of the map indicates the astrometric correction to be applied to the a priori coordinates of AB Dor A (those used for the data correlation)

for further reduction in DIFMAP [14]. The same calibration procedure was repeated for the pair AB Dor Ba/AB Dor Bb; the analysis of the map of this latter pair will be reported elsewhere.

2.1 Phase-Referenced Map of AB Dor A

The phase-referenced map of AB Dor A is shown in Fig. 1. The map was obtained by Fourier inversion of the naturally-weighted visibilities, and after performing a CLEAN deconvolution. Since the source is partially resolved, we fitted a one-component uniform disk model (uniform intensity) to the AB Dor A data. For an elliptical disk, we obtained a size of $(1.4 \times 0.8) \pm 0.3$ mas (1σ errors). This measurement does not mean a strong evidence of ellipticity and, therefore we repeated the fit using a circular disk. The best fit for this simpler model, of similar quality to the fit to the elliptical disk, provided a size of 1.0 ± 0.3 mas (1σ errors).

The astrometric information of AB Dor A is extracted from Fig. 1. The coordinates of the phase-referenced map are referred to the PKS 0516-621 position. The relative position of AB Dor A is then found measuring the coordinates of the brightness peak. The resulting position of AB Dor A at the epoch of the observations (2007.863) is:

$$\alpha = 5^h\,28^m\,44^s\!.91605 \pm 0^s\!.00025$$
$$\delta = -65^\circ\,26'\,53''\!.8683 \pm 0''\!.0010$$

where the quoted uncertainties are overall standard errors.

2.2 *Reflex Orbit*

We calculated the reflex motion of AB Dor A, consequence of the gravitational interaction with the low mass companion AB Dor C (see Fig. 2), using the same procedure and data set reported in [8], the latter augmented with the new LBA position of AB Dor A and the infrared relative positions AB Dor A/AB Dor C from [13] and [4]. This new fit provides very similar orbital elements to those quoted in [8], thus confirming the mass of AB Dor C to be $0.090 \pm 0.005\,M_\odot$. One important point of this new orbital analysis is that the new data set (remarkably more complete than the previous one, with more absolute positions of AB Dor A and more relative positions AB Dor A/ C) allows an *independent* estimate of the dynamical mass of the primary star, AB Dor A of $0.86 \pm 0.09\,M_\odot$. This first measurement of the dynamical mass of AB Dor A removes a subtle ambiguity in previous estimates of the mass of the companion AB Dor C, coupled to an empirical, but not dynamical, mass of AB Dor A (see [3] for details).

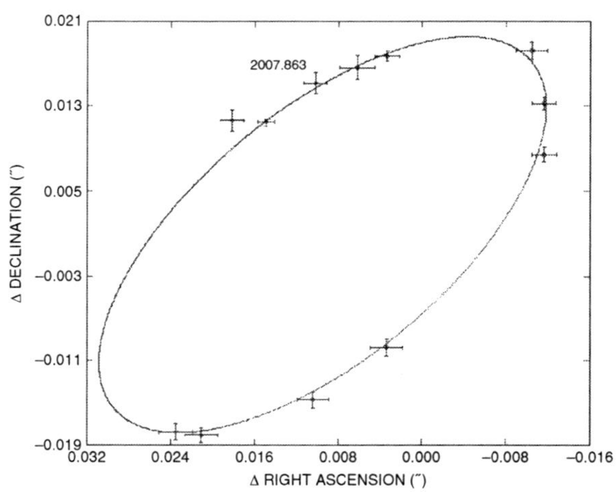

Fig. 2 Orbit of the reflex motion of AB Dor A corresponding to the presence of the companion AB Dor C. The new position at epoch 2007.863 is indicated. The rest of the positions marked on this revised orbit corresponds to the data (VLBI and HIPPARCOS) taken during the period 1992–1996 [7]. The orbital period is 11.82 ± 0.06 years

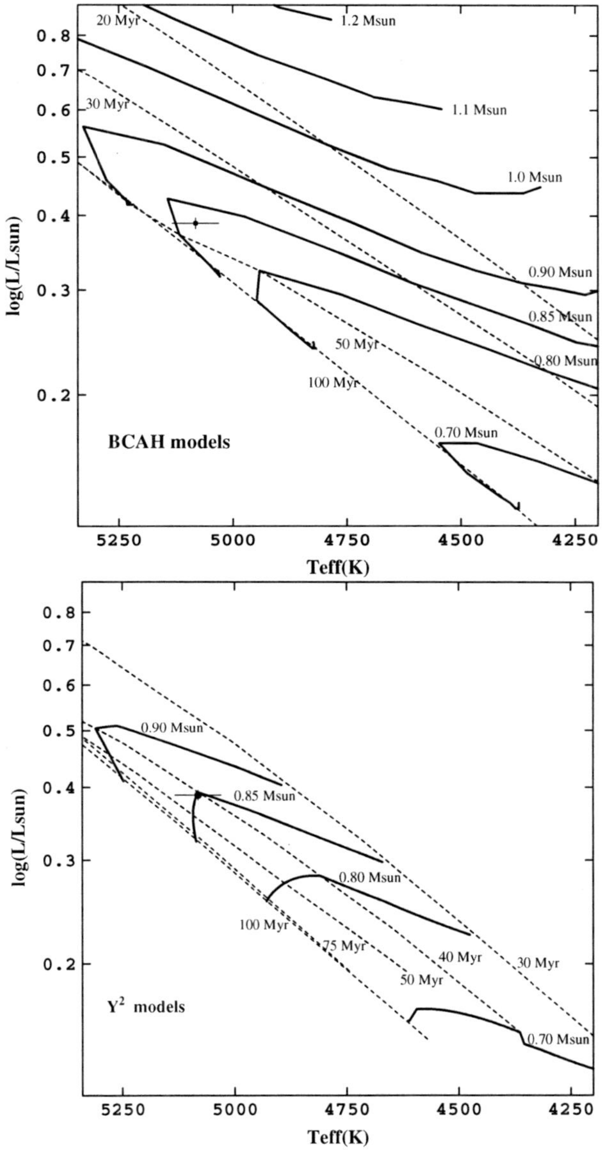

Fig. 3 AB Dor A compared with two PMS models, the BCAH models (*top*; [1]) and the Y^2 models (*bottom*; [6]). These HR diagrams show a good agreement between measurements and predictions: the 0.85 $M_\odot$ isomass is compatible with the luminosity and temperature measured for this star (the dynamical mass derived from this work is $0.86 \pm 0.09\,M_\odot$)

3 Comparison with PMS Models

The dynamical mass of AB Dor A, $0.86 \pm 0.09\,M_\odot$, can be compared to theoretical masses from PMS stellar evolutionary models. In this contribution we have used two PMS models, the BCAH models [1], and the more recent Y^2 models [6]. As seen in Fig. 3, both models offer good predictions for the mass of AB Dor A: the Y^2 models are in excellent agreement with the measurements, meanwhile a slight underprediction is shown in the BCAH models, but certainly, well within the quoted uncertainties of the dynamical estimate. Are these results reasonable? References [9] and [12] have shown that for all PMS models there is a good agreement between theoretical and dynamical masses above $1.2\,M_\odot$; these authors have also shown that, between 1.2 and $0.5\,M_\odot$, most of the models tend underestimate the dynamical masses around 10–30%. This may explain the underprediction of the BCAH models in Fig. 3. Still, the extraordinary agreement of the dynamical mass with the Y^2 predictions is remarkable.

An important outcome from these dynamical vs. theoretical mass comparisons is related to the age of AB Doradus. The age of this system is still a matter of discussion among some authors, with different estimates in the literature ranging from 50 Myr [4, 15] to ages coeval with the Pleiades (120 Myr [11]). The placement of AB Dor A in the HR diagrams, along with the isochrones of different PMS models in Fig. 3, seems to favor an early age for AB Dor A (40–50 Myr), substantially younger than the Pleiades.

We have presented new radio interferometry data of the AB Doradus system, in particular those referred to the main star AB Dor A. We have recalculated the reflex orbit of this star, consequence of the presence of the low-mass companion AB Dor C, which is fully compatible with previous estimated. This new calculation provides the first estimate of the *dynamical* mass of AB Dor A, $0.86 \pm 0.09\,M_\odot$. The comparison of the dynamical mass of AB Dor A with theoretical masses from PMS stellar evolutionary models show a reasonable agreement, which, in turn, place constraints on the age of AB Dor A.

Acknowledgements This work has been partially founded by grant AYA2006-14986-CO2-02 of the Spanish DGICYT. The Long Baseline Array is part of the Australia Telescope which is funded by the Commonwealth of Australia for operation as a National Facility managed by CSIRO. A portion of this work was carried out at the Jet Propulsion Laboratory, California Institute of TEchnology, under contract with the National Aeronautics and Space Administration. IMV is a fellow of the Alexander von Humboldt-Stiftung.

References

1. Barrafe, I., et al., A&A 337, 403 (1998)
2. Chabrier, G., et al., ApJ 542, 464 (2000)
3. Close, L.M., et al., Nature 433, 286 (2005)
4. Close, L.M., et al., ApJ 665, 736 (2007)
5. Deller, A.T., et al., PASP 119, 318 (2007)

6. Demarque, P., et al., ApJS 155, 667 (2004)
7. Guirado, J.C., et al., ApJ 490, 835 (1997)
8. Guirado, J.C., et al., A&A 446, 733 (2006)
9. Hillenbrand, L.A., White, R.J., ApJ 604, 741 (2004)
10. Janson, M., et al., A&A 462, 615 (2007)
11. Luhman, K.L., et al., ApJ 628, L69 (2005)
12. Mathieu, R.D., et al., in *Protostars and Planets V*, 951, 411 (2007)
13. Nielsen, E.L., et al., AN 326, 1033 (2005)
14. Shepherd, M.C., et al., BAAS 26, 987 (1995)
15. Zuckerman, B., Song, I., ARA&A 42, 685 (2004)

Spectrophotometry with Gaia

C. Jordi, J.M. Carrasco, C. Fabricius, F. Figueras, and H. Voss

Abstract Gaia is an all-sky survey satellite, to be launched by ESA in late 2011, to obtain parallaxes and proper motions to microarcsecond precision, radial velocities and astrophysical parameters for about 10^9 objects down to a limiting magnitude of 20 mag, which means about 10^4 times more stars than observed with Hipparcos mission. The astrophysical information for all sources will be derived from broad-band photometry and low-resolution spectrophotometry complemented with astrometric and high-resolution spectroscopy measurements around CaII triplet. This paper describes the instrument and its capabilities in terms of stellar parameters determination, as well as the current status of the mission.

1 Gaia: An Enormous Step Forward

The ESA space astrometry mission Gaia will scan the sky continuously during 5 years. The principle for obtaining accurate absolute parallaxes is the same of the precursor Hipparcos mission: two telescopes pointing at two viewing directions separated a basic angle spinning around a perpendicular axis, which at the same time precesses. Gaia will be located at the Lagrangian L2 point of the Sun–Earth system, spin with a period of 6 h and precess around the line L2–Earth–Sun with a period of 70 days, allowing a full coverage of the sky several times (80 in average) along the mission duration.

The science case has been fully described several times (see, for instance, [9, 10, 14]) and so, we will not repeat it again. Just to remind, it covers an extremely wide range of topics in galactic and stellar astrophysics, solar system and exoplanet science, as well as the establishment of a very accurate, dense and faint optical reference frame. This is accomplished because Gaia will be monitoring the 1 billion

C. Jordi, J.M. Carrasco, C. Fabricius, F. Figueras, and H. Voss
University of Barcelona, ICC-IEEC, Martí i Franquès 1, E-08028 Barcelona, Spain
e-mail: carme.jordi@ub.edu

J.M. Diego et al. (eds.), *Highlights of Spanish Astrophysics V*, Astrophysics and Space Science Proceedings, DOI 10.1007/978-3-642-11250-8_14,

brightest sources on the sky. Expected accuracies are in the 7–25 μ as range down to 15 mag and sub-mas accuracies at the faint limit (20 mag).

From the astrometric measurements of unfiltered (white) light, Gaia will yield G magnitudes in a very broad band covering the 350–1,000 nm wavelength range, while the spectral energy distribution of each source will be sampled by a dedicated spectrophotometric instrument, with a dispersion between 4 and 32 nm/pixel. Gaia also features, for the brighter stars, a high-resolution ($R = 11,500$) integral field spectrograph, in the range 847–874 nm around the CaII triplet, the so-called Radial Velocity Spectrometer (RVS) instrument. The total flux in RVS instrument will provide the G_{RVS} magnitude.

There are in total six mirrors per light path. The two primary mirrors are $1.45{\times}0.5\,\mathrm{m}^2$ in size. The two fields of view are combined in a unique focal plane of about $1{\times}0.5\,\mathrm{m}^2$ in size with 106 CCDs. Figure 1 shows the focal plane with its different sections. Gaia's observations are made with high-quality, large-format CCDs operated in integration (TDI) mode, with charge images being transported in synchrony with optical images moving across the field due to the rotation of the satellite. The sources are (a) detected in the sky mappers, which only see one field of view each, (b) further confirmed as a source or rejected in the first strip of CCDs in the astrometric section and, if confirmed, (c) an image window around the position of the source on every transiting CCD is downloaded to the ground. If the source is bright enough (brighter than about 16.5 mag) it will be observed with the radial-velocity spectrometer as well.

The mission operation is fully funded by ESA. On February 2006, EADS-Astrium was selected as prime contractor by the Science Programme Committee of ESA. Currently, the mission is in phase C/D after the Preliminary Design Review was passed in July 2007. The major current concerns are on the basic angle stability, the mitigation of the radiation damage and the performances of the radial velocity spectrometer. Efforts are being made to improve these topics. With a launch foreseen for December 2011 (see Fig. 2), transfer to L2, commissioning phase, routine

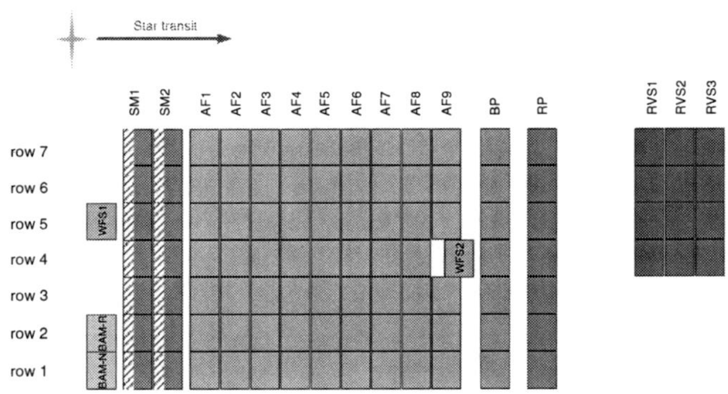

Fig. 1 Combined focal plane. The sources are detected in the sky mappers and observed on nine astrometric CCDs, *blue* and *red* photometers and if bright enough in the radial-velocity spectrometer. Figure courtesy of J. de Bruijne

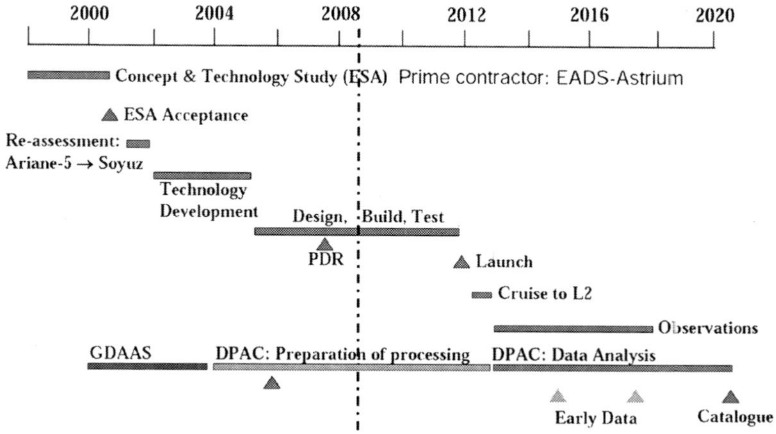

Fig. 2 Gaia schedule. The mission is in phase C/D and DPAC is preparing the data processing

science operations lasting 5 years, with a possible 1 year extension, and data processing, the final catalogue is expected around 2021. The ingestion of the large amount of data that Gaia will obtain is analyzed in [7, 12] in this volume.

The list of persons currently involved in the Gaia Data Processing and Analysis Consortium (DPAC) includes some 370 people, not counting the industrial contractors. The community of scientists interested in Gaia is much larger, due to the wide science case.

2 Aim of Spectrophotometry

Gaia measurements will provide the basic diagnostics for classifying all sources as stars, quasars, solar-system objects, or otherwise, and for parameterizing them according to their nature. Stellar classification and parameterization across the entire Hertzsprung–Russell diagram is required as well as the identification of peculiar objects. This demands observations in a wide wavelength range, extending from the very blue to the infrared. The photometric data must determine: (a) effective temperatures and reddening at least for O–B–A stars (needed both as tracers of Galactic spiral arms and as reddening probes), (b) at least effective temperatures and abundances for F–G–K–M giants and dwarfs, (c) luminosities (gravities) for stars having large relative parallax errors, (d) indications of unresolved multiplicity and peculiarity, and (e) a map of the interstellar extinction in the Galaxy. All of this has to be done with an accuracy sufficient for stellar age determination in order to allow a quantitative description of the chemical and dynamical evolution of the Galaxy over all galactocentric distances. Separate determination of Fe and α-elements abundances is essential for mapping Galactic chemical evolution and understanding the formation of the Galaxy.

Photometry is also crucial to identify and characterize the set of ~500,000 quasars that the mission will detect. Apart from being astrophysically interesting in their own right, quasars are key-objects for defining the fixed, non-rotating Gaia Celestial Reference Frame, the optical equivalent of the International Celestial Reference Frame [9]. On the other hand, Gaia will identify about 900 quasars (QSO) with multiple images produced by macrolensing. Since this number is sensitive to cosmological parameters, the Gaia observations will be able to constrain the latter.

Due to diffraction and optical aberrations of the instrument, the position of the center of the stellar images is wavelength dependent. To achieve the microarcsec accuracy level, astrometry has to be corrected for this chromatic aberration through the knowledge of the spectral energy distribution of the observed objects. Photometry is indispensable for this. If uncorrected, chromatic errors could reach several milliarcsec.

3 Spectrophotometry Instrument

The spectrophotometric instrument consists on two low-resolution slitless spectrographs made of fused silica and named BP (*blue photometer*) and RP (*red photometer*). They cover the wavelength intervals 330–680 nm and 650–1050 nm, respectively, and their resolution was designed in such a way to provide the same science outputs that the set of photometric passbands proposed by the scientific community [5]. Both spectra have similar lengths in order to provide similar angular resolution, which is able to deal with stellar densities of up to 750, 000 sources deg^{-2}.

The total flux of these BP and RP spectra will yield G_{BP} and G_{RP} magnitudes as two broad passbands. Therefore, four passbands and their corresponding magnitudes, can be associated with the Gaia instruments: G, G_{BP}, G_{RP} and G_{RVS}. Figures 3 (left) and 4 show the four passbands and colour–colour diagrams, respectively. The late involves Johnson–Cousins B, V and I_c passbands and provide relationships computed using 141 spectra from Pickles' library [11].

The white light observations, G band photometry, will yield the best signal-to-noise ratio and hence is the most suitable for variability detection. One source will be observed by the two fields of view with a separation of 1.6 h and repeated again after a revolution of the satellite after 6 h.

The spectral resolution is a function of wavelength as a result of the natural dispersion curve of fused silica; the dispersion is higher at short wavelengths, and ranges from 4 to 32 nm $pixel^{-1}$ for BP and from 7 to 15 nm $pixel^{-1}$ for RP. The variation across-scan does not exceed ±6% for BP and ±4% for RP. Figure 5, based on simulated spectra provided by [2], shows the observations for two stars (T_{eff} = 8,500 and 3,500 K), which SEDs have been taken from BaSeL-2.2 spectral library [6]. End-of-mission error bars for $G = 15$ and 18 are shown. Different spectra displayed correspond to different surface gravities and different metallicities. The differences are relevant compared to the error bars.

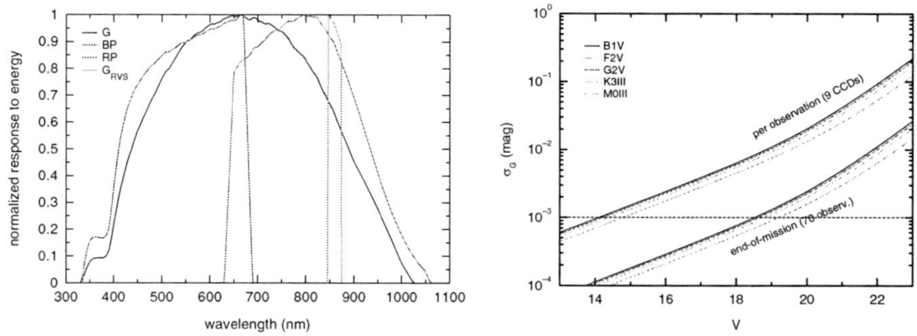

Fig. 3 *Left*: Response curves yielding G, G_{BP}, G_{RP} and G_{RVS}. *Right*: Expected precisions for G fluxes as a function of magnitude for a single transit and at the end-of-mission. A threshold of 1 mmag has been marked

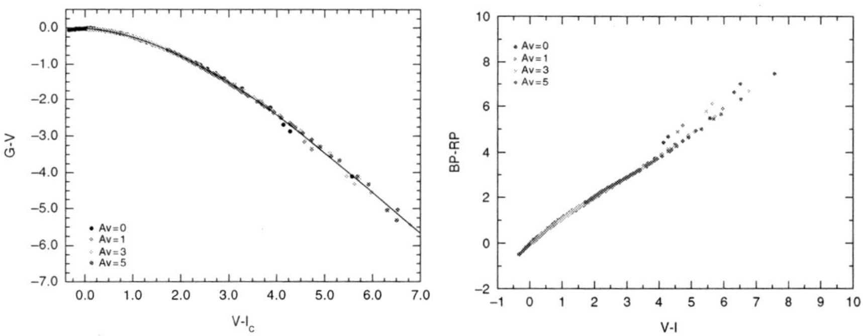

Fig. 4 Colour–colour diagrams involving Gaia passbands and Johnson–Cousins B, V and I_c passbands. Different colors are used for different absorption values

4 Calibration Principle

Gaia is essentially a self-calibrated mission with a strong inter-relationship among astrometric, photometric and spectroscopic data. The calibration parameters and the fluxes and spectra of the sources are solved for in an iterative procedure that updates the solutions as new data come in. The procedure is based on a large set of internally defined *well-behaved* sources.

The variations in response and wavelength scale across the focal plane, and with time, will be monitored taking advantage of the scanning mode in which Gaia will be operated. For average stellar densities on the sky about 60 stars per CCD per second will cross the focal plane. This translates to about 660 stars per pixel column per 6 h (Gaia spin period). The same stars are observed repeatedly in different parts of the focal plane on time scales varying from hours to weeks to months, up to the mission life time of 5 years. There will thus be plenty of measurements to perform detailed CCD calibrations on scales from pixel columns to CCDs.

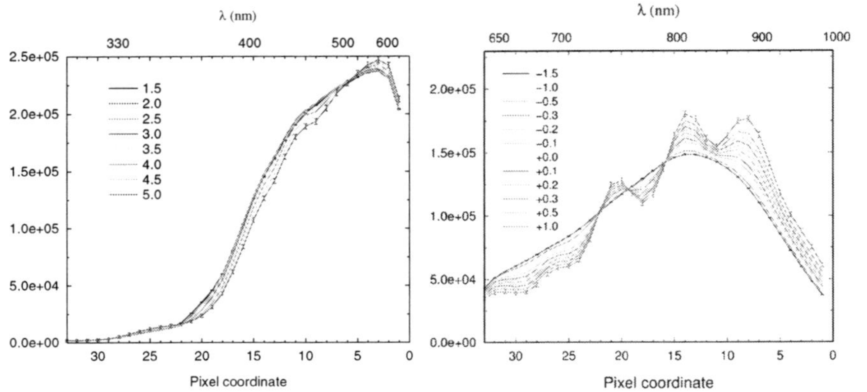

Fig. 5 1D spectra based on simulations from [2] for a $T_{\rm eff} = 8,500$ K star with several $\log g$ observed in BP (*left panel*) and for a $T_{\rm eff} = 3,500$ K star with several [M/H] observed in RP (*right panel*). The end-of-mission error bars in the violet line correspond to $G = 18$ and in the black line to $G = 15$

The photometric system will be defined by the average response for all the CCDs over the duration of the mission. The conversion to absolute fluxes will be based on a set of 100–150 SpectroPhotometric Standard Stars (SPSS). For this, we are conducting an specific ground-based observational project at 2.2 m at CAHA, TNG at La Palma, NTT at La Silla, REM at La Silla, 1.5 m at San Pedro and Cassini at Loiano. We wish to achieve 1–2% precision based on Vega for a set of 100–150 SPSS. Medium resolution spectra and photometry are being obtained, both for absolute calibrating the spectra and for monitoring variability at small and large time baseline.

To know more details about Gaia photometric calibration see [3] in this volume.

5 Results

The Besançon Galaxy Model [13] has been used to simulate the content of the Galaxy and the stellar population that Gaia will observe (see Fig. 6 and [4, 8], in this volume). In total, the simulations predict about 300 million stars with parallax errors better than 10% and 30 million stars with precisions better than 1%. All parts of the HR diagram and all type of variable stars will be well covered, so absolute luminosities for all of them will be derived.

The simulations have been used to establish internal calibration models and to evaluate precisions and systematics. The first results show that the calibration errors could be below 1 mmag. Figure 3 (right) shows the expected precisions for G fluxes as a function of magnitude for a single focal plane crossing (9 CCDs) and at the end-of-mission (assumed 70 observations for safety).

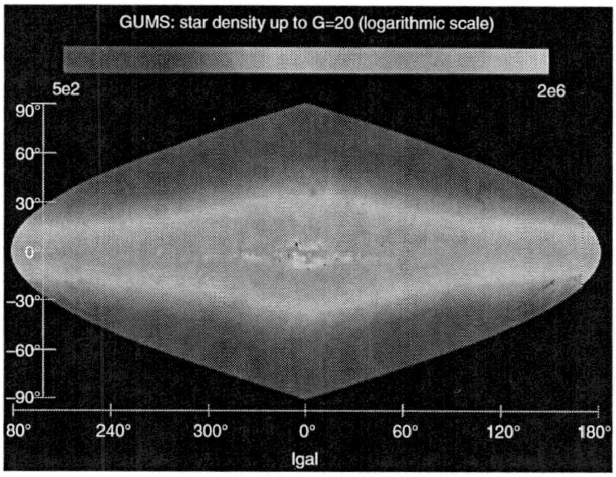

Fig. 6 Sky density from Gaia Universe Model Snapshot (GUMS) up to $G = 20$

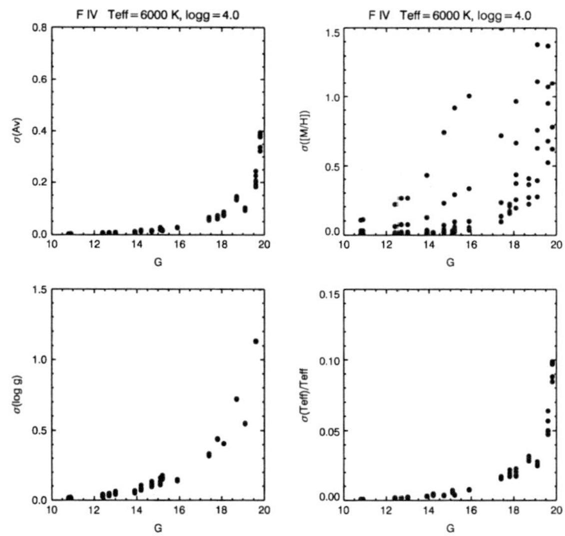

Fig. 7 Precision of A_v, [M/H], $\log g$ and $T_{\rm eff}$ determination for an F IV star at different distances as a function of G. The largest errors in $\sigma_{[M/H]}$ corresponds to very metal poor stars, with [M/H] $= -4.0$

Precision of the astrophysical parameters has been preliminary estimated taking into account poisson, background and readout noises. Parallax value has been introduced as an additional information into the procedure. As an example, Fig. 7 shows the estimated precision of A_v, [M/H], $\log g$ and $T_{\rm eff}$ determination for F-type dwarfs as a function of G magnitude. These stars are key for the halo and old disk age determination.

According to [1], at $G = 15$, the uncertainties derived are 1% in $T_{\rm eff}$ for a wide range types (OB to K stars); A_V, to 0.05–0.1 mag for hot stars; lower than 0.2 dex in [Fe/H] for spectral types later than F down to ~2.0 dex metallicity; $\log g$ to between

0.1 and 0.4 dex for all luminosity classes, but to better than 0.1 dex for OBA stars. The performance varies a lot with the physical parameters themselves, as well as with G magnitude (i.e. signal-to-noise ratio).

Neural Network algorithms have been applied to parametrize the stars, see for instance [15]. The systematic trends have been evaluated as a function of A_v, [M/H], $\log g$ and $T_{\rm eff}$. The largest ones are for the determination of [M/H] for hot stars and very metal poor, as it can be anticipated. In fact, in Fig. 7, the largest errors correspond to very metal poor stars, with [M/H]$= -4.0$. The correlations between [M/H] and $\log g$ determinations will be improved by introducing the parallax information into the classification algorithms.

Acknowledgements This research has been granted by MCyT under contracts ESP2006-26356-E, ESP2006-13855-C02-01.

References

1. Bailer-Jones, C.A.L., in *The Galaxy disk in a cosmological context*, Proc. IAU Symp. No. 254 (2008)
2. Brown, A., Internal Gaia report GAIA-CA-TN-LEI-AB-005-7 (2006)
3. Carrasco, J.M., Voss, H., Jordi, C., Fabricius, C., Figueras, F., this volume (2009)
4. Isasi, Y., Luri, X., Robin, A., this volume (2009)
5. Jordi, C., Høg, E., Brown, A.G.A., Lindegren, L., Bailer-Jones, C.A.L., Carrasco, J.M., Knude, J., Straižys, V., de Bruijne, J.H.J, Claeskens, J.-F., et al., MNRAS 367, 290 (2006)
6. Lejeune, Th., Cuisinier, F., Buser, R., A&AS 125, 229 (1997)
7. Luri, X., et al., this volume (2009)
8. Masana, E., Isasi, Y., Luri, X., Peralta, J., this volume (2009)
9. Mignard, F., in *The Three Dimensional Universe with Gaia*, ESA SP-576, p. 5 (2005)
10. Perryman, M.A.C., de Boer, K.S., Gilmore, G., Høg, E., Lattanzi, M.G., Lindegren, L., Luri, X., Mignard, F., Pace, O., de Zeeuw, P.T., A&A 369, 339 (2001)
11. Pickles, A.J., PASP 110, 863 (1998)
12. Portell, J., Castañeda, J., Isasi, Y., Fabricius, C., Luri, X., Torra, J., Vicente, D., this volume (2009)
13. Robin A.C., Reyl, C., Derrière, S., Picaud, S., A&A 409, 523 (2003)
14. Torra, J., 2007, in *Highlights of Spanish Astrophysics IV*, Proc. of the VII Scientific Meeting of the Spanish Astronomical Society, Figueras, F., Guirart, J.M., Hernanz, M., & Jordi, C., eds., Springer, p. 255 (2007)
15. Willemsen, P.G., Kaempf, T.A., de Boer, K.S., Bailer-Jones, C.A.L., 2005, Gaia technical report ICAP-PW-005 (2005)

The Least Massive (Sub)Stellar Component of the Milky Way

E.L. Martín, V.J.S. Béjar, H. Bouy, J. Licandro, B. Riaz, F. Rodler, L. Valdivielso, R. Deshpande, and R. Tata

Abstract This review presents a panorama of the research topics that are currently being developed by our *strategic* research group at the IAC in the field of very low-mass stars, brown dwarfs, extrasolar giant planets and the solar system. Our main goal is to investigate the cosmogony of the least massive stellar and sub-stellar component of the Milky Way. We are using multiwavelength observations and theoretical modeling to provide constrains to different scenarios of star and planet formation. We present summaries of the following results: (1) the study of a deeply-embedded low-mass protostar in the B59 molecular cloud; (2) the discovery of accreting very low-mass objects in the IPHAS survey; (3) the identification of faint planetary-mass candidates in the cores of young open clusters using multi-conjugate adaptive optics; (4) the discovery of a widely separated companion of a young brown dwarf; (5) the search for reflected light from hot Jupiters; and (6) the spectroscopic study of relics of the formation of our Solar System.

1 Introduction

Thirteen years ago the first unambiguous discoveries of brown dwarfs (BDs) and extrasolar giant planets (EGPs) [11, 12, 16] opened a new window into the elusive very low-mass population of the Milky Way. Observational and theoretical progress on the study of these objects has been enormous but, of course, a lot remains to be done. One of the most controversial subjects is the origins of BDs and EGPs. This

---

E.L. Martín, V.J.S. Béjar, H. Bouy, B. Riaz, F. Rodler, and L. Valdivielso
Instituto de Astrofísica de Canarias, Calle Vía Láctea s/n, E-38200, La Laguna, Tenerife, Spain
e-mail: ege@iac.es

R. Deshpande and R. Tata
University of Central Florida, Physics Department, 4000 Central Florida Blvd, Orlando, USA

J.M. Diego et al. (eds.), *Highlights of Spanish Astrophysics V*, Astrophysics and Space Science Proceedings, DOI 10.1007/978-3-642-11250-8_15,

bears upon our basic understanding of these objects and the difficulties to separate them with a clear and universally accepted definition [10].

This year, at the Instituto de Astrofísica de Canarias (IAC), a strategic project has started to foster the research on BDs and EGPs for the next 5 years. In this review paper we provide a panorama of some of the research topics that are underway in our group. These projects attack a common theme from different angles. They are diverse but connected by the general framework of the cosmogony of BDs and EGPs.

2 2MASS 171123: A Deeply-Embedded Low-Mass Protostar

Basmah Riaz is currently a Marie Curie postdoctoral fellow at IAC. She is a young researcher in the framework of the CONSTELLATION network (http://www.constellation-rtn.eu). She is studying a protostellar system (2MASS 171123) in the molecular cloud B59. This cloud is known to be the only star-forming region in the Pipe nebula (see [15] and references therein). An outflow has been detected in B59 and this protostar is considered a prime candidate to be the driving source of this outflow. We have detected scattered nebulosity near 2M171123 in our Ks-band observations (obtained with the Blanco 4 m telescope by R. Tata) that seems to trace the edges of this outflow cavity. Another interesting feature observed in the Ks-image is a dark lane offset from the protostar, which we interpret as the shadow cast by this system onto an optically thick background cloud located close to it. Basmah has been working on modeling of this system to determine the basic envelope and disk parameters, as well as obtain model images of it. We have been able to explain the observed Ks-band variability for this protostar by variations in its mass infall rate.

We have also been analyzing the different absorption features observed in 2M 171123 mid-IR spectrum obtained with Spitzer IRS. This protostar exhibits a rarely observed 11.3 μm absorption feature within its 10 μm silicate band. We have found a strong correlation between the strength in this feature and the water-ice column densities, indicating an origin of this feature in the thickness of ice mantle over the silicate grains (see Fig. 1). This rare feature has been reported before for a few other protostars in the literature [4], but no clear conclusions have been presented on its origin, other than hinting at the possible presence of crystalline forsterites. Such a scenario though seems improbable considering the very young age of these systems, and the cool temperatures that prevail in their environments. The strong correlation that we have found between the inclusion of a thicker ice coating over the grains, and the strengthening of the 11.3 μm shoulder provides a more plausible explanation, in line with 2M171123 being a deeply-embedded source.

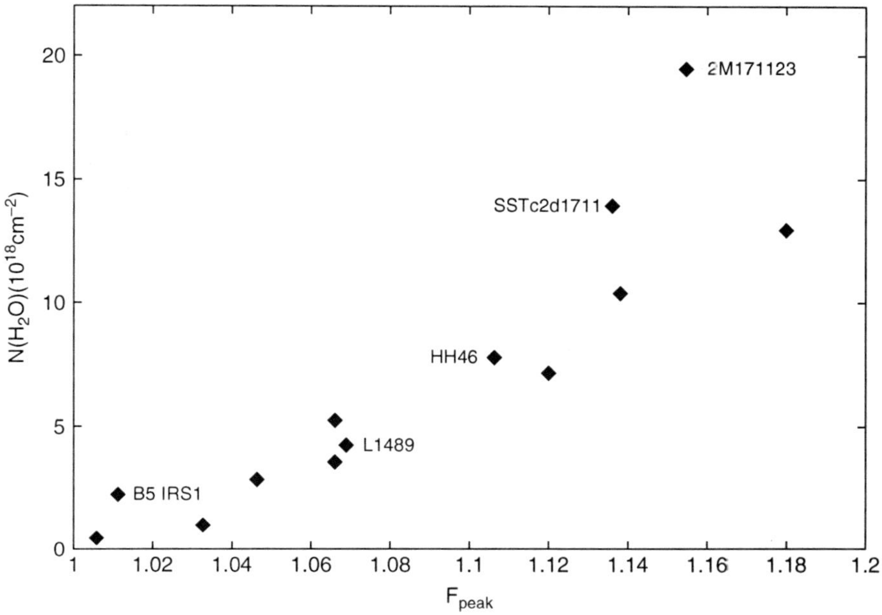

Fig. 1 Direct correlation between the strength in the 11.3 μm shoulder, and the water-ice column densities. The thicker the ice mantle over silicate grains, the stronger is the absorption in the 11.3 μm feature

3 Very Low-Mass Accreting Objects Scattered Around the Milky Way

Luisa Valdivelso is currently a PhD student at IAC. She is expected to graduate in 2009. Her work has proved that accreting very low-mass stars and brown dwarfs can be identified in IPHAS, a Hα emission survey of the northern Milky Way (http://www.iphas.org/).

To carry out this project, we have used Virtual Observatory tools to cross-match the IPHAS catalogue with the 2MASS catalogue. We defined photometric criteria to indentify Hα emission sources with near-infrared colors similar to known young very low-mass stars and brown dwarfs. 4,000 candidates were initially identified that met our criteria over an area of 1,600 square degrees.

Low-resolution optical spectra were obtained for 113 candidates using the Nordic Optical Telescope and the William Herschel Telescope in La Palma. Spectral types have been derived for the 33 candidates that have spectroscopically confirmed Hα emission, negligible reddening and M spectral class. We have also measured Hα emission and NaI doublet (818.3 nm, 819.5 nm) equivalent widths in these 33 objects. We confirm that these 33 IPHAS candidates have strong Hα indicative of disk accretion for their spectral type (see Fig. 2). Twenty-three of them have spectral class M4 or later, of which 10 have classes in the range M5.5–M7.0 and thus could

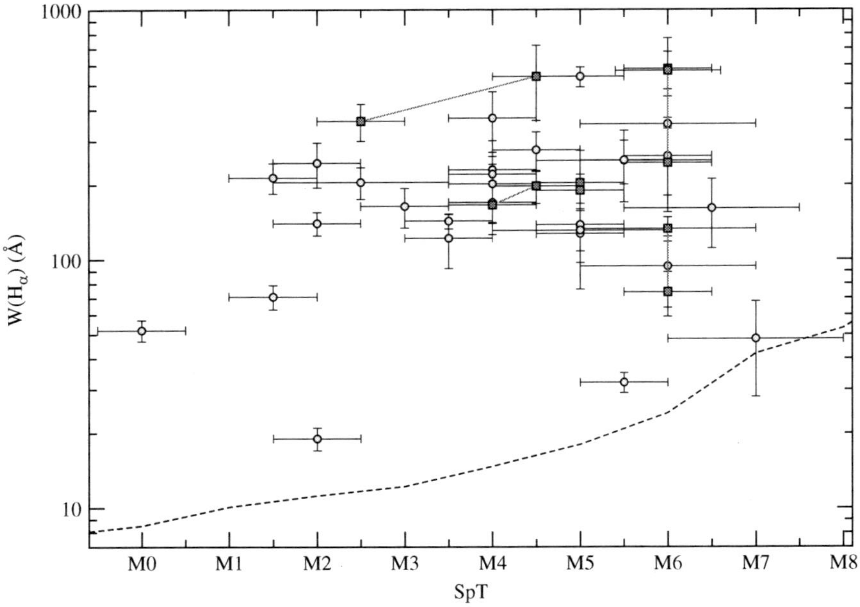

Fig. 2 Hα equivalent width against adopted spectral type for our objects. The *dashed line* denotes the dividing line between chromospheric activity and disk accretion [2]. Our objects are clearly above the *dashed line* and hence they are likely undergoing mass accretion

be very young brown dwarfs. Also 23 objects have weak NaI doublet, an indication of low surface gravity.

We conclude that IPHAS provides a very valuable database to identify accreting very low-mass stars and brown dwarfs, and that Virtual Observatory tools provide an efficient method for identifying these objects over large areas of the sky. Based on our success rate of 23 Hα emission objects with spectral type in the range M4–M7 out of 113 candidates with spectroscopic follow-up, we estimate that there could be hundreds of such objects in the full IPHAS survey.

4 Probing the Very Low-Mass Population of the Cores of Nearby Open Clusters

Hervé Bouy is a Marie Curie IAC–Berkeley outgoing fellow. After spending 2 years at the University of California in Berkeley, he came back to IAC this year. One of the focus points of his research has been the study of the very low mass end of the (sub)stellar population in the core of young clusters (see Fig. 3) using multi-conjugated adaptive optics at the Very Large Telescope in Paranal. This allowed us to indentify a number of new BD and isolated EGP candidates in the Sigma Orionis and Trapezium clusters [5]. Additional panchromatic data allowed us to confirm some of them because their spectral energy distribution were consistent with the

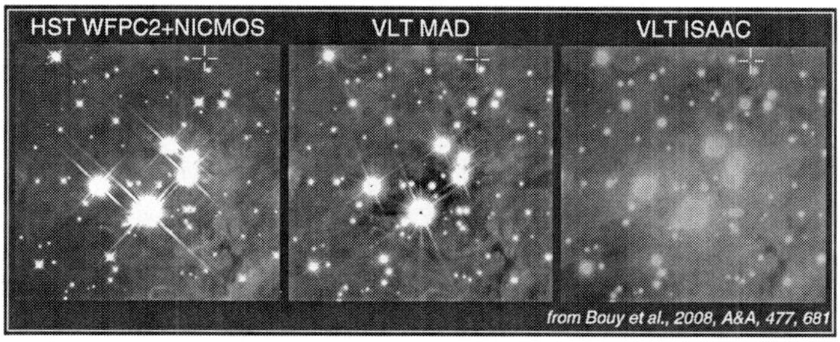

Fig. 3 Comparison of HST, ISAAC and MAD images of the Trapezium. The new MCAO facility MAD allows to reach diffraction limited images with a resolution comparable to HST and a depth better than ISAAC

age and distance of the clusters. Masses in the range 8–20 Jupiters were obtained from theoretical models. This project shows that the deeper we probe the murky end of the stellar population, we keep uncovering fainter and fainter objects.

5 A Young Very Low-Mass Wide Couple

Víctor Sánchez Béjar is a Ramón y Cajal fellow at IAC who has led the discovery of a $14^{+2}_{-8}\,M_{Jup}$ companion [3] located at an angular separation of 4.6 ± 0.1 arcsec (projected distance of ~670 AU) from UScoCTIO 108, a $60 \pm 20\,M_{Jup}$ brown dwarf of the very young Upper Scorpius association. Optical and near-infrared photometry and spectroscopy confirm the cool nature of both objects, with spectral types of M7 and L1–L3, respectively (corresponding to T_{eff} of 2,700 and 2,100 K, respectively) and that they are bona fide members of the association, showing low gravity and features of youth. This is one of the least luminous and massive companions imaged to date, and the widest substellar system with the smallest gravitational bound energy. The existence of this object around a BD at this wide orbit suggests that the companion is unlikely to be formed in a disk based on current planet formation models. Because this system is rather weakly bound, they did not probably form through dynamical ejection of stellar embryos. If the formation of these wide and very low mass systems in the denser central part of clusters is relatively frequent, this could explain the existence of isolated planetary mass objects as planets that became unbound from their primary stars or brown dwarfs.

6 The Search for Starlight Reflected from Extrasolar Planets

Florian Rodler is a postdoctoral fellow at IAC who works for the NAHUAL project (http://www.iac.es/proyecto/nahual/). He also continues to work on a project about the search for starlight reflected from extrasolar planets. Once detected, we will be able to determine the albedo of the planetary atmosphere, which is a key parameter for atmospheric models of BDs and EGPs.

In 2000, high-precision RV measurements revealed the existence of a hot Jupiter orbiting the G0 main-sequence star HD 75289A, indicated by the characteristic wobbling of the star. The planet was found to be at an average distance of only 0.048 AU, and its mass was determined to be larger (or equal) than $M_{\mathrm{p}} \sin i = 0.46\,\mathrm{M_{Jup}}$, which indicated a Jupiter-sized planet [19]. Hence, a campaign for the search for reflected starlight was launched, but resulted in a non-detection [7]. Instead, a stringent upper limit to the geometric albedo was placed ($p < 0.12$). This upper limit was the tightest limit to the geometric albedo of an EGP, confirming theoretical models which predicted a very low albedo for hot Jupiters at optical wavelengths. We reanalyzed these UVES-data, which we had retrieved from the ESO science archive. However, we realized that the results of that research group were based on erroneous orbital phase information of the planet. Adopting the corrected orbital phase information, it turned out that this planet was largely observed at new phase, where it was at its faintest. It was not possible to find any evidence for reflected light. However, this non-detection allowed us to establish new upper limits on the planet-to-star flux ratio as a function of the orbital inclination for different confidence levels. The upper limit to the geometric albedo of the planet HD 75289Ab was found to be $p = 0.46$ for the grey albedo model and $p = 0.57$ for the isolated Class IV model [18]. Jupiter's albedo is $\approx$0.43 in the observed wavelength regime ($\lambda = 400$–530 nm). Although no detection of reflection light could be achieved, this result is still important: the previously upper limit to the geometrical albedo of 0.12 was by far the deepest one. This is no longer valid. The deepest upper limits to the albedos of hot Jupiters were found just to be $\approx$0.30 [17]. We therefore do not have any observational evidence so far that hot Jupiters are really as opaque as predicted by some theoretical models.

7 Relics of the Formation of the Solar System

Javier Licandro is a Ramón y Cajal fellow at IAC who leads a very active group that concentrates in spectroscopic studies of asteroids and comets. In Licandro et al. [9] we studied the composition of the surface of asteroid (3,200) Phaethon which is likely the parent of the Geminid meteor stream and it is a paradigmatic case of asteroid-comet transition object. Phaethon has not displayed cometary activity but models favor the ejection of meteoroids during cometary activity. For this reason, Phaethon has been considered a *dead* or *dormant* comet. The study of asteroid-comet objects is of fundamental importance to address several astronomical problems including the end states of comet nuclei, the abundance of water in main belt asteroids, and its role as a possible source of terrestrial water.

The spectrum of 3,200 Phaethon in the range 0.35–2.4 μm was obtained by us using the 4.2 m William Herschel Telescope, the 2.5 m Nordic Optical Telescope, and the Italian 3.58 m Telescopio Nazionale Galileo at "El Roque de los

Muchachos" Observatory (ORM, La Palma, Spain). It was compared with those of meteorite samples and man-made mineral mixtures to determine possible components, modeled using multiple scattering formulations, and also compared with the spectra of comet nuclei and other comet-asteroid transitional objects. Phaethon's spectrum shows important differences with the few comet nuclei properly observed at these wavelengths and is similar to the spectra of other peculiar comet-asteroid transition objects. The spectral shape is similar to that of aqueously altered CI/CM meteorites and of hydrated minerals. A surface composition with hydrated silicates is also suggested by the models. The spectral properties and dynamical properties of (3,200) Phaethon support an asteroidal nature rather than a cometary one. Phaethon is more likely an *activated* asteroid, similar to the population of activated asteroids in the Main Belt Comets [6], than an extinct comet.

In [13] we showed that there is a group of trans-Neptunian objects (TNOs) (2003 EL_{61} being the biggest member), with surfaces composed of almost pure water ice and with very similar orbital elements. We have already studied another member of the population, 2002 TX_{300} [8]. These objects provide exciting laboratories for the study of the processes that prevent the formation of an evolved mantle of organics on the surfaces of the bodies in the trans-Neptunian belt (TNb). We studied the surface composition of one of the members of the family, 2005 RR_{43}, by modeling its spectrum in the 0.53–2.4 μm spectral range, obtained with the WHT and the TNG telescopes at the ORM. Scattering models show that its surface is covered by water ice, a significant fraction in crystalline state with no trace (5% upper limit) of complex organics. We discussed three possible scenarios to explain the existence of this population of TNOs: a giant collision, an originally carbon depleted composition, or a common process of continuous resurfacing. Assuming that this population is the product of a giant collision, [14] identified a group of TNOs that are potentially candidates to belong to this collisional family.

Acknowledgements This paper makes use of results obtained as part of the INT Photometric Hα Survey of the Northern Galactic Plane (IPHAS) carried out at the Isaac Newton Telescope. The INT and WHT are operated on the island of La Palma by the Isaac Newton Group in the ORM. All IPHAS data are processed by the Cambridge Astronomical Survey Unit, at the Institute of Astronomy in Cambridge. This publication makes use of data products from the Two Micron All Sky Survey, which is a joint project of the University of Massachusetts. This work makes use of EURO-VO software, tools or services. The EURO-VO has been funded by the European Commission through contract numbers RI031675 (DCA) and 011892 (VO-TECH) under the 6th Framework Programme. This research has made use of the Spanish Virtual Observatory supported from the Spanish MEC through grants AyA2005-04286, AyA2005-24102-E. Support for our research group has been provided by the MEC grant AyA2007-67458. H. Bouy acknowledges the funding from the European Commission Sixth Framework Program as a Marie Curie Outgoing International Fellow (MOIF-CT-2005-8389).

References

1. A'Hearn, M.F., Millis, R.L., Schleicher, D.G., et al., Icarus 118, 223 (1995)
2. Barrado y Navascués, D., Martín, E.L. , AJ 126, 2997 (2003)
3. Béjar, V.J.S., et al., ApJ, 673, L185 (2008)

4. Boogert, A.C.A., et al., ApJ 678, 985 (2008)
5. Bouy, H., et al., A&A 477, 681 (2008)
6. Hsieh, H., Jewitt, D., Science 312, 561 (2006)
7. Leigh, C., Collier Cameron, A., Guillot, T., MNRAS 346, 890 (2003)
8. Licandro, J., di Fabrizio, L., Pinilla-ALonso, N., et al., A&A 457, 323 (2006)
9. Licandro, J., Campins, H., Mothé-Diniz, T., Pinilla-ALonso, N., De León, J., A&A 461, 751 (2007)
10. Martín, E.L., ed., *Brown Dwarfs*, IAU Symp. 211 (2002)
11. Mayor, M., Queloz, D., Nature 378, 355 (1995)
12. Nakajima, T., Oppenheimer, B.R., Kulkarni, S.R., Golimowski, D.A., Matthews, K., Durrance, S.T., Nature 378, 463 (1995)
13. Pinilla-Alonso, N., Brunetto, R., Licandro, J., Gil-Hutton, R., A&A 468, L25 (2007)
14. Ragozzine, D., Brown, M.E., AJ 134, 2160 (2007)
15. Rathborne, J.M., Lada, C.J., Muench, A.A., Alves, J.F., Lombardi, M., ApJS 174, 396 (2008)
16. Rebolo, R., Zapatero-Osorio, M.R., Martín, E.L., Nature 377, 129 (1995)
17. Rodler, F., Kürster, M., Henning, T., A&A 485, 859 (2008)
18. Sudarsky, D., Burrows, A., Hubeny, I., ApJ 588, 1121 (2003)
19. Udry, S., Mayor, M., Naef, D., et al., A&A 356, 590 (2000)

A Pilot Survey of Stellar Tidal Streams in Nearby Spiral Galaxies

David Martínez–Delgado, R. Jay Gabany, Jorge Peñarrubia, Hans-Walter Rix, Steven R. Majewski, Ignacio Trujillo, and Michael Pohlen

Abstract Within the hierarchical framework for galaxy formation, merging and tidal interactions are expected to shape large galaxies to this day. While major mergers are quite rare at present, minor mergers and satellite disruptions–which result in stellar streams–should be common, and are indeed seen in both the Milky Way and the Andromeda Galaxy. As a pilot study, we have carried out ultra-deep, wide-field imaging of some spiral galaxies in the Local Volume, which has revealed external views of such stellar tidal streams at unprecedented detail, with data taken at small robotic telescopes ($0.1 - 0.5$ m) that provide exquisite surface brightness sensitivity. The goal of this project is to undertake the first systematic and comprehensive imaging survey of stellar tidal streams, from a sample of ~ 50 nearby Milky Way-like spiral galaxies within 15 Mpc, that features a surface brightness sensitivity of ~ 30 mag/arcsec^2. The survey will result in estimates of the incidence, size/geometry and stellar luminosity/mass distribution of such streams. This will not only put our Milky Way and M31 in context but, for the first time, also provide an extensive statistical basis for comparison with state-of-the-art, self-consistent cosmological simulations of this phenomenon.

D. Martínez–Delgado and I. Trujillo
Instituto de Astrofisica de Canarias, La Laguna, Spain
e-mail: ddelgado@iac.es

D. Martínez–Delgado and H.-W. Rix
Max-Planck-Institute fur Astronomie, Heidelberg, Germany

R. Jay Gabany
Black Bird Observatory, NM, USA

J. Peñarrubia
Institute of Astronomy, Cambridge, UK

S. R. Majewski
Department of Astronomy, University of Virginia, USA

M. Pohlen
Cardiff University, UK

J.M. Diego et al. (eds.), *Highlights of Spanish Astrophysics V*, Astrophysics and Space Science Proceedings, DOI 10.1007/978-3-642-11250-8_16,

1 Introduction

Within the hierarchical framework for galaxy formation (e.g. [29]), the stellar bodies of galaxies are expected to form and evolve through dark-matter-driven mass infall and successive coalescence of smaller, distinct sub-units that span a wide mass range. Mergers of initially bound sub-halos (which we refer to as *satellites*; they consist of dark matter, gas, and, in most cases, stars) are effected by dynamical friction, either through gradual orbital decay or by a single encounter (depending on the initial orbit), its eccentricity and the satellite-to-main-galaxy mass ratio. It is likely that a satellite becomes disrupted by the tidal forces of the larger companion before its orbit spirals all the way to the center. If such a tidal disruption is complete, and no bound satellite is left, dynamical friction ceases to act. If the disruption is only partial at this epoch, the surviving satellite fragment displays extensive tidal tails, leading and trailing its current position in the galactic halo.

While in Λ-Cold Dark Matter (ΛCDM) the interaction rate is expected to drop to the present-day epoch, such disruption of satellites should still occur around normal spiral galaxies. The fossil records of these merger events may be detected nowadays in the form of distinct coherent stellar structures in the outer regions of massive systems. The most spectacular cases of tidal debris are long, dynamically cold, stellar streams from a disrupted dwarf satellite, which have wrapped around the host galaxy's disk and roughly trace the orbit of the progenitor satellite. The now well-studied Sagittarius tidal stream surrounding the Milky Way [12] and the giant stream in Andromeda galaxy [8] are archetypes of these satellite galaxy merger *fossils* in the Local Group. They provide sound qualitative support for the scenario that tidally disrupted dwarf galaxies are important contributors to stellar halo formation in the Local Group spirals.

State-of-the-art, high-resolution numerical simulations of galaxy formation, built within the ΛCMD context (e.g. [21, 27]), can guide the quest for observational signatures of such star-streams (e.g. [4, 9]). Recent simulations have demonstrated that the characteristics of substructure currently visible in the stellar halos are sensitive to the last (0–8 Gyr ago) merger histories of galaxies, a timescale that corresponds to the last few to tens of percent of mass accretion for a spiral galaxy like the Milky Way. While stellar streams in the Milky Way and Andromeda can be studied in detail, comparison with cosmological models is limited by *cosmic variance*. However, the current models imply that a survey of 50–100 parent galaxies reaching to a surface brightness of 30 mag/arcsec^2 would reveal many tens of tidal features, perhaps nearly one detectable stream per galaxy [9]. However, a specific comparison of these simulations with observations is missing because no suitable data sets exist. Such a comparison, which could quantify the present sub-halo merger rate, is not only important as a test of ΛCDM models, but also as a more direct probe of how resilient disks are to minor mergers.

2 Stellar Tidal Streams in External Galaxies

Galactic archeology by looking at tidal remnants is a relatively new field of research that so far has been primarily focused on the Local Group spiral galaxies. The first known tidal stream surrounding the Milky Way (the Sagittarius tidal stream) was discovered less than one decade ago [15, 20]. In recent years, studies have focused on the formation and evolution of our Galaxy have been revolutionized by the first generation of wide-field, digital imaging surveys. The resulting extensive photometric databases have provided, for the first time, spectacular panoramic views of Milky Way tidal streams (e.g. the *Field of Streams* [2]) and have revealed the existence of large stellar sub-structures in the halo [10,22], which have been interpreted as observational evidence of our home Galaxy's hierarchical formation. The discovery of the Monoceros tidal stream [30] and the possible Canis Major dwarf galaxy [14, 16], located close to the Galactic plane, indicates that minor mergers might play a relevant role in the formation of the outer regions of spiral disks [24]. A multitude of tidal streams, arcs, shells and other irregular structures that are possibly related to ancient merger events can be seen in deep panoramic views of the Andromeda halo [8]. These pictures show in detail the level of stellar sub-structure that might be present in the halos of nearby external spiral galaxies.

Our current understanding of the Local Group spirals has provided a couple of firsthand examples of individual minor mergers and their link to the current state of massive galaxy building [9]. The search for analogues to these galactic fossils beyond the Local Group is required not only to see whether the Milky Way and Andromeda galaxies are *typical* with regard to substructure formation, but to estimate the fractional contribution of accreted mass and the mass spectrum of accreted bodies in the life of these massive systems, an issue that remains unresolved.

Unfortunately, over the past decade only a few cases of confirmed stellar tidal streams have been detected in spiral galaxies outside the Local Group (e.g. [25] and references therein). The first cases of extragalactic tidal streams were reported a decade ago by [13]. Using special contrast enhancement techniques on deep photographic plates, they were able to highlight two possible tidal streams surrounding the galaxies M83 and M104. Then, deep CCD images of the nearby, edge-on, galaxy NGC 5907 by [26] revealed an elliptically-shaped loop in the halo of this galaxy. This was the most compelling example of a non-Local-Group tidal stream up to now. More recently, very deep images [17, 18] have clearly revealed large scale, complex structures of arcing loops in the halos of several nearby galaxies (see Fig. 1). These detailed observations provide an elegant example of how a single, current epoch, low-mass satellite accretion can produce a very complex, rosette-like structure of debris dispersed in the halo of its host galaxy.

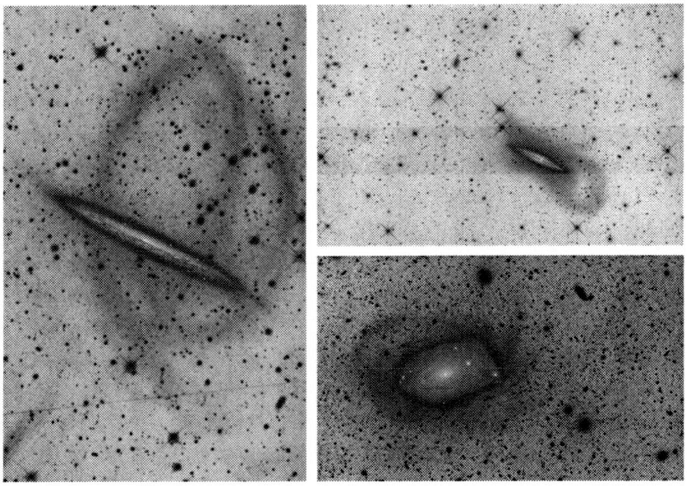

Fig. 1 *Left:* Deep image of the stellar tidal stream around NGC 5907 obtained with the 0.5 m Black Bird Observatory (BBO) telescope [17]. A N-body model of this structure is shown in Fig. 2. *Right, top:* A low-galactic latitude stellar tidal stream of NGC 4013, discovered by our team from deep images taken with the BBO telescope. *Right, bottom:* Deep images taken with a FSQ-106ED telescope of only 10 cm aperture allowed the discovery of a giant tidal stream in the halo of the spiral galaxy Messier 63 [19]. A color inset of the disk of each galaxy has been inserted with reference purpose

3 A Pilot Survey of Stellar Tidal Stream in Nearby Galaxies (2006–2008)

We have initiated a pilot survey of stellar tidal streams in a select number of nearby, edge-on spiral systems using modest (0.1 − 0.5 m), robotic telescopes operating under very dark skies. The main results of this first observational effort are given below.

3.1 The Tidal Stream of NGC 5907

In summer 2006, we re-observed the tidal stream of NGC 5907 as a commissioning target to demonstrate the sensitivity of our small aperture telescope for detecting extragalactic tidal streams. Our deep observations showed for first time an interwoven, rosette-like structure of debris dispersed in the halo of this spiral galaxy (Fig. 1; [17]). Its presence provides confirmation that these tidal remnants can survive several Gigayears, as predicted by N-body simulations of tidally disrupted stellar systems around the Milky Way (e.g. [11, 23]).

Our N-body simulations (Fig. 2) of the tidal disruption of a dwarf satellite by a disk galaxy and its dark halo potential suggest that most of the tidal features

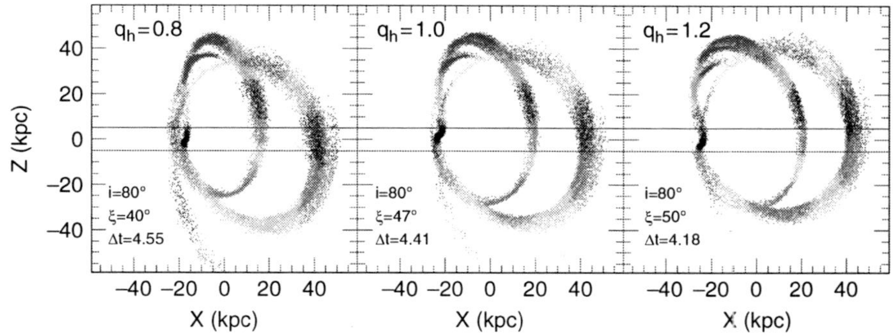

Fig. 2 *N*-body model of the stellar stream detected in NGC 5907. The satellite is realized as a King model with an initial mass, King core, and tidal radii of $M = 2 \times 10^8\,M_\odot$, $r_c = 0.39$ kpc and $r_t = 2.7$ kpc, respectively (see [17] for details). For this particular model the orbital period is $T_r = 0.9$ Gyr. Each panel adopts a different halo minor-to-major axis-ratio (q_h). Different colors denote different ranges of projected radial velocities, namely yellow (−210, −150) km/s; blue (−150, −90) km/s; green (−90, −30) km/s; black (−30, +30) km/s; red (+30, +90) km/s; cyan (+90, +150) km/s, and magenta (+150, +210) km/s. Note that the three models show similar variations of the projected radial velocity along the stream. This result suggests that additional kinematical data may provide insufficient information to break the model degeneracies discussed in [17]

observed in NGC 5907 can be explained by a single accretion event. Interestingly, this model finds that the stellar stream may be relatively old with the fainter, outer loop material becoming unbound at least 3.6 Gyrs ago. The stellar stream around NGC 5907 may therefore represent one of the most ancient tidal debris ever reported in the halo of a spiral galaxy. It also confirms that spiral galaxy halos in the Local Universe still contain a significant number of galactic fossils from their hierarchical formation and that they can be detected with modest instruments.

3.2 Discovery of Stellar Tidal Streams in Warped Disk Galaxies

Promising galaxies in the hunt for extragalactic tidal streams are those that display outstanding asymmetries in optical or HI images. It has been long suggested that these perturbations are a result of gravitational interaction with nearby companions. The most striking case is NGC 4013, an isolated spiral galaxy famous for having one of the most prominent HI warps detected so far [3]. Our deep images of this galaxy [18] revealed a faint loop-like structure that appears to be part of a gigantic, low-inclination stellar tidal stream (Fig. 1). Although its true three dimensional geometry is unknown, the sky-projected morphology of this structure displays a remarkable resemblance to the theoretical predictions for an edge-on view of the Milky Way's Monoceros tidal stream [23]. This suggests that the progenitor system (whose current position and final fate remains unknown) may have been a galaxy with an initial mass $\sim 10^8\,M_\odot$ moving on a low-inclination ($\simeq 25°$), nearly-circular orbit. Using this model as a template, the tidal stream may be approximately ~ 3 Gyr of age.

We have also discovered a stellar tidal stream in the halo of the nearby spiral Messier 63 [19]. Our data, collected from different telescopes, reveals an enormous, arc-like structure around this galaxy's disk extending ~ 29 kpc from its center (Fig. 1), apparently tilted with respect to its strong gaseous warp. This unassailable indication of a past merger event provides an additional example of apparently isolated galaxies with significantly warped gaseous disks that also show evidence indicating the ongoing tidal disruption of a dwarf companion.

These results highlight the fact that disks that appear to be undisturbed in high surface brightness optical images but warped in HI maps can reveal complex signatures of recent accretion events when viewed in deep optical surveys. Additionally, with the growing number of examples of spirals (NGC 5907, NGC 4013, M 63) showing a connection between warped disks and evidence of mergers, the origin of galactic warps via interactions with minor mergers is a possibility worth further investigation.

3.3 Diffuse Light Structures in Nearby Spiral Galaxies

Over the last two years, we have also obtained wide field, follow-up images of a dozen familiar, nearby spiral galaxies that are widely known to have diffuse light features in their outer regions. Although our current data for them is not yet deep enough yet (with an estimated surface brightness limit of ~ 28 mag/arcsec^2), our panoramic views of their halos have revealed their fossil structures in detail. The most conspicuous cases detected during this observational effort are shown in Fig. 3. In addition, we have also discovered striking stellar tidal streams of different morphological types in several neighboring galaxies [19]. Thus, our deep images are very effective in revealing a plethora of very faint morphological perturbations and dynamical features in the external regions of nearby galactic disks, presumably all signposts of minor gravitational interactions.

4 Future Work

The promising results of our foray into a more systematic look for tidal streams in the nearby Universe encourage a more aggressive attention to this new way of understanding galaxy formation. The overall goal of this project is to conduct the first systematic survey of stellar tidal streams for $\sim$ 50 nearby spiral galaxies ($D < 15$ Mpc) to a (large area) surface brightness sensitivity of ~ 30 mag/arcsec^2, in order to obtain the first comprehensive census of such structures in the Local Volume. Based on theoretical predictions on stream counts (one stream expected per galaxy; see Sect. 1), we expect this observational effort to yield a tidal stream sample with sufficient statistical significance to undertake a direct comparison with high

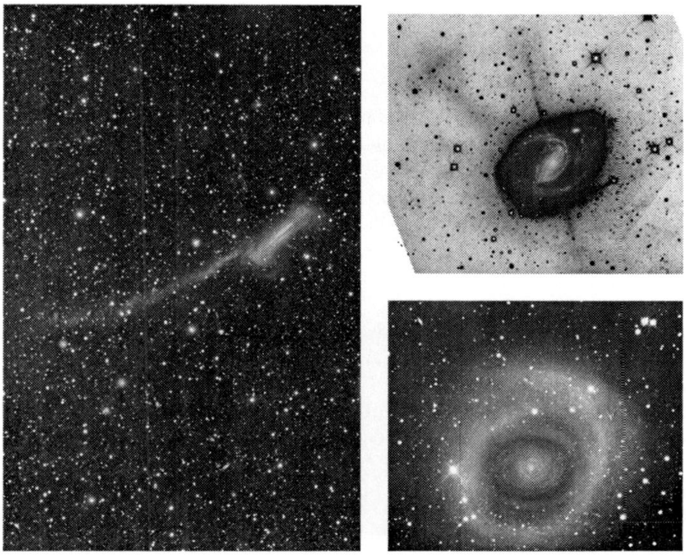

Fig. 3 Diffuse light, giant structures detected in the outskirts of several nearby galaxies in our pilot survey. *Left:* panoramic view of the halo of NGC 3628 obtained with a 10 cm telescope. In addition to the known giant filament [5], our deep images reveal for first time new arc-like features in the edge of its perturbed disk, which suggest that this galaxy could be suffering a tidal encounter with a dwarf satellite. *Bottom, right:* a striking, very faint ring structure in M94 detected with the BBO telescope [28], consistent with being the optical counterpart of the ultraviolet extended disk discovered from GALEX observations [6]. *Top, right:* first wide-field CCD panoramic image of the mysterious jet-like features in the halo of the active galaxy NGC 1097, previous reported in photographic plates [1]. The tidal origin of these features is still controversial [7]

resolution cosmological simulations of hierarchical galaxy formation and satellite dark matter halo dynamics.

A number of astrophysical problems can be tackled with the data produced from this survey and covering a large variety of classical topics (stellar populations, formation of galactic disk, galactic dynamics, globular cluster formation, dark matter halo flattening, cosmology; see a discussion in [17]). The survey will result in estimates of the incidence, size/geometry and stellar luminosity/mass distribution of such streams. The results of the project will provide a direct and stringent test of hierarchical structure formation on this scale, will constrain the present-epoch (minor) interaction rate and probe the minor-merger resilience of stellar disks.

Acknowledgements The authors gratefully thank the following astrophotographers whose collaboration helped make our initial survey possible: Ray Gralak (Fig. 2, M63), Steve Mandel (Fig. 3, NGC 3628), Kenneth Crawford and Mischa Schirmer.

References

1. Arp, A., ApJ 207, 147 (1976)
2. Belokurov, V., et al., ApJ 642, L137 (2006)
3. Bottema, R., Shostak, G.S., van der Kruit, P.C., Nature 328, 401 (1987)
4. Bullock, J.S., Johnston, K.V., ApJ 635, 931 (2005)
5. Chromey, F.R., Elmegreen, D.M., Mandell, A., McDermott, J., AJ 115, 2331 (1998)
6. Gil de Paz, A., et al., ApJS 173, 185 (2007)
7. Higdon, J.L., Wallim, J.F., ApJ 585, 281 (2003)
8. Ibata, R., Martin, N.F., Irwin, M., Chapman, S., Ferguson, A.M.N., Lewis, G.F., McConnachie, A.W., ApJ 671, 1591 (2007)
9. Johnston, K.V., Bullock, J.S., Sharma, S., Font, A., Robertson, B.E., Leitner, S N., ApJ 689, 936 (2008)
10. Juric, M., et al., ApJ 673, 864 (2008)
11. Law, D. R., Johnston, K.V., Majewski, S.R., ApJ 619, 807 (2005)
12. Majewski, S.R., Skrutskie, M.F., Weinberg, M.D., Ostheimer, J.C., ApJ 599, 1082 (2003)
13. Malin, D., Hadley, B., PASA 14,52 (1997)
14. Martin, N.F., Ibata, R.A., Bellazzini, M., Irwin, M.J., Lewis, G.F., Dehnen, W., MNRAS 348, 12 (2004)
15. Martínez-Delgado, D., Aparicio, A., Gómez-Flechoso, M.Á., Carrera, R., ApJ 549, L199 (2001)
16. Martínez-Delgado, D., Butler, D., Rix, H.W., Franco, I.V., Peñarrubia, J., Alfaro, E.J., Dinescu, D.I., ApJ 633, 205 (2005)
17. Martínez-Delgado, D., Peñarrubia, J., Gabany, R.J., Trujillo, I., Majewski, S.R., Pohlen, M., ApJ 689, 184 (2008)
18. Martínez-Delgado, D., Pohlen, M., Gabany, R.J., Majewski, S.R., Palma, C., ApJ 692, 955 (2009)
19. Martínez-Delgado, D., et al., in preparation (2009)
20. Mateo, M., Olszewski, E.W., Morrison, H.L., ApJ 508, 55 (1998)
21. Moore, B., Ghigna, S., Governato, F., Lake, G., Quinn, T., Stadel, J., Tozzi, P., MNRAS 304, 465 (1999)
22. Newberg, H.J., et al., ApJ 569, 245 (2002)
23. Peñarrubia, J., et al., ApJ 626, 128 (2005)
24. Peñarrubia, J., McConnachie, A., Babul, A., ApJ 650, L33 (2006)
25. Pohlen, M., Martínez-Delgado, D., Majewski, S., Palma, C., Prada, F., Balcells, M., in *Satellites and Tidal Streams*, PASP 327, 288 (2004)
26. Shang, Z., et al., ApJ 504, L23 (1998)
27. Springel, V., Wang, J., Vogelsberger, M., Ludlow, A., Jenkins, A., Helmi, A., Navarro, J.F., Frenk, C.S., White, S.D., MNRAS 391, 1685 (2008)
28. Trujillo, I., in preparation (2009)
29. White, S.D.M., Rees, M.J., MNRAS 183, 341 (1978)
30. Yanny, B., et al., ApJ 540, 825 (2000)

Massive Young Stellar Clusters in the Milky Way

Ignacio Negueruela

Abstract Compact young open clusters with very high masses ($M_{\rm cl} \gtrsim 10^5\,\rm M_\odot$) have been observed in many galaxies, and their connection to globular clusters is a matter of discussion. However, until very recently, no clusters with masses $M_{\rm cl} \gtrsim 10^4\,\rm M_\odot$ were known in the Milky Way. Their absence was considered a natural consequence of the mild star formation rate in the Milky Way. The development of new infrared observational techniques has completely changed our perception. At present, almost a dozen young massive clusters are known in the Milky Way, and there are reasons to believe that many more wait to be found. In this paper, I briefly review our knowledge of some of these objects: the extended star-forming region Cygnus OB2, the compact young massive cluster Westerlund 1, currently believed to be the most massive young cluster in the Galaxy, and the new massive clusters hosting a large population of red supergiants found at the base of the Scutum Arm.

1 Introduction

The development of infrared detectors has allowed us to realize that the majority of stars form in clusters inside giant molecular clouds, even if many of these clusters disperse before becoming visible [24]. Clusters in the solar neighborhood are typically small, with masses up to $\sim 10^3\,\rm M_\odot$ [4], though perhaps their natal masses are somewhat higher [39]. More massive clusters were thought scarce in the Milky Way. The double cluster in Perseus (h & χ Per) is a good example of rather young (double) cluster, with a combined initial mass $M_{\rm cl} \gtrsim 10^4\,\rm M_\odot$ [40]. Another example is the very young and compact cluster NGC 3603, which has a large population of very massive stars [34] and $M_{\rm cl} \gtrsim 10^4\,\rm M_\odot$ [20].

I. Negueruela
Departamento de Física, Ingeniería de Sistemas y Teoría de la Señal. Escuela Politécnica Superior. Universidad of Alicante. Apdo.99 E-03080 Alicante, Spain
e-mail: ignacio@dfists.ua.es

J.M. Diego et al. (eds.), *Highlights of Spanish Astrophysics V*, Astrophysics and Space Science Proceedings, DOI 10.1007/978-3-642-11250-8_17,

In contrast, globular clusters are testimony to an earlier epoch when enormous clusters were formed. Low-mass stars lost in the course of their dynamical evolution could explain the majority of Population II field stars found in galaxy halos [30]. The formation of massive cluster, however, is not restricted to the distant past. Indeed, hundreds of young clusters with estimated masses comparable to or greater than Galactic globular clusters ($\geq 10^5\,M_\odot$) have been identified in starburst galaxies like, e.g., the Antennae galaxies (e.g. [41]). Since the size of the clusters formed seems to be correlated to the intensity of the starburst and star formation was more intense in the past, these Young Massive Clusters (YMCs) might represent the dominant mode of star formation in the Universe.

Unfortunately, at extragalactic distances, even the brightest individual cluster members are unresolved. The physical properties of YMCs must be estimated from their integrated spectra and photometry, dominated by the most massive and luminous stars, or from their dynamical masses, estimated from their half mass radius and velocity dispersion. These methods require strong assumptions, which may or not be justified (e.g. [2, 7]). We do not know if the observed YMCs will evolve into globular clusters. The fate of a cluster depends on how much mass is contributed by low-mass stars [14], i.e. on the initial mass function (IMF), and the low-mass portion of the IMF cannot be observed in YMCs. Studies of YMCs by different groups have come to entirely contradictory conclusions about the shape of their IMF. Some find a truncation of the IMF towards lower stellar masses (e.g. [33]). Others find that a standard IMF [29] describes well the whole range of stellar masses (e.g. [31]). Because of the complications involved in studying unresolved populations (e.g. [1]), the study of the most massive young clusters in the Milky Way provides a useful testbed for comparison, at a distance where individual stars may be observed. In recent years, important efforts have been carried out to locate the most massive clusters in the Galaxy (e.g. [15]). The presence of a population of YMCs with $M_{\rm cl} \gtrsim 10^5\,M_\odot$ in the Milky Way had not been considered until recently, but is not ruled out at present [19, 39].

2 Cygnus OB2

Cyg OB2 was considered an interesting target since a population of very massive stars was discovered, but early studies were hampered by high obscuration [22]. Advances in detector technology allowed comprehensive investigations, which identified ~ 60 stars more massive than $15\,M_\odot$ [32]. Cyg OB2 is very compact for an OB association and seems to occupy a more or less unique position somewhat intermediate between an open cluster and a normal OB association (c.f. [26]). The central region contains two cluster-like stellar concentrations [5]. They form an elongated figure $\sim 4' \times 10'$ ($\sim 2 \times 5$ pc at the distance of Cyg OB2) and contain 18 O-type stars (not counting binaries, which could perhaps almost double this figure).

Star counts in the 2MASS observations of this region suggested a much richer population of early type stars, leading to the proposal that the number of O-type stars in Cyg OB2 could exceed 100 [26]. Several studies have followed [10, 18, 38],

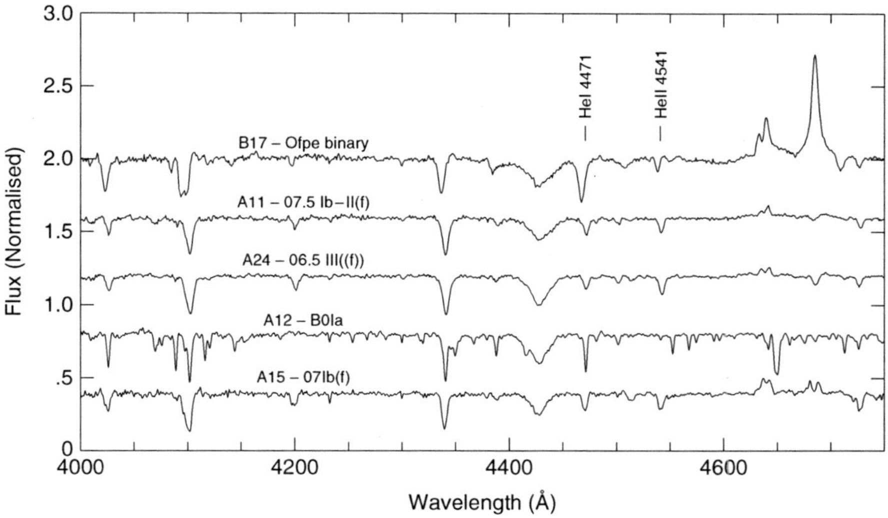

Fig. 1 New evolved very massive stars in Cyg OB2 found by [38] amongst the candidates of [10]

resulting in the discovery of many new members (see Fig. 1). The whole association contains > 50 (and likely ~70–80) O-type stars [27, 38], including another small cluster [38]. The exact mass of the association is very difficult to determine. The number of O-type stars is highly uncertain due to their propensity to appear in binaries or groups. A high binary fraction in Cyg OB2 has been confirmed observationally [28]. In addition, Cyg OB2 is projected along the line of sight spanning the Local Spur and so suffers from high fore- and background contamination, especially stars connected with the Cygnus Superbubble [9].

The properties of Cyg OB2 represent clear demonstration that massive star clusters can exist without being conspicuous. It is unlikely that Cyg OB2 would stand out if it were at 5 kpc or projected against a denser background. Another region which may contain a very high number of massive stars spread over a moderately wide area, resulting in a low apparent density, is the massive star forming region W49A [21]. Subclustering is seen in both Cyg OB2 and W49A, as well as in other large Galactic star forming regions, such as W51 [36]. Study of large stellar complexes in M51 [3] shows that the age spread within different clusters is $\lesssim$ 10 Myr. It is thus possible that the largest stellar aggregates in the Galaxy have escaped detection until now.

3 Westerlund 1

Our perception of the cluster population in the Milky Way has changed since the discovery that Westerlund 1 (Wd 1) is much more massive than other Galactic clusters [8]. Three independent mass estimates place its mass around $10^5\,M_\odot$ [6, 8, 35], making it the only Galactic young cluster approaching the size of YMCs in starburst regions. With a half-mass radius of ~ 0.5 pc, it is as dense and compact as extragalactic YMCs [8]. These properties provide us with an excellent opportunity to measure the physical properties of a YMC at a distance that allows the observation of individual stars down to subsolar masses. Direct measurement of its stellar content finds an IMF consistent with standard down to $\sim 2\,M_\odot$ [6].

Moreover, at an age ~ 4.5 Myr, Wd 1 is an ideal laboratory to study the post-MS evolution of massive stars. The cluster contains 23 Wolf–Rayet stars [11] and a large ($\gtrsim 100$) population of luminous supergiants with spectral types from O to M [8, 37]. This huge population of evolved massive stars contains several examples of rare evolutionary phases and peculiar objects, such as the strange sgB[e] star 9, which may be a common envelope binary, or the red supergiant stars 20 & 26, which show bipolar nebulae in radio and Hα. Such rare stars are frequent in Wd 1 simply because of the size of the population. A recent report on the state of the study of this population is presented in [37]. A detailed investigation of its evolutionary context will be crucial to understand the evolutionary paths of massive stars and the way in which they lose a significant fraction of their masses before becoming Wolf–Rayet stars. The exact physical processes producing this mass loss will determine the nature of the final remnant left by a massive star and the contribution made by such stars to the enrichment of the ISM.

Wd 1 provides us with a wonderful example of a compact cluster, most likely formed monolithically, and so an excellent analogue to the huge young globular clusters seen in external galaxies.

4 The Red Supergiant Clusters

The realization that Wd 1 is a YMC has led to a search for further examples of very massive young open clusters [15, 19]. As they are likely to be very reddened, a sensible strategy may be looking for them in the near-infrared. The brightest massive stars in the near-infrared will be those in the red supergiant (RSG) phase. A very obscured cluster containing many luminous red stars was found by [16]. Further work suggests that this cluster (known as Red Supergiant Cluster 1; RSGC1) contains at least 15 yellow and red supergiants and has a mass $M_{\rm cl} \approx 3 \times 10^4\,M_\odot$ at an age of ~ 12 Myr [13]. Its dynamical distance is 6.6 ± 0.6 kpc. RSGC1 is located only $\sim 1°$ away from Stephenson 2 (= RSGC2). This cluster contains ~ 25 RSGs and has a mass estimate $M_{\rm cl} \approx 4 \times 10^4\,M_\odot$ at an age in the 10–17 Myr range [12]. Its distance is estimated at 6 ± 2 kpc.

The presence of these two clusters in a small area is unlikely to be a coincidence. As a matter of fact, a third RSG cluster candidate is being investigated by our group in the neighborhood of the other two. The three clusters are located in a region (around Galactic latitude $l = 26°$) where [17] had already discovered a very high density of field RSGs. This overdensity had been attributed to a concentration of sources at distances ~ 6.5 kpc in an area of only a couple of squared degrees, and interpreted as evidence for the presence of a prominent star formation region possibly associated with the end of the Galactic Bar, where it meets the Scutum spiral arm [25].

This area may represent the first fully blown starburst region discovered in the Milky Way. Its location close to the base of the Scutum Arm is almost certainly not casual and suggests that this region may be almost unique in the Galaxy.

5 Future Work

In order to check the possibility that more RSG clusters may exist outside the region at the base of the Scutum Arm, we have made a shallow survey of optically visible clusters in the $l = 30°–100°$ range, using 2MASS photometry. We have searched for the presence of a large number of RSGs, which should be seen in the $K_S/(J - K_S)$ diagrams as a clump of very bright stars with high $(J - K_S)$ and values of the Q_{IR} parameter clearly separating them from field red giants. We have not found any cluster resembling the known RSGCs, but we have come across some interesting cases.

A good example is the open cluster Berkeley 55. I-band spectra of its brightest members (Fig. 2) reveal a population of K–M stars with luminosities close to the

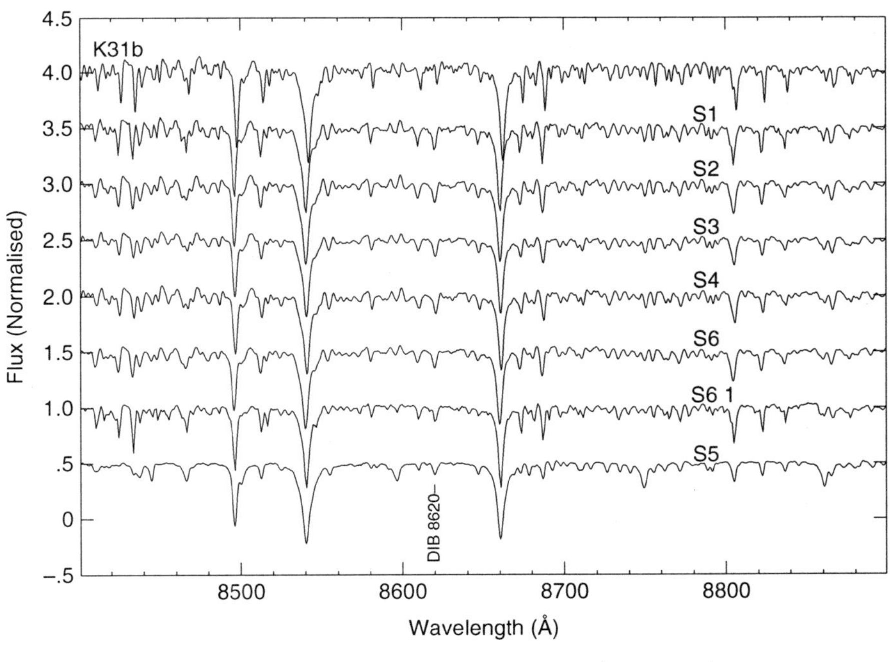

Fig. 2 Seven low-luminosity supergiants in the previously unstudied cluster Berkeley 55. The top spectrum is a K3 Ib supergiant used as comparison. The six stars immediately below are luminosity class Ib or II K-type members of Berkeley 55. The bottom spectrum corresponds to a late-F supergiant member

borderline between Ib supergiants and II bright giants and suggest a moderately massive cluster with an age $\sim$ 50 Myr. Therefore this method looks a very effective way of finding obscured compact clusters with masses above the average.

If we take into account that the area that we have surveyed is unlikely to be the richest in obscured clusters (most lines-of-sight are towards the outer reaches of the Galaxy and extinction is much lower than along line of sights towards the Galactic Bulge), we expect many more interesting clusters to lie hidden in other regions of the Galaxy. Given its closeness to the Sun, $\sim$ 4 kpc [23], it is unlikely that Wd 1 will turn out to be unique. Very young massive clusters may, however, be relatively inconspicuous. Only clusters older than $\sim$ 4 Myr will have a population of yellow and red supergiants that will make them stand out in infrared surveys. Likewise, less compact clusters, like Cyg OB2 may be more difficult to detect. Therefore future detections may be biased towards compact, moderately evolved massive clusters. Current and future wide-field surveys will provide the mining ground for our searches.

Acknowledgements I would like to thank my collaborators in cluster research and especially Dr. Simon Clark for many years of fruitful collaboration. This research is partially supported by the Spanish Ministerio de Ciencia e Innovación under grants AYA2005-00095 and CSD2006-70.

References

1. Bastian, N., in *Pathways Through an Eclectic Universe* , Knapen, J.H., et al., eds., ASP Conference Series, San Francisco, Astronomical Society of the Pacific, Vol. 390, p. 47 (2008)
2. Bastian, N., Goodwin, S.P., MNRAS 369, L9 (2006)
3. Bastian, N., Gieles, M., Efremov, Yu.N., Lamers, H.J.G.L.M., A&A 443, 79 (2005)
4. Battinelli, P., Capuzzo-Dolcetta, R., MNRAS 249, 76 (1991)
5. Bica, E., Bonatto, Ch., Dutra, C.M., A&A 405, 991 (2003)
6. Brandner, W., et al., A&A 478, 137 (2008)
7. Cerviño, M., Valls-Gabaud, D., MNRAS 338, 481 (2003)
8. Clark, J.S., Negueruela, I., Crowther, P.A., Goodwin, S.P., A&A 434, 949 (2005)
9. Comerón, F., Torra, J., ApJ 423, 652 (1994)
10. Comerón, F., et al., A&A 389, 874 (2002)
11. Crowther, P.A., et al., MNRAS 372, 1407 (2006)
12. Davies, B., et al., ApJ 671, 781 (2007)
13. Davies, B., et al., ApJ 676, 1016 (2008)
14. de Grijs, R., Parmentier, G., ChJAA 7, 155 (2007)
15. Figer, D.F., in *Massive Stars as Cosmic Engines*, Bresolin, F., et al., eds., IAU Symposium, Cambridge, Cambridge University Press, Vol. 250, p. 247 (2008)
16. Figer, D.F., et al., ApJ 643, 1166 (2006)
17. Garzón, F., et al., ApJ 491, L31 (1997)
18. Hanson, M.M., ApJ 597, 957 (2003)
19. Hanson, M.M., Popescu, B, IAU Symposium, Cambridge, Cambridge University Press, Vol. 250, p. 307 (2008)
20. Harayama, Y., Eisenhauer, F., Martins, F., ApJ 675, 1319 (2008)
21. Homeier, N.L., Alves, J., A&A 430, 481 (2005)
22. Johnson, H.L., Morgan, W.W., ApJ 119, 344 (1954)
23. Kothes, R., Dougherty, S.M., A&A 468, 993 (2007)

24. Lada, J., Lada, E.A., ARA&A 41, 57 (2003)
25. López-Corredoira, M., et al., AJ 118, 381 (1999)
26. Knödlseder, J., A&A 360, 539 (2000)
27. Kiminki, D.C., et al., ApJ 664, 1102 (2007)
28. Kiminki, D.C., McSwain, M.V., Kobulnicky, H.A., ApJ 679, 1478 (2008)
29. Kroupa, P., Science 295, 82 (2002)
30. Kroupa, P., Boily, C.M., MNRAS 336, 1188 (2002)
31. Larsen, S.S., Richtler, T., A&A 427, 495 (2004)
32. Massey, P., Thompson, A.B., AJ 101, 1408 (1991)
33. McCrady, N., Graham, J.R., Vacca, W.D., ApJ 621, 278 (2005)
34. Melena, N.W., Massey, P., Morrell, N.I., Zangari, A.M., AJ 135, 878 (2008)
35. Mengel, S., Tacconi-Garman, L.E., in *Dynamical Evolution of Dense Stellar Systems*, Vesperini, E., et al., eds., IAU Symposium, Cambridge, Cambridge University Press, Vol. 246, p. 113 (2008)
36. Nanda-Kumar, M.S., Kamath, U.S., Davis, C.J., MNRAS 353, 1025 (2004)
37. Negueruela, I., Clark, J.S., Hadfield, L.J., Crowther, P.A., in *Massive Stars as Cosmic Engines*, Bresolin, F., et al., eds., IAU Symposium, Cambridge, Cambridge University Press, Vol. 250, p. 301 (2008)
38. Negueruela, I., Marco, A., Herrero, A., Clark, J.S., A&A 487, 575 (2008)
39. Piskunov, A.E., et al., A&A 487, 557 (2008)
40. Slesnick, C.L., Massey, P., Hillenbrand, L.A., ApJ 576, 880 (2002)
41. Whitmore, B.C., Schweizer, F., AJ 109, 960 (1995)

Part V
The Sun and the Solar System

The Impact of Energetic Particle Precipitation on the Earth's Atmosphere

B. Funke, M. López-Puertas, M. García-Comas, D. Bermejo-Pantaleón, G.P. Stiller, and T. von Clarmann

Abstract Energetic particle precipitation (EPP) represents an important Sun–Earth coupling mechanism with important implications on polar stratospheric ozone chemistry. Solar protons generated during solar storms cause sporadically in situ production of stratospheric NO_x and HO_x radicals involved in catalytic ozone destruction. Further, NO produced continuously in the mesosphere and lower thermosphere by medium energy electron precipitation (EEP) descends to the stratosphere during the polar winter, where it represents an additional, though variable source of NO_x. The capability of the Michelson Interferometer for Passive Atmospheric Sounding (MIPAS) to measure all important NO_y species, as well as ClO and HOCl with global coverage including the polar night regions make it an ideal instrument for studying EPP effects on stratospheric chemistry. We present a quantitative assessment of EPP-induced composition changes as observed by MIPAS during 2002–2004, including the unusually strong solar proton event in October/November 2003. The impact of EPP on the stratospheric ozone budget has been studied with chemical models. The stratospheric ozone loss in the polar regions reached 18 DU and lasted over months to years.

1 Introduction

During polar winter, large amounts of upper atmospheric NO_x (= NO + NO_2) can be transported down to the stratosphere by the meridional circulation without being photochemically destroyed. Once transported into the stratosphere, NO_x is photo-chemically stable and thus may contribute to the stratospheric NO_y budget,

B. Funke, M. López-Puertas, M. García-Comas, and D. Bermejo-Pantaleón
Instituto de Astrofísica de Andalucía, CSIC, Granada, Spain
e-mail: bernd@iaa.es, puertas@iaa.es, maya@iaa.es, diego@iaa.es

G.P. Stiller and T. von Clarmann
Institut für Meteorologie und Klimaforschung, University and Forschungszentrum Karlsruhe, Karlsruhe, Germany
e-mail: gabriele.stiller@imk.fzk.de, thomas.clarmann@imk.fzk.de

J.M. Diego et al. (eds.), *Highlights of Spanish Astrophysics V*, Astrophysics and Space Science Proceedings, DOI 10.1007/978-3-642-11250-8_18,

which largely controls the O_3 abundance. This solar–terrestrial coupling mechanism was already proposed in the early 1980s [3,5,30], and its first experimental evidence was given by the Limb Infrared Monitor of the Stratosphere (LIMS) which measured enhanced upper stratospheric and mesospheric NO_2 in the 1978–1979 polar night Northern hemisphere [27]. Later, several satellite occultation measurements have corroborated this mechanism and helped in understanding it [1,20,25,29].

NO_x is produced in the polar regions by different processes at different altitudes. In the lower thermosphere, the main production mechanism is energetic electron precipitation (EEP) dissociating N_2, and then N reacting with O_2 to form NO. This is a continuous source although highly variable since it is modulated by the electron fluxes at different energies in short (days) an long (solar cycle) scales [1,2,23].

In addition, NO and HO_x is produced at lower altitudes, from the upper mesosphere down to the stratosphere, during solar protons events (SPEs). Protons enter these atmospheric regions during solar eruptions when large fluxes of high-energy solar protons reach the Earth [11]. This is a rather sporadic source, although also affected by the long-term solar cycle.

The potential impact of energetic particle precipitation (EPP) on stratospheric ozone by modulating catalytic O_3 destruction cycles represents an important uncertainty in the evaluation of present and future ozone trends, particularly due to its long-term solar cycle dependence. The major interest in the past was to determine the amount of NO_x that is produced in the upper atmosphere and injected in the polar stratosphere. The major factors controlling the NO_x stratospheric deposition are twofold: (1) the source strength of EPP-induced NO production, largely modulated by the solar and geomagnetical activity, and (2) the factors controlling the dynamics of this region.

Despite the numerous measurements confirming the downward transport of NO_x in polar winter, only few studies deal with the quantitative assessment of the injection of NO_x in the polar winter stratosphere, partially because of the lack of wintertime NO_x data poleward of 50° [29]. The Michelson Interferometer for Passive Atmospheric Sounding (MIPAS) measures most of the NO_y constituents, including NO, NO_2, HNO_3 and N_2O_5, as well as dynamical tracers as CH_4 and CO, with global coverage and independent of illumination conditions. It is thus perfectly suited for studying EPP-related composition changes in the mesosphere and stratosphere, the downward transport of NO_x, and its effect on the stratospheric NO_y budget. Here, we summarize scientific key results obtained from the analysis of MIPAS data taken during September 2002 to March 2004, including the unusually strong solar proton event in October/November 2003.

2 MIPAS Observations

MIPAS is a mid-infrared Fourier transform limb emission spectrometer designed and operated for measurement of atmospheric trace species from space [4]. It is part of the instrumentation of the Environmental Satellite (ENVISAT) which was launched into its sun-synchronous polar orbit on 1 March 2002. MIPAS observes the atmosphere during day and night with global coverage from pole to pole. The instrument's field of view is 30 km in horizontal and approximately 3 km in vertical direction. MIPAS operated from July 2002 to March 2004 at full spectral resolution of 0.035 cm^{-1} (unapodized) in terms of full width at half maximum. During this period, MIPAS recorded a limb sequence of 17 spectra each 90 s, corresponding to an along track sampling of approximately 500 km and providing about 1,000 vertical profiles per day in its standard observation mode. Tangent heights covered the altitude range from 68 down to 6 km with tangent altitudes at 68, 60, 52, 47, and then at 3 km steps from 42 to 6 km. The raw signal is processed by the European Space Agency (ESA) to produce calibrated geolocated limb emission spectra.

Species abundance profiles are retrieved with the scientific MIPAS level 2 processor developed and operated by the Institute of Meteorology and Climate Research (IMK) in Karlsruhe together with the Instituto de Astrofísica de Andalucía (IAA) in Granada. The general retrieval strategy, which is a constrained multi-parameter nonlinear least squares fitting of measured and modeled spectra, is described in detail in [35, 36]. Its extension to retrievals under consideration of non-LTE as required for the abundances of NO, NO_2, and CO is described in [6].

3 Composition Changes During the October/November 2003 Solar Proton Event

In recent years, there have been two large solar proton events (SPEs) (October-November 2003 and January 2005) [14] which have been intensively observed by several instruments on different satellite platforms, including, for example, NOAA 16 SBUV/2 and HALOE data [12, 13, 21], MIPAS, GOMOS and SCIAMACHY on Envisat [16, 17, 19, 26, 28, 37], and MLS on AURA [32]. In particular, during late October and early November 2003, three active solar regions produced solar flares and solar energetic particles of extremely large intensity, the fourth largest event observed in the past forty years [12, 14]. During and after this SPE, the MIPAS instrument observed global changes (e.g. in both the Northern and Southern polar regions, during day and nighttime) in the stratospheric and lower mesospheric composition. This includes enormous enhancements in NO_x, and large depletions in O_3 [16] as well as significant changes in other NO_y species, such as HNO_3, N_2O_5, $ClONO_2$ [17], and N_2O [9]. In addition, there also have been observed changes in ClO and HOCl, as evidence of perturbations by solar protons on the HO_x and chlorine species abundances [37].

MIPAS observations of NO_2, O_3, HNO_3, N_2O_5, $ClONO_2$, ClO and HOCl during this SPE have been compared to Whole Atmosphere Community Climate Model (WACCM3) simulations [14] which includes a proton-driven ionization source based on GOES proton flux observations. Generally, there is reasonable agreement between the WACCM3 predictions and the observations, especially for SPE-caused NO_x enhancements and ozone depletion (see Fig. 1). Polar mesospheric

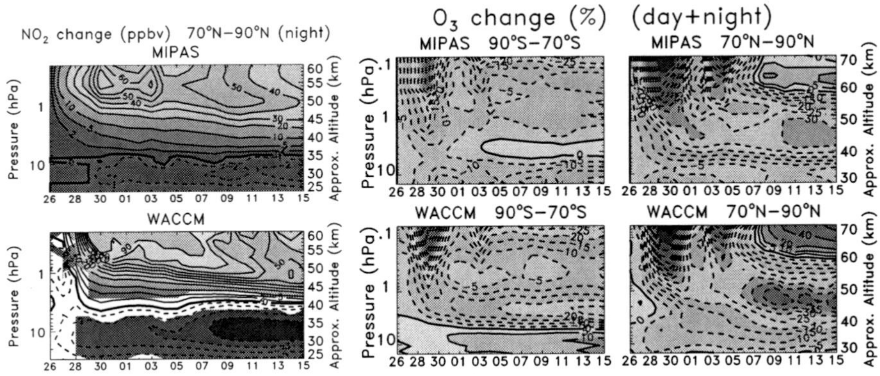

Fig. 1 *Left*: Temporal evolution of observed MIPAS (*top*) and modeled WACCM3 (*bottom*) nighttime NO_2 abundance changes relative to 26 October for the polar Northern Hemisphere (70–90° N). *Middle* and *right*: Same as *left panels*, but for ozone changes for 70–90° S and 70–90° N, respectively. The *colored areas in the bottom panels* demonstrate the 2σ statistically significant regions. Reproduced from [14]

NO_x increased by over 50 ppbv and mesospheric ozone decreased by over 30% during these very large SPE periods. Also, observed HOCl enhancements in the order of 0.3 ppbv are reproduced well by the model (see Fig. 2, upper panel). There are, however, some disagreements between WACCM3 and satellite instrument observed enhancements of HNO_3, N_2O_5, $ClONO_2$, and ClO. The underestimation of the observed HNO_3 enhancements by WACCM3 (see Fig. 2, lower panel) could be explained by missing ion-ion recombination between NO_3^- and H^+ cluster ions in the chemical scheme [33]. Canadian Middle Atmospheric Model (CMAM) simulations have shown that observed N_2O enhancements of 6–7 ppbv were generated by the reaction of NO_2 with atomic nitrogen produced by proton-induced ionization [9] (see Fig. 3).

4 Polar Winter Descent of NO_x Generated by Energetic Electron Precipitation During 2002–2004

Polar winter descent of NO_x generated by energetic EEP was detected by MIPAS during all polar winters in 2002–2004, although with variable magnitude (see Fig. 4). This inter-annual variability with stronger NO_x enhancements during 2003–2004 compared to 2002–2003 could be explained partly by variations of the solar and geomagnetic activity [7, 18] in agreement with other studies [2, 20, 22, 29]. The MIPAS measurements further suggest that the NO_x descent is well confined to the polar vortex [7]. In consequence, descent velocities related to the vortex strength and vortex extension are important dynamical factors controlling the efficiency of the downward transport. Dynamical modulations by planetary wave activity, resulting

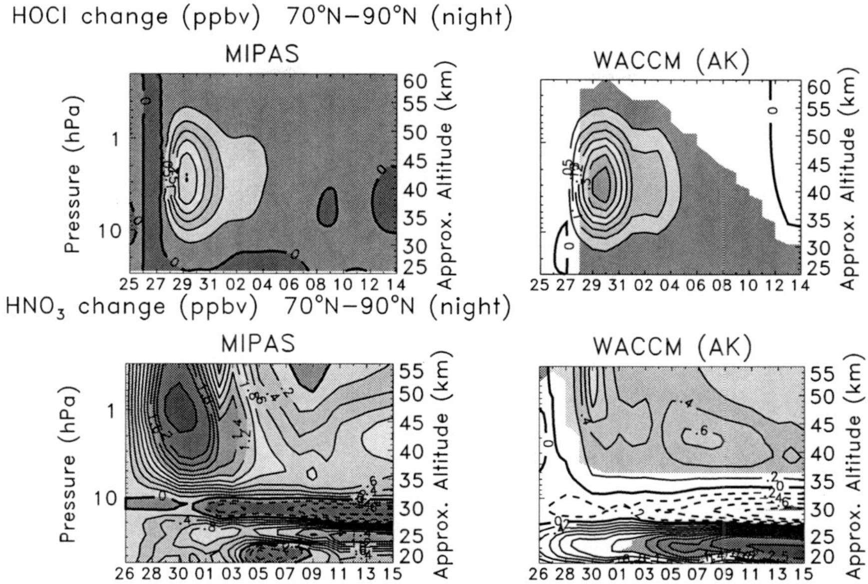

Fig. 2 *Top*: Temporal evolution of observed MIPAS (*left*) and modeled WACCM3 (*right*) nighttime HOCl abundance changes relative to 26 October for the polar Northern Hemisphere (70–90° N). *Bottom*: Same as *top panels*, but for HNO_3. MIPAS averaging kernels have been applied to model data. The *colored areas in the right panels* demonstrate the 2σ statistically significant regions. Reproduced from [14]

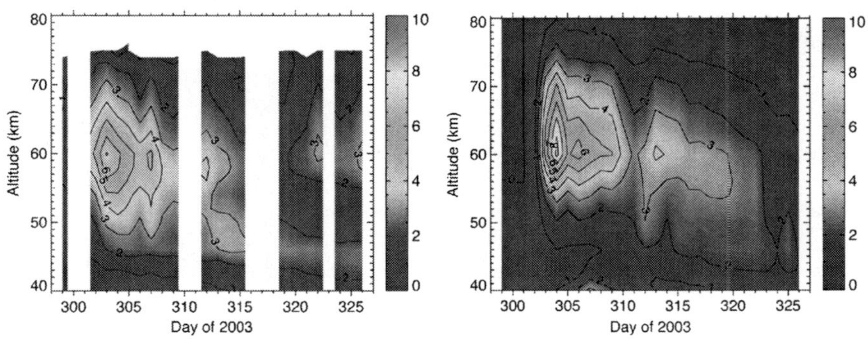

Fig. 3 Temporal evolution of observed MIPAS (*left*) and modeled CMAM (*right*) nighttime N_2O abundances (ppbv) for the polar Northern Hemisphere (70–90° N). MIPAS averaging kernels have been applied to model data. Reproduced from [9] with kind permission of Copernicus

in stratospheric warming events, have been observed in the Arctic winters. During these events, the NO_x descent was disrupted by shifting the polar vortex to sunlit mid-latitudes where NO_x was photochemically destroyed [7]. This disruption was particularly pronounced during the December 2003 – January 2004 warming event (see Fig. 4, right panel). The origin of the subsequent quasi-instantaneous

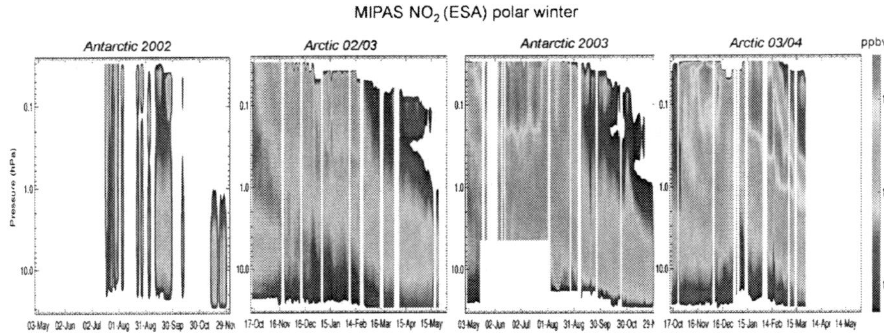

Fig. 4 Temporal evolution of nighttime NO_2 volume mixing ratio (vmr) (ESA reprocessed data) averaged over equivalent latitudes 60–90° N during the polar 2002–2004 polar winters

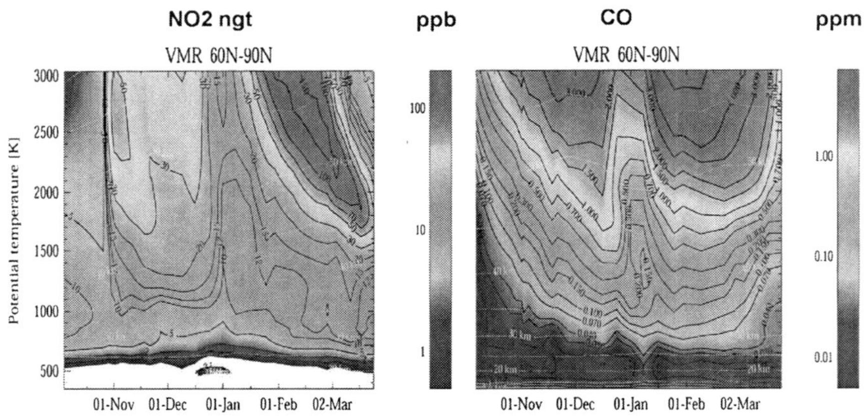

Fig. 5 Temporal evolution of nighttime NO_2 (*left*) and CO (*right*) vmr retrieved by IMK/IAA averaged over equivalent latitudes 60–90° N from September 2003 to March 2004

increase of NO_x around 15 January, resulting in enhancements of several hundred ppbv, raised a controversial scientific discussion. Renard et al. [24] proposed that these enhancements were produced in situ by a particular nighttime ion chemistry related to an high-energy electron event on 22 January. The simultaneous increase of CO and NO_x observed by MIPAS demonstrated, however, that these enhancements were caused by extraordinary efficient descent of upper atmospheric air masses with vertical velocities around 1,200 m day^{-1} [8] (see Fig. 5).

Thanks to the spatial coverage of MIPAS observations including the polar night regions, a quantitative assessment of EEP-related NO_x deposition into the stratosphere could be performed for the first time. During the Antarctic winter 2003, a total amount of 2.4 Gmole was released into the stratosphere [7] which makes up 9% of the local source due to N_2O oxidation in the SH, twice as much as estimated in previous studies. This additional NO_x was subsequently converted to HNO_3 at

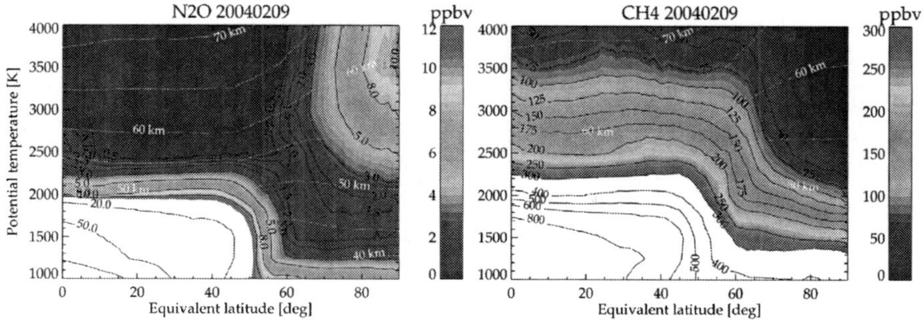

Fig. 6 Potential temperature-equivalent latitude daily mean cross sections of N_2O (*left*) and CH_4 (*right*) vmr retrieved by IMK/IAA for the NH for on 9 February 2004. Mean geometric heights are indicated by *dotted white lines*. Reproduced from [10]

altitudes around 30–35 km producing a secondary polar winter HNO_3 maximum [31].

Also, N_2O enhancements ranging from 0.5 to 6 ppbv were observed in the polar upper stratosphere/lower mesosphere by MIPAS during the 2002–2004 Arctic and Antarctic winters [10]. A detailed study of the observed N_2O–CH_4 correlations showed that such enhancements cannot be explained by dynamics without invoking an upper atmospheric chemical source of N_2O (see Fig. 6). The analysis of possible chemical production mechanisms showed that the major part of the observed N_2O enhancements is most likely generated under dark conditions by the reaction of NO_2 with atomic nitrogen at altitudes around 70–75 km in the presence of EPP, although an additional contribution by the reaction of $N_2(A^3\Sigma_u^+)$ with O_2 cannot be excluded.

The impact of the downward transport of enhanced upper atmospheric NO_x caused by mesospheric NO_x intrusions in early 2004 produced throughout the winter by auroral and precipitating electrons on the stratospheric ozone budget was studied by performing model simulations with the chemical transport model CLaMS [34]. Upper boundary conditions were taken from the results of a long-term simulation conducted with KASIMA [15], where increased NO_x concentration in the mesosphere was derived from MIPAS measurements. For the Arctic polar region (equivalent lat. >70°N) we found that mesospheric NO_x intrusion due to downward transport of upper atmospheric NO_x produced by EEP affected NO_x mixing ratios down to about 700 K potential temperature (27 km) until March 2004. A comparison with a simulation without an additional NO_x source at the upper boundary showed that O_3 mixing ratios were affected by transporting high burdens of NO_x down to about 400 K (17–18 km) during the winter. Locally, an additional ozone loss of the order of 1 ppmv was simulated for January between 850–1,200 K potential temperature during the period of the strong polar vortex in the middle stratosphere. We found that the additional ozone loss caused by transport of mesospheric NO_x-rich air into the stratosphere was about 10–18 DU (3–6%) at the end of March 2004.

5 Conclusions

The impact of EPP on the Earth's atmosphere has been studied by means of MIPAS observations of various trace gases during 2002–2004. During the SPE in October/November 2003, important atmospheric composition changes were observed, in many species for the first time. Comparison to model simulations have shown that SPE-induced NO_x, HNO_3, and HOCl increases, as well as related O_3 losses are well understood. Uncertainties remain, however, with respect to the chemical processes leading to increases of N_2O_5, $ClONO_2$, and ClO. Polar winter descent of NO_x was observed by MIPAS in all polar winters during 2002–2003. Inter-annual differences could be explained by variations in the EEP source related to solar and geomagnetical activity, as well as dynamical factors. An EEP-related NO_x deposition into the stratosphere of 2.4 GMole was determined from the observations during the Antarctic winter 2003. Apart of NO_x enhancements, MIPAS observed also EEP-related increases of mesospheric N_2O and stratospheric HNO_3. The additional ozone loss induced by EPP during the 2003/2004 Arctic winter was estimated by means of ClaMS model simulations to as much as 18 DU.

Acknowledgements The IAA team was supported by the Spanish MICINN under contract AYA2008-03498 and EC FEDER funds. The IMK team was supported by the Priority Program CAWSES of the German science foundation (DFG) under the project MANOXUVA.

References

1. Callis, L.B., Lambeth, J.D., Geophys. Res. Lett. 25(11), 1875 (1998)
2. Callis, L.B., Natarajan, M., Lambeth, J.D., J. Geophys. Res. 106(D7), 7523 (2001)
3. Crutzen, P.J., Heidt, L.E., Krasnec, J.P., Pollock, W.H., Seiler, W., Nature 282, 253 (1979)
4. Fischer, H., Birk, M., Blom, C., Carli, B., Carlotti, M., von Clarmann, T., Delbouille, L., Dudhia, A., Ehhalt, D., Endemann, M., Flaud, J.M., Gessner, R., Kleinert, A., Koopmann, R., Langen, J., López-Puertas, M., Mosner, P., Nett, H., Oelhaf, H., Perron, G., Remedios, J., Ridolfi, M., Stiller, G., Zander, R., Atmos. Chem. Phys. 8, 2151 (2008)
5. Frederick, J.E., Orsini, N., J. Atm. Solar-Terr. Phys. 44, 479 (1982)
6. Funke, B., López-Puertas, M., Stiller, G.P., von Clarmann, T., Höpfner, M., Adv. Space Res. 27(6-7), 1099 (2001)
7. Funke, B., López-Puertas, M., Gil-López, S., von Clarmann, T., Stiller, G.P., Fischer, H., Kellmann, S., J. Geophys. Res. 110, D24308 (2005)
8. Funke, B., López-Puertas, M., Fischer, H., Stiller, G.P., von Clarmann, T., Wetzel, G., Carli, B., Belotti, C., Geophys. Res. Lett. 34(7), L07813 (2007)
9. Funke, B., García-Comas, M., López-Puertas, M., Glatthor, N., Stiller, G.P., von Clarmann, T., Semeniuk, K., McConnell, J.C., Atmos. Chem. Phys. 8, 3805 (2008)
10. Funke, B., López-Puertas, M., García-Comas, M., Stiller, G.P., von Clarmann, T., Glatthor, N., Atmos. Chem. Phys. 8, 5787 (2008)
11. Jackman, C.H., McPeters, R.D., Geophys. Monogr. 141, 305 (2004)
12. Jackman, C.H., DeLand, M.T., Labow, G.J., Fleming, E.L., Weisenstein, D.K., Ko, M.K.W., Sinnhuber, M., Anderson, J., Russell, J.M., Adv. Space Res. 35(3), 445 (2005)
13. Jackman, C.H., DeLand, M.T., Labow, G.J., Fleming, E.L., Weisenstein, D.K., Ko, M.K.W., Sinnhuber, M., Russell, J.M., J. Geophys. Res. 110, A09S27 (2005)

14. Jackman, C.H., Marsh, D.R., Vitt, F.M., Garcia, R.R., Fleming, E.L., Labow, G.J., Randall, C.E., López-Puertas, M., Funke, B., von Clarmann, T., Stiller, G.P., Atmos. Chem. Phys. 8, 765 (2008)
15. Kouker, W., Offermann, D., Küll, V., Reddmann, T., Ruhnke, R., Franzen, A., J. Geophys. Res. 104(D13), 16,405 (1999)
16. López-Puertas, M., Funke, B., Gil-López, S., von Clarmann, T., Stiller, G.P., Höpfner, M., Kellmann, S., Fischer, H., Jackman, C.H., J. Geophys. Res. 110(A9), A09S43 (2005)
17. López-Puertas, M., Funke, B., Gil-López, S., von Clarmann, T., Stiller, G.P., Höpfner, M., Kellmann, S., Mengistu Tsidu, G., Fischer, H., Jackman, C.H., J. Geophys. Res. 110(A9), A09S44 (2005)
18. López-Puertas, M., Funke, B., Bermejo-Pantaleón, D., von Clarmann, T., Stiller, G.P., Grabowski, U., Höpfner, M., Geophys. Res. Lett. 34(2), L02825 (2007)
19. Orsolini, Y.J., Manney, G.L., Santee, M.L., Randall, C.E., Geophys. Res. Lett. 32, L12S01 (2005)
20. Randall, C.E., Rusch, D.W., Bevilacqua, R.M., Hoppel, K.W., Lumpe, J.D., J. Geophys. Res. 103(D21), 28,361 (1998)
21. Randall, C.E., Harvey, V.L., Manney, G.L., Orsolini, Y.J., Codrescu, M., Sioris, C., Brohede, S., Haley, C.S., Gordley, L.L., Zawodny, J.M., Russell III, J.M., Geophys. Res. Lett. 32, L05802 (2005)
22. Randall, C.E., Harvey, V.L., Singleton, C.S., Bernath, P.F., Boone, C.D., Kozyra, J.U., Geophys. Res. Lett. 22, L18811 (2006)
23. Randall, C.E., Harvey, V.L., Singleton, C.S., Bailey, S.M., Bernath, P.F., Codrescu, M., Nakajima, H., Russell III, J.M., J. Geophys. Res. 112(D8), D08308 (2007)
24. Renard, J.B., Blelly, P.L., Bourgeois, Q., Chartier, M., Goutail, F., Orsolini, Y.J., Geophys. Res. Lett. 33, L11801 (2006)
25. Rinsland, C.P., Goldman, A., Murcray, F.J., Stephen, T.M., Pougatchev, N.S., Fishman, J., David, S.J., Blatherwick, R.D., Novelli, P.C., Jones, N.B., Connor, B.J., J. Geophys. Res. 104(D15), 18,667 (1999)
26. Rohen, G.J., von Savigny, C., Sinnhuber, M., Llewellyn, E.J., Kaiser, J.W., Jackman, C.H., Kallenrode, M.B., Schröter, J., Eichmann, K.U., Bovensmann, H., Burrows, J.P., J. Geophys. Res. 110, A09S39 (2005)
27. Russell III, J.M., Farmer, C., Rinsland, C., Zander, R., Froidevaux, L., Toon, G., Gao, B., Shaw, J., Gunson, M., J. Geophys. Res. 93(D2), 1718 (1988)
28. Seppälä, A., Verronen, P.T., Kyrölä, E., Hassinen, S., Backman, L., Hauchecorne, A., Bertaux, J.L., Fussen, D., Geophys. Res. Lett. 31, L19107 (2004)
29. Siskind, D.E., Geophys. Monogr. 123, 101 (2000)
30. Solomon, S., Crutzen, P.J., Roble, R.G., J. Geophys. Res. 87, 7206 (1982)
31. Stiller, G.P., Mengistu Tsidu, G., von Clarmann, T., Glatthor, N., Höpfner, M., Kellmann, S., Linden, A., Ruhnke, R., Fischer, H., López-Puertas, M., Funke, B., Gil-López, S., J. Geophys. Res. 110(D20), D20303 (2005)
32. Verronen, P.T., Seppälä, A., Kyrölä, E., Tamminen, J., Pickett, H.M., Turunen, E., Geophys. Res. Lett. 33, L24811 (2006)
33. Verronen, P.T., Funke, B., López-Puertas, M., Stiller, G.P., von Clarmann, T., Enell, C.F., Turunen, E., Tamminen, J., Geophys. Res. Lett. 35, L20809 (2008)
34. Vogel, B., Konopka, P., Grooß, J.U., Müller, R., Funke, B., López-Puertas, M., Reddmann, T., Stiller, G., von Clarmann, T., Riese, M., Atmos. Chem. Phys. 8, 5279 (2008)
35. von Clarmann, T., Ceccherini, S., Doicu, A., Dudhia, A., Funke, B., Grabowski, U., Hilgers, S., Jay, V., Linden, A., López-Puertas, M., Martín-Torres, F.J., Payne, V., Reburn, J., Ridolfi, M., Schreier, F., Schwarz, G., Siddans, R., Steck, T., J. Geophys. Res. 108(D23), 4746 (2003)
36. von Clarmann, T., Glatthor, N., Grabowski, U., Höpfner, M., Kellmann, S., Kiefer, M., Linden, A., Mengistu Tsidu, G., Milz, M., Steck, T., Stiller, G.P., Wang, D.Y., Fischer, H., Funke, B., Gil-López, S., López-Puertas, M., J. Geophys. Res. 108(D23), 4736 (2003)
37. von Clarmann, T., Glatthor, N., Höpfner, M., Kellmann, S., Ruhnke, R., Stiller, G.P., Fischer, H., Funke, B., Gil-López, S., López-Puertas, M., J. Geophys. Res. 110(A9), A09S45 (2005)

Marco Polo: Hunting and Capture of Material from a Primitive Asteroid

Javier Licandro

Abstract A description of the Spanish contribution to the Marco Polo mission and of the mission itself is presented. Marco Polo is a joint European–Japanese mission of sample return from a Near Earth Object (NEO). Submitted to ESA on July 2007 in the framework of the Cosmic Vision 2015–2025, Marco Polo passed the first evaluation process on October 2007. Seventeen Spanish researchers belonging to six Spanish institutes signed the proposal. The mission is planned to visit a primitive NEO, belonging to a class that cannot be related to known meteorite types, to characterize it at multiple scales, and to bring samples back to Earth. Marco Polo will give us the first opportunity for detailed laboratory study of the most primitive materials that formed the planets. This will allow us to improve our knowledge on the processes which governed the origin and early evolution of the Solar System, and possibly of the life on Earth. Three Spanish institutes are involved in the feasibility studies of two instruments: the THERmal MAPper (THERMAP) and the Marco Polo Camera System (MPCS).

1 Introduction

Minor Bodies, in particular asteroids and comets, are the relic of the building blocks of the planets [5]. Their analysis will bring us crucial information on the nature of the protoplanetary disk [6–8]. The Near Earth Object (NEO) population comprises both asteroids and comet nuclei, in orbits with perihelion distances $q < 1.3$ AU. It's one of the most interesting populations, considering also that they constitute a potential hazard for the Earth. NEOs must be continuously replenished from major small bodies reservoirs because of the short dynamical lifetimes of their orbits, These reservoirs are identified mainly in the asteroid main belt, with a possible significant contribution of extinct cometary nuclei [2]. The knowledge of the structure and

J. Licandro
Instituto de Astrofísica de Canarias, La Laguna, Spain
e-mail: jlicandr@iac.es

J.M. Diego et al. (eds.), *Highlights of Spanish Astrophysics V*, Astrophysics and Space Science Proceedings, DOI 10.1007/978-3-642-11250-8_19,

composition of NEOs is still rather limited, since only 10% of the known NEOs (∼5,500 objects) have spectral types determined from observations. NEOs present a high degree of diversity in terms of physical properties (shapes, rotational and spectroscopical properties). All the taxonomic spectral classes present among main belt asteroids are recognized among NEOs. The taxonomic classification can give some hints about the regions of the Solar System where these objects come from: S-, Q- and E-types seem to come from the inner asteroid belt, C-types from the mid-to-outer belt, while D- and P-types from the outer belt and/or are dormant Jupiter family comets [10]. The knowledge of the chemical and mineralogical composition of asteroid surfaces gives us an insight into the processes that driven the formation and evolution of our planetary system, and of the materials that formed the protoplanetary nebula at different solar distances. It can also provide information on the evolution of small bodies [4]. NEOs have the advantage of being much more accessible for space missions than small bodies in other populations providing a major opportunity for advancement in our understanding of some of the fundamental issues on the origin and early evolution of the Solar System. With the exception of asteroid Matilde, all the asteroids already visited by spacecrafts are not primitive, belonging to the S-type. The study of the physical nature of NEOs is also very important to define successful mitigation strategies in the case of possible impactors. NEOs play an important role also in exobiological scenarios: in fact the delivery of exogeneous material from primitive NEOs is invoked by current theories for the triggering of life on Earth [3]. Planets of the inner Solar System experienced an intense influx of cometary and asteroidal material for several hundred million years after they formed. The earliest evidence for life on Earth coincides with the decline of this enhanced bombardment. The fact that the influx contained vast amounts of complex organic material and water offers a possibility that it may be related to the origin of life on the Earth.

In this context, a joint European–Japanese consortium proposed a mission of sample return from a primitive C-, D-, P-type NEO. The Marco Polo (hereafter MP) proposal [1] was submitted to ESA on July 2007 in the framework of the Cosmic Vision 2015–2025 context. On October 2007 it passed the first evaluation process and it is now in its feasibility study phase.

2 Scientific Objectives

Asteroids and comets are relics of the building blocks of the planets. Their study will bring us crucial information on the nature of the protoplanetary disk. Returned primitive material from a small body offers the possibility to identify nebula effects and resolve them from asteroidal processes. Primitive objects include material originating from, or modified by stellar outflows, the interstellar medium, the solar protoplanetary disk, and the parent-body processing. The isotopic composition of various elements, the nature of the organics and the mineralogy of the rocky elements in primitive Solar System bodies, are requisite data to obtain information on the great variety of processes that took place during Solar System history. Primitive asteroids, e.g. C-, D-, P-types, have low albedo probably due to abundant organic matter present on their surface, suggesting that these bodies have experienced little or no thermal processing. The surface of D and P type asteroids is probably composed of anhydrous minerals and organic matter. A sample from a primitive NEO would give some definitive answers on the formation processes of carbonaceous matter in interplanetary material, including key biological compounds like the aminoacids.

Most of our knowledge of the composition of the asteroids is a product of the study of meteoritic material, that, e.g., allowed the isolation and detailed analyses of a wide range of different pre-solar grains found in the primitive meteorites (e.g. [11]). But the analysis of meteoritic material is rather ambiguous due to terrestrial contamination. By returning samples free from contamination, any ambiguity is eliminated. Also, mechanical weak material should exist in significant quantities within the inner Solar System. But the existing meteorite collection is strongly biased towards more heavily processed material, because it better survives atmospheric entry than the weak one. Meteorites are very tough, coherent rocks, result of metamorphism, shock and/or aqueous alteration on the parent asteroids that mobilise elements and isotopic ratios within and between minerals, re-set radioisotope chronometers, destroy and modify primitive materials, and synthesize and mobilise organic compounds. The materials that form primitive asteroids rarely survive passage through the Earth's atmosphere as meteorites and should be some of the most primitive materials available to study the early Solar System formation processes. Primitive material from near the surface of an asteroid should contain abundant presolar grains. MP will provide a unique opportunity to investigate the abundance of such grains accreted to the parent body and to search for new grains which have not survived the meteorite formation processes, and to study the complex organic material without interference from terrestrial contamination.

Why do we need to return a sample to Earth? Many of the science questions we are attempting to resolve stem from detailed knowledge obtained from high precision and sensitivity measurements of meteorite properties. To advance the science with the new samples anticipated from a primitive asteroid will require comparable levels of analytical capability. The ability of "in situ" or remote sensing instruments to emulate lab-based instruments is compromised by constraints due to limitations of size, mass, power, data rate, and reliability imposed by the practical aspects of space missions. Many of the answers to the scientific questions will only be obtained through sophisticated analysis chains and integrated studies.

In summary, a mission to a primitive asteroid will provide crucial elements to answer the following key questions:

- What were the processes occurring in the primitive Solar System and accompanying planet formation?
- Do NEOs of primitive classes contain presolar material yet unknown in meteoritic samples?
- How did asteroid and meteorite classes form and acquire their present properties?
- How do asteroids and meteoritic classes relate to each other?

- What is the link between the vast array of spectral information on asteroids and the detailed knowledge available from meteorites?
- What are the main characteristics of the internal structure of a NEO – both physically and chemically?
- What is the nature and origin of organic compounds on a NEO?
- How do NEO organics shed light on the origin of molecules necessary for life?
- What is the role of NEO impacts in the origin of life on Earth?
- Why are the existing meteorite specimens not suitable?

3 Possible Targets

Several hundreds of NEOs are easily accessible with $\delta_v < 6\,\mathrm{km\,s^{-1}}$, where δ_v is the velocity change needed to realize a rendezvous mission (e.g. orbiting around an object). The current number of easily accessible objects (see Fig. 1) will certainly increase considering the high rate of NEOs discoveries. Dedicated observational campaigns will characterize among them the scientifically interesting targets for space missions. This is particularly interesting for MP. A number of possible targets of high scientific interest have been selected covering a wide range of launch windows in the time span 2017–2019. Each target has flexible launch windows and short mission durations (from about 4 to 8.5 years). Between them the most interesting cases are:

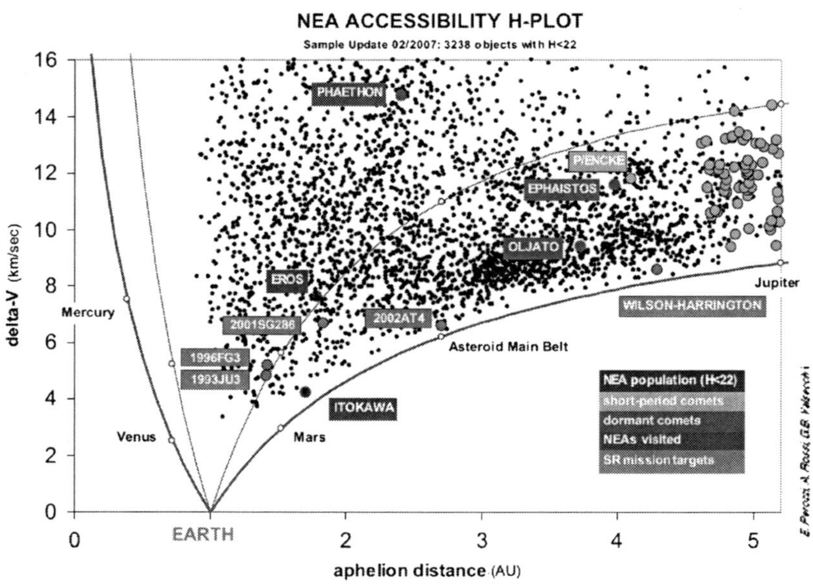

Fig. 1 The accessibility of NEOs versus required δ_v

- The comet–asteroid transition object 4,015 Wilson–Harrington, which can provide insights into the already unknown link between asteroids and comets [9].
- Asteroids belonging to the D-type class, such as 2002 AT4, or 2001 SG286.
- The C-type double asteroid 1996 FG3, another type of primitive asteroid that also offers the opportunity to provide insight into binary formation processes.

4 Mission Scenario

MP will use a Soyuz–Fregat launcher (ESA) to be inserted into an initial heliocentric orbit, and a EDVEGA (Electric Delta-V Gravity Assistant) to the interplanetary cruise phase. The baseline mission scenario includes a launch of a Mother Spacecraft (MSC, provided by JAXA), sampling devices (ESA and JAXA), a re-entry capsule (ESA and JAXA) and scientific payloads (shared between Japan and Europe consortia). The MSC consists of the body and the solar array panel. The Lander, re-entry capsule, the ion engines, the sampling devices, the antennas, and the scientific equipments are attached to the body (Fig. 2). The MSC benefits from the successful development and operation of the Hayabusa spacecraft. The MSC leaves Earth orbit, cruises toward the target with ion engines, rendezvous with the target, and conducts a global characterization of the target. It will characterize the global properties of the target with onboard remote sensing scientific instruments in order to determine in particular a three dimensional shape model and its gravitational

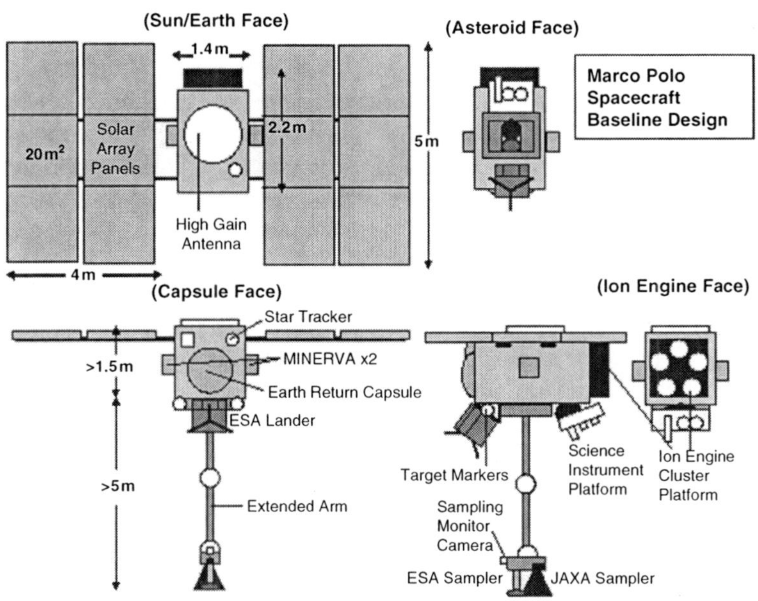

Fig. 2 Draw of the MSC

Fig. 3 Artistic view of the MSC with the deployed arm as in the touch and go maneuver, the Lander Sifnos, and a Minerva hopper

field. Moreover, surface morphology at global scale will be investigated to select the preferred landing and sampling sites.

During the descent sequence, small hoppers (MINERVA-type) will be released while a larger Lander called SIFNOS (provided by ESA), if added, will be deployed to the selected site (Fig. 3). The Lander itself profits considerably from the Rosetta Lander (Philae) heritage. SIFNOS will perform a soft landing, anchor to the asteroid surface and make various "in situ" measurements of surface/subsurface material near the sampling site with on board instruments within several Earth days.

For the samples collected on the surface by a *touch and go* strategy of the MSC, two different sampling techniques will be mounted on a retractable extension arm. One sampling technique developed in Japan will use the heritage from the Hayabusa technique. The other concept from Europe will probably be based on the use of a sticky pad which is a very light, robust and simple mechanism. Once the samples have been collected, the arm will be retracted to transfer the sample containers into the MSC and the samples will be inserted into the re-entry capsule (ERC). Then the latch and ERC sealing is performed. The MSC will return towards the Earth and release the capsule for the high-speed reentry into Earth's atmosphere. The capsule will be retrieved on the ground and samples will be taken out of the capsule in a dedicated curation facility to conduct initial sample characterization, prior to their distribution to designated scientists for detailed analyses. All the scientific data acquired and analytical results of the returned samples will be jointly archived by ESA and JAXA for public release after the proprietary period.

For the proposed baseline (and most challenging) target, Wilson–Harrington, we have designed a mission scenario with launch in April 2018 (possible back up in late 2018 and 2019, see Table 1).

Table 1 A mission scenario for (4015) Wilson–Harrington

Dates	Events	Notes
2018/04/25	Soyuz-Fregat launch MSC (1,320 kg)	Inserted to interplanetary orbit for EDVEGA acceleration (first back-up Window in 18/10)
2019/10/10	Earth gravity assist (Swing-by)	IES acceleration
2022/09/24	Rendezvous with Wilson–Harrington	At 1.04 AU after perihelion passage
2022/09/12	Global characterization	Staying at the target for ∼100 days
	Sampling/landing site selection	
	Touchdown rehearsals	
	Lander deployment and measurements	Few days
	Touch-and-Go sampling	
	IES restart tests	
2023/01/01	Departure from Wilson–Harrington	IES operation restart at 1.7 AU
2026/10/01	Earth return and capsule retrieval	Re-entry velocity at 14 km s^{-1}

The mission scenario depends on the orbital properties of the target of the mission. As those properties are different for the three targets, our mission can offer different launch opportunities and mission durations, depending on which target is eventually chosen. This is a big advantage of such a space mission, which offers considerable flexibility in case some unexpected changes in the programme occur (as happened for the Rosetta mission). Depending on the target, the mission can last from 4 to 8 years. The overall spacecraft mass is also constrained by the target choice.

5 Payload

A multi-scale approach is proposed to reach the scientific goals: macroscopic global scale (kilometer, meter) by the MSC, local scale (centimeter, millimeter) by the Lander, and microscopic scale (micrometer, nanometer) by laboratory analyses of returned samples.

To link the collected samples with the origin and evolution of the parent object a global characterization of the asteroid is crucial. Decisions on the sample return and the landing site need characterization of the object obtained both by remote measurements with instruments on board the MSC and the Lander. The scientific payload instruments of the MSC and the Lander, and the science that will be done with them are summarized in Figs. 4 and 5, respectively.

The Global-scale key measurements to be performed are: surface topography and morphology; overall characteristics: orbit, rotation, size, shape, mass, gravity and density; mineralogical and chemical compositions; main characteristics of the internal structure; and dust conditions around the object.

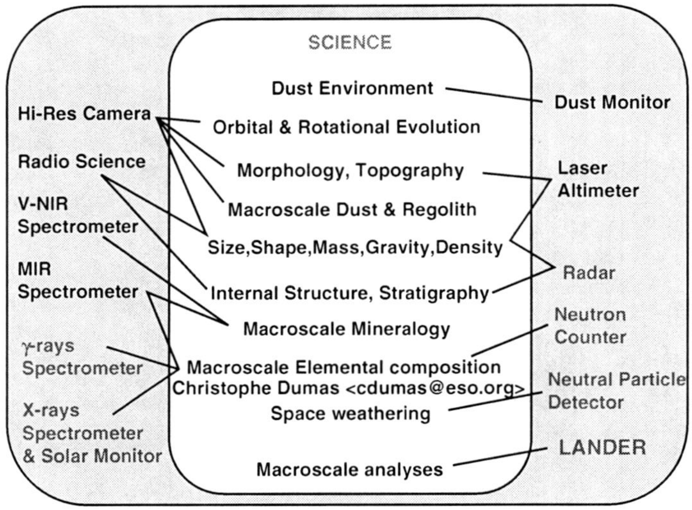

Fig. 4 The payload MSC proposed instruments to meet the scientific objectives shown. In *black* the primary payload, in *blue* the secondary priority

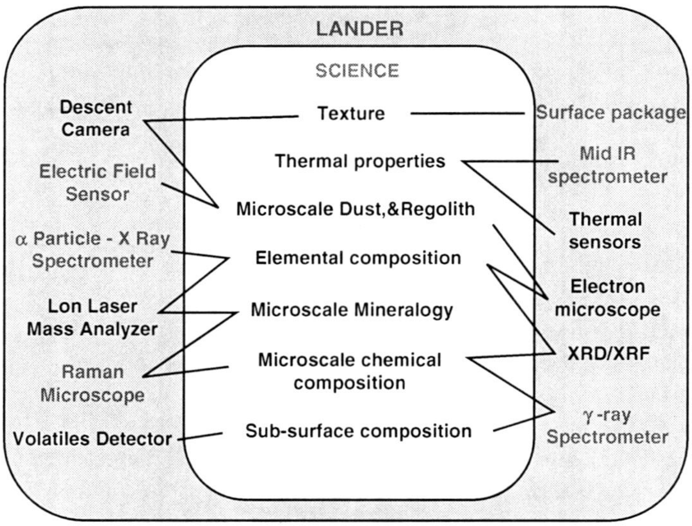

Fig. 5 The payload Lander proposed instruments to meet the scientific objectives shown. In *black* the primary payload, in *blue* the secondary priority

The Local-scale key measurements are: surface thermal properties; mineralogical composition and crystal structure of surface minerals; out-gassing volatiles (e.g. H_2O, CO_2,...); and complex organic molecules.

The "in situ" investigation by the lander SIFNOS is considered particularly important to identify: the context where the samples are taken; the heterogeneity of the sampling area; and any contamination/modification due to the sample transfer to Earth.

6 The Spanish Contribution

Seventeen Spanish researchers from the following institutes were included in the MP proposal: Instituto de Astrofísica de Canarias, Instituto de Astrofísica de Andalucía–CSIC, Universidad de Valladolid, Universidad Politécnica de Valencia, Universidad de Alicante, and IEEC–Universidad Autónoma de Barcelona–CSIC.

There is also Spanish participation in two consortia for instruments proposed for feasibility study, both for the MSC: the MP Camera Systems (MPCS; P.I. Hermann Boehnardt; MPS, KLatemburg–Lindau, Germany), and the mid-infrared camera and spectrograph THERmal MAPper (THERMAP; P.I. Olivier Groussin; LAM, France)

The MPCS consist of three different cameras, i.e. the wide-angle camera WAC, the narrow-angle camera NAC, and the close-up camera CUC. The scientific goals for the WAC rely on the global coverage and mapping of the asteroid surface and of the body as a whole (order decimeter). The WAC may also image during descent and at the surface. The science goals of the NAC target – besides global mapping – for more detailed investigations of the surface at higher spatial resolution (order millimeter), in particular also of the anticipated landing sites. A major common goal for both cameras is the provision of stereo imaging of the landforms. The CUC is the camera for close-up imaging of the sampling sites of Marco Polo mission (order 20–50 μm). All three cameras shall work in the visible wavelength range from about 350 to 1,000 nm. L. Lara is Co-Leader of the project and member of the Science Team. J. Licandro (IAC) is also member of the science team. The IAA and the Universidad Politécnica de Madrid (UPM) are also part of the engineering team. In this stage the team will perform the design and engineering analysis of the camera system and evaluate the scientific requirements. The IAA team (L. Lara, J.M. Castro-Marín, M. Herranz, J.L. Ramos, and J. Rodrigo) is working in the main electronics, while the UPM team (I. Pérez Grande, G. Alonso, and A. Sanz Andrés) is working on the thermal analysis of the device.

THERMAP is a thermal infrared imaging spectrometer based on the MERTIS instrument already flying in the Bepi–Colombo mission. It works in the 7–14 μm region. The scientific goal of THERMAP is to provide detailed information about the mineralogical composition and determine the surface temperature of the asteroid. It will allow to study silicate features and determine the thermal inertia and thermal skin depth. The heart of the instrument is a large 2D microbolometer array. Our best candidate for the mid-infrared is an uncooled microbolometer array with 640×480 pixels and a pixel size of 25 μm available from the ULIS company (Grenoble, France). J. Licandro (IAC) is Co-Leader of this consortium and part of the Science Team. A group of the IAC (J. Licandro, M. Serra Ricart, A. Oscoz Abad, G. Herrera and E. Joven) will develop the electronics and will do a full characterization of the ULIS (UL 04171) detector in the IAC laboratories in order to reach a TRL of 5 for this detector at the end of the feasibility study phase.

7 Conclusions

The Spanish community of astrophysicists that study NEOs is strongly involved in the scientific and technological aspects of the most ambitious ESA–JAXA mission proposed to study a primitive asteroid. NEOs are representative of the less-evolved populations of small bodies of the Solar System (asteroids and comets), and have the advantage to be more accessible for space missions. A space mission to a primitive NEO provides major opportunities to have some hints on the origin and early evolution of the Solar System. The Marco Polo mission has the potential to revolutionize our knowledge of primitive materials, essential to understand the conditions for planetary formation and emergence of life, and can provide important information to develop strategies to protect the Earth from potential hazards. A robotic sample return mission to a NEO is also innovative. It will allow us to test new challenging technologies, to develop new microanalysis techniques and to prepare laboratory facilities for the next generation analysis of extraterrestrial samples and will be the precursor of future sample return missions to high surface gravity bodies (e.g. Mars).

Acknowledgement This research has been supported by the Spanish Ministry of Science and Innovation (MICINN) under the grant AYA2008-06202-C03-02.

References

1. Barucci, M.A., Yoshikawa, M., Michel, P., Kawagushi, J., Yano, H., Brucato, J.R., Franchi, I.A., Dotto, E., Fulchignoni, M., Ulamec, S., Exp. Astron. 23, 785 (2009)
2. Bottke, W.F., Morbidelli, A., Jedicke, R., Petit, J.M., Levison, H.F., Michel, P., Metcalfe, T.S., Icarus 156, 399 (2002)
3. Chyba C.F., Owen T.C., Ip W.-H., in *Hazards Due to Comets and Asteroids*, ed. T. Gehrels, University of Arizona Press, Tucson, 9 (1994)
4. Ciesla F.J., Charnley S.B., in *Meteorites and the Early Solar System II*, eds. D.S. Lauretta & H.Y. McSween H.Y., University of Arizona Press, Tucson, 209 (2006)
5. Drake, M.J., Righter, K., Nature 416, 39 (2002)
6. Jacobsen, S.B., Ann. Rev. Earth Planet. Sci. 33, 531 (2005)
7. Kleine, T., Munker, C., Metzger, K., Palme, H., Nature 418, 952 (2002)
8. Levison, H.F., Morbidelli, A., Gomes, R., Backman, D., in *Protostars and Planets V*, eds. B. Reipurth, D. Jewitt & K. Keil, University of Arizona Press, Tucson, 669 (2006)
9. Licandro, J., Campins, H., Mothé-Diniz, T., Pinilla-Alonso, N., de León, J., A&A 461, 751 (2007)
10. Licandro, J., Alvarez-Candal, A., de León, J., Pinilla-Alonso, N., Lazzaro, D., Campins, H., A&A 481, 861 (2008)
11. Meyer, B.S., Zinner, E., in *Meteorites and the Early Solar System*, eds. D.S. Lauretta & H.Y. McSween, University of Arizona Press, Tucson, 69 (2006)

Part VI
Observatories and Instrumentation

The DUNE Mission

F.J. Castander

Abstract The Dark UNiverse Explorer (DUNE) is a wide-field imaging mission concept whose primary goal is the study of dark energy and dark matter with unprecedented precision. To this end, DUNE is optimised for weak gravitational lensing, and also uses complementary cosmological probes, such as baryonic oscillations, the integrated Sachs–Wolf effect, and cluster counts. Besides its observational cosmology goals, the mission capabilities of DUNE allow the study of galaxy evolution, galactic structure and the demographics of Earth-mass planets. DUNE is a medium class mission consisting of a 1.2 m telescope designed to carry out an all-sky survey in one visible and three NIR bands. The final data of the DUNE mission will form a unique legacy for the astronomy community. DUNE has been selected jointly with SPACE for an ESA Assessment phase which has led to the Euclid merged mission concept which combines wide-field deep imaging with low resolution multi-object spectroscopy.

1 Introduction

Current observations indicate that the energy–matter content of the Universe is composed of approximately 5% of baryons, 20% of dark matter and 75% of dark energy. While the observational effects of both dark components can be measured, their nature is unknown and remain a mystery. Especially the existence of dark energy challenges our current understanding of Fundamental Physics.

Dark energy is probed through how it influences the expansion rate of the Universe $H(z)$ and the rate growth of structure $g(z)$. One can study the expansion rate and geometry of the Universe measuring distances and volumes and one can probe the growth rate measuring how the structure of the Universe evolves with

F.J. Castander (for the DUNE and EIC collaborations)
Facultat de Ciències, Institut de Ciències de l'Espai, Campus UAB, 08192 Bellaterra, Barcelona, Spain
e-mail: fjc@ieec.fcr.es

J.M. Diego et al. (eds.), *Highlights of Spanish Astrophysics V*, Astrophysics and Space Science Proceedings, DOI 10.1007/978-3-642-11250-8_20,

time. During the last years there have been two influential committees that were appointed to advice agencies on the strategy to attack the study of dark energy: the Dark Energy Task Force (DETF) in the USA and the ESA/ESO Working Group on Fundamental Cosmology (WGFC) in Europe. The DETF highlighted four observational probes as the most promising to study dark energy: weak gravitational lensing, baryon acoustic oscillations, supernovae and clusters of galaxies. Baryon acoustic oscillations and supernovae are pure geometrical tests while weak lensing and clusters probe both the geometry and the growth of structure. The recommendations of the ESA/ESO working group on cosmology suggested several avenues to advance in our understanding of the Universe. Amongst them there was a wide field optical and NIR imaging survey from space and the ground.

In 2005 ESA presented its Cosmic Vision 2015–2025 Themes Programme. One of the fundamental questions posed was: *How did the Universe originate and what is made of?* Among the themes to address this questions was the investigation of the nature and origin of dark energy that is accelerating the expansion of the Universe. A wide-field optical–infrared imager was proposed as a possible tool to pursue this investigation.

2 The Dark Universe Explorer Concept

The Dark Universe Explorer (DUNE) is a wide-field space imager whose primary goal is the study of dark energy and dark matter with unprecedented precision using weak gravitational lensing along with other cosmological probes.

The DUNE mission concept was presented to ESA in June 2007 in response to its Cosmic Vision 2015–2025 programme. DUNE was an M-class mission with limited risks and consisting of a 1.2 m telescope with a combined optical and near-infrared focal plane of 1 deg^2. DUNE will carry out an all sky survey in one wide optical band and three near infrared bands that will constitute a unique legacy for astronomy. DUNE addresses multiple themes of the Cosmic Vision Programme and is a realization of the wide-field imaging survey recommended by the ESO/ESA Working Group on Fundamental Astronomy.

DUNE main science objective is to investigate the nature of dark energy using weak gravitational lensing as its main observational tool. Weak lensing has been recognized by the DETF and the ESO/ESA WGFC as the most promising tool to study dark energy. The weak lensing technique samples the distribution of matter in the Universe probing both the geometry and the grow of structure of the Universe. For this purpose DUNE will measure galaxy shapes of millions of galaxies over the whole extragalactic sky (20,000 deg^2). The precision required in these measurements cannot be achieved from the ground and is only feasible from space. DUNE will complement its weak lensing measurements with other independent cosmological probes: baryon acoustic oscillations, integrated Sachs–Wolfe effects and galaxy clusters. In addition to studying dark energy, DUNE will also measure

Table 1 DUNE baseline summary

Science objectives	Must: Cosmology and Dark Energy
	Should: Galaxy formation
	Could: Exoplanets
Surveys	Must: 20,000 deg^2 extragalactic
	Should: Full sky (20,000 deg^2 galactic), 100 deg^2 medium-deep
	Could: 4 deg^2 exoplanet finding
Requirements	1 visible band (R + I + Z) for high-precision shape measurements
	3 NIR bands (Y, J, H) for photometry
Resolution	0.23″ PSF FWHM, 0.10″ pixels (vis)
Payload instrument	⊘ 1.2 m telescope, Visible & NIR cameras each with 0.5 deg^2 FOV
Service module	Mars/Venus Express, GAIA heritage
Spacecraft	2,013 kg launch mass
Orbit	Geosynchronous
Launch	Soyuz S-T Fregat
Operations	4-year survey mission

with exquisite accuracy the distribution of dark matter and yield information on the initial conditions of structure formation and test of General Relativity.

DUNE's main extragalactic survey of 20,000 deg^2 will be complemented by a medium-deep survey reaching fainter in a smaller area (~100 deg^2), a galactic survey covering the rest of the sky, and a microlensing survey to search for Earth-mass planets.

DUNE's baseline concept evolved from a phase 0 study conducted by the CNES and is summarized in Table 1. It consists in a 1.2 m mirror telescope with two focal planes, one in the optical and the other one in the near infrared covering 0.5 deg^2 each. The telescope is a Korsch-type design with three mirrors and a dichroic separating the optical and near infrared focal planes (see Fig. 1). DUNE will operate in time-delay-and-integrate mode (TDI or drift scanning). There will be a de-scanning mechanism for the near infrared focal plane. The visible focal plane will have 36 CCD detectors of a pixel scale of 0.10″/pix. The near infrared focal plane will have 60 detectors with a scale of 0.15″/pix (see Table 2). DUNE will image in a wide band (550–920 nm) in the optical and three near infrared bands in the 920–1,600 nm range.

The spacecraft will have a 2,013 kg wet mass and will be launched with a Soyouz S-T Fregat. It will be placed in a geosynchronous orbit, providing thermal stability to the mission necessary for the required stable Point-Spread Function (PSF). The baseline spacecraft module is based on the Mars/Venus express and Gaia missions. DUNE has a low technological risk as it uses components of high reliability coming from the Gaia mission heritage.

DUNE was presented to the ESA Cosmic Vision programme by a consortium of institutes from different countries. Spain participates with three institutions: the Institut de Ciències de l'Espai (IEEC-CSIC), the Institut de Física d'Altes Energies (IFAE) and the Centro de Investigaciones Energéticas, Medioambientales y Tecnológicas (CIEMAT).

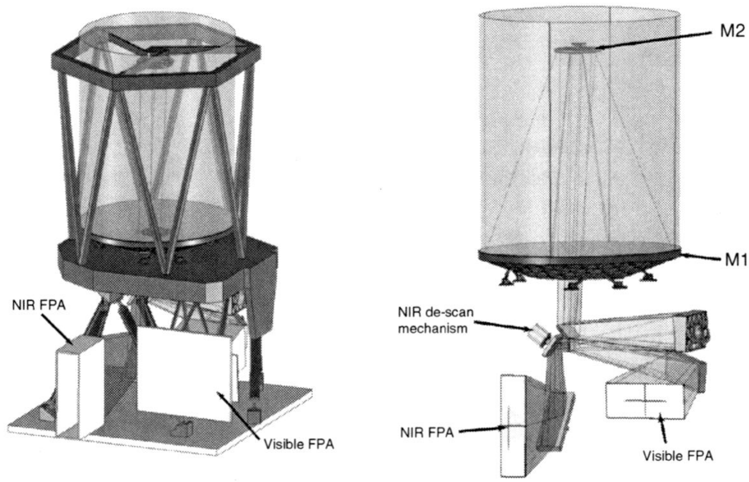

Fig. 1 DUNE optical configuration

Table 2 DUNE payload parameters

Optical configuration	Off-axis Three Mirror Anastigmat (TMA)
Pupil diameter	1.2 m
Focal length	24.2 m
f-Number	20
Spectral range	VIS: 550–920 nm NIR: 920–1,600 nm (3 bands)
Spectral separation	Dichroic mirror
Resolution (pixel size)	0.102 arcsec 0.15 arcsec
Visible FOV	$1.09° \times 0.52°$ (array of 4 AL × 9 AC matrices / 4,096 × 4,096 pixels per matrix / 12 μm pitch)
NIR FOV	$1.04° \times 0.44°$ (array of 5 AL × 12 AC matrices / 2,048 × 2,048 pixels per matrix / 18 μm pitch)

3 The Euclid Mission Concept

In October 2007 ESA selected the DUNE and SPACE concept missions for a joint assessment study, given their similarities in scientific goals, in objects to be studied, the wavelengths used and the area coverage on the sky. The technical appraisal of the proposals stated that the DUNE mission was of reasonable size but that the NIR focal plane was too large and dependent on US technology and that the PSF degradation due to radiation damage should be further investigated. Regarding the SPACE mission, it expressed concerns about the risks involved in the micro-mirror technology readiness level used for spectroscopy and the very complex optical design.

In May 2008 the Concept Advisory Team (CAT) formed reported their conclusions of the joint mission to the ESA advisory structure. The new merged mission concept combines the wide-field deep imaging inherited from DUNE with the low resolution multi-object spectroscopy adopted from SPACE. The new mission concept was named Euclid in honour of the pioneer of geometry. It was chosen to continue as one of the pre-selected missions in the ESA Cosmic Vision programme.

Euclid is an observational cosmology mission whose primary science objectives are: (1) The study of the nature of dark energy, for which it will measure the dark energy equation of state parameter ($w = P/\rho \simeq w_0 + w_a\ (1 - a)$, where a is the scale factor of the Universe relative to its current scale), zeroth order (w_0) and first derivative (w_a) to 2% and 10% precision respectively, using both the expansion rate of the universe and the growth of structure. (2) The study of the nature of dark matter, testing the structure formation paradigm and measure the sum of the neutrino masses in combination with Planck to 0.04 eV. (3) The study of the seed of structure formation, improving the determination of the initial conditions parameters by a factor of 20 compared to Planck alone. (4) Testing general relativity and distinguish it from the simplest modified gravity theories by measuring the growth factor exponent γ to a 2% precision.

To reach its science goals, Euclid will carry out an entire extra-galactic survey of 20,000 $\deg^2$ both in imaging and spectroscopy to measure weak lensing and baryon acoustic oscillations. On the one hand, it will measure diffraction limited galaxy shapes in one broad band and obtain photometric redshifts using three near infrared bands down to $H_{AB} = 24$. On the other, it will obtain spectroscopic redshifts for one-third of the galaxies down to $H_{AB} = 22$. Besides its two main cosmology probes, the data gathered will also probe cosmology using galaxy clustering using the whole information of the power spectrum, redshift space distortions, the number density of clusters and the integrated Sachs–Wolfe (ISW) effect. Figure 2 summarizes the constraints on the redshift dependence of w that can be achieved with Euclid's cosmological probes.

Complementary to the wide extra-galactic survey, Euclid will also carry out two additional surveys. A deep extragalactic survey that will sample of the order of 30 $\deg^2$ two magnitudes deeper than the wide survey, reaching an $H_{AB} = 26$ magnitude limit in optical and near infrared imaging and taking spectra of galaxies down to $H_{AB} = 24$. The main science driver of this deep survey is the study of galaxy evolution. Euclid is also expected to carry out a Galactic survey whose characteristics are still to be determined.

In addition to determining cosmological parameters with great accuracy, Euclid will also have a tremendous legacy value. It will provide morphology and deep visible and near infrared photometry for thousands of millions of galaxies up to redshifts $z \sim 2$. It will also take the near infrared spectra of hundreds of millions of galaxies up to $z \sim 2$. In the deep survey, it will reach magnitude limits impossible to achieve from the ground, allowing the study of very high redshift galaxies up to $z \sim 10$.

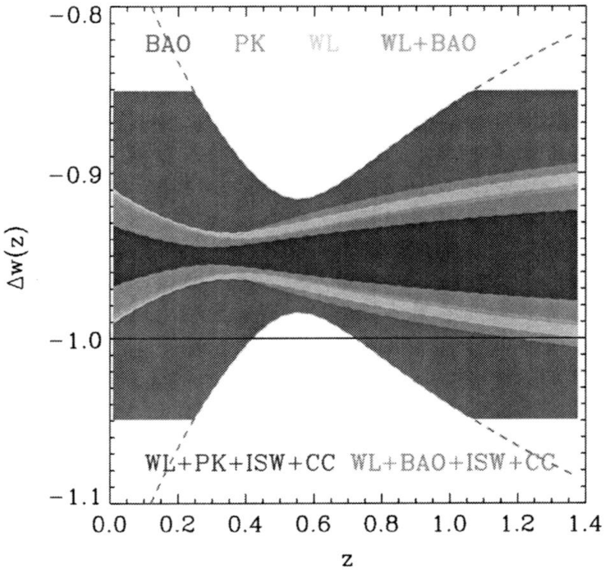

Fig. 2 Predicted redshift dependence of $w(z)$ errors taking into account Planck priors using Euclid's cosmology probes. Figure courtesy of J. Weller and the Euclid Cosmology Working Group

3.1 The Euclid Mission Implementation

The implementation of the merged mission recommended by the Concept Advisory Team has the following elements. Euclid will be launched by a SOYUZ ST from Kourou. It will operate at a large amplitude Lissajous orbit around the L2 Lagrange point, reached with free insertion in 30 days of transfer time. The spacecraft will have a solar panel attached to the body and will have a three-axis stabilization. The attitude control will be done with gas propellers. The science telemetry will be performed in the K band with an expected rate of 700 Gbits/day after compression. The mission duration is expected to be five years.

The Euclid payload will be composed of a Korsch type three mirror anastigmat (TMA) telescope with a 1.2 m primary mirror (see Fig. 3). It will be passively cooled. The light beam coming from the secondary mirror will feed three instruments. The optical design to provide the optimal solution for these different channels is under study. The NIR spectroscopic channel will have a 0.5 deg^2 field of view. It will perform multi-slit spectroscopy using Digital Micro-Devices (DMD). It will have a resolution $R \sim 400$ and a wavelength range coverage 0.9–1.7 μm. As the DMD technology has to demonstrate a high enough readiness level for space qualification, a slitless spectrograph is a backup option. Euclid will also have two imaging instruments fed with the same beam split with a dichroic. The near infrared imaging channel will have a 0.5 deg^2 field of view covered with near-infrared detectors sensitive in the wavelength region from 1.0 to 1.7 μm and with a scale of 0.3″/pixel. It will have a filter wheel with at least three filters Y, J and H. The visible channel

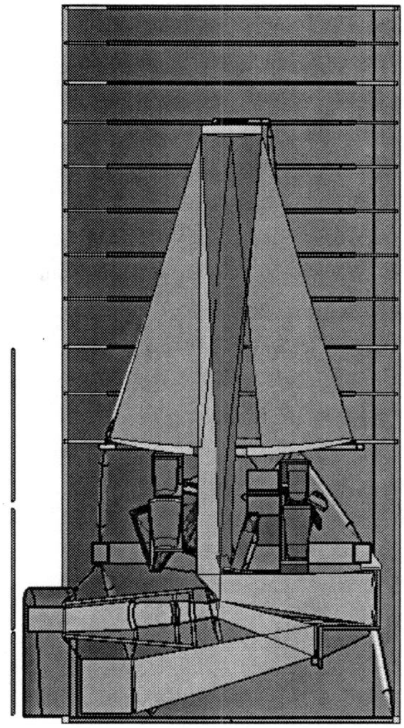

Fig. 3 Euclid configuration sketch. The spectroscopic beam is represented in *dark pink-red*, while the imaging channels in *light violet-pink*. The NIR channel focal plane is the smaller field of view to the *left*. The visible focal plane is the larger one at the *bottom left corner*

will also have a field o view of 0.5 $\deg^2$. The pixel scale will be of 0.10″/pixel to be able to properly sample the instrument PSF of 0.23″ FWHM. The instrument will contain only one wide filter (equivalent to R + I + z) covering the wavelength range from 5,500 to 9,200 Å. The payload is expected to weigh ~660 kg, including 300 kg coming from the instruments. The power consumption will be lower than 200 W. Euclid will contain an on-board data handling unit to provide spectroscopic targets and compress the data volume before transmission.

3.2 The Euclid Mission Assessment Phase

In May 2008 ESA released a call for industry (ITT) to carry out assessment studies of the Euclid concept. Two industrial consortia applied and were selected. One led by Astrium GmbH at Friedrichshafen and the other led by TAS Italia at Torino. The industrial assessment phase started in September 2008. At the same time in June 2008 ESA released a call for Declarations of Interests to study the Euclid payload. Two consortia applied and were selected. One, the Euclid Imaging Channels (EIC) consortium, the former DUNE team, applied to study the visible and near infrared imaging channels. The other was the Euclid Near Infrared Spectrograph (ENIS) consortium, formerly the Space team, that applied to study the spectrograph. The assessment study (or Phase A) for the payload started on October 2008. The study consisted of three stages: a fist one, to evaluate requirements and make trade off studies between different solutions; a second one, to perform preliminary design and cost estimates; and a third one, to report the conclusions. At the end of this phase, scheduled for Autumn 2009, the Euclid mission, in conjunction with the other M-class missions, has to be reviewed by the ESA advisory bodies to pass a down selection process for selection for phase B.

Meanwhile, at the end of 2008 ESA and NASA started negotiations to study the possibility of merging their corresponding cosmology missions (Euclid and JDEM). In March 2009 they announced a joint mission called International Dark Energy Cosmology Survey (IDECS), very similar to the Euclid concept. However, in April 2009 NASA reported a delay in their schedule. As the European cosmology mission (either IDECS or Euclid) has to pass the Cosmic Vision Programme selection process, ESA decided to leave the IDECS concept on hold and continue with the Euclid concept but with the modification of using slitless spectroscopy as the default spectroscopic mode.

Acknowledgements FJC acknowledges the DUNE and EIC consortia as well as the Euclid Cosmology Working Group for useful discussions and providing material for these proceedings.

The Nordic Optical Telescope

Anlaug Amanda Djupvik and Johannes Andersen

Abstract An overview of the Nordic Optical Telescope (NOT) is presented. Emphasis is on current capabilities of direct interest to the scientific user community, including instruments. Educational services and prospects and strategies for the future are discussed briefly as well.

1 Introduction

The Nordic Optical Telescope (NOT) is a modern 2.6 m alt-azimuth telescope, operating at Roque de los Muchachos Observatory, La Palma, since 1989 as the main northern-hemisphere optical facility for Nordic astronomers (Fig. 1). Its f/11 Ritchey–Chretien optical system, the telescope, and the enclosure were carefully designed to deliver the best image quality at the site [2].

Structure. The Nordic Optical Telescope Scientific Association (NOTSA) was formed in 1984 to construct and operate the NOT on behalf of the Research Councils of Denmark, Finland, Norway, and Sweden; the University of Iceland joined in 1997. The budget is shared roughly in the ratios 20:30:20:30:1%.

A Director has overall legal, financial, and scientific responsibility for operations and strategic planning, under the authority of the NOTSA Council, which represents the owners. On-site operations are ensured by a team of 13 staff members headed by an Astronomer-in-Charge; 5 research students complement the team.

User Community. Seventy-five percent of the science time on the telescope is at the disposal of the Nordic community through competitive proposals peer-reviewed by the NOT Observing Programmes Committee. Twenty percent of the time is allocated to Spanish astronomers through the Comisión de Asignación de Tiempo (CAT), while 5% goes to international teams through the CCI International Time Progamme (ITP).

A.A. Djupvik and J. Andersen
Nordic Optical Telescope, Apdo 474, 38700 Santa Cruz de La Palma, Spain
e-mail: amanda@not.iac.es, ja@not.iac.es

J.M. Diego et al. (eds.), *Highlights of Spanish Astrophysics V*, Astrophysics and Space Science Proceedings, DOI 10.1007/978-3-642-11250-8_21,

Fig. 1 The NOT at sunset with the dome lights on while turning (Photo: Jacob Clasen, NOT)

Not only Nordic astronomers, but *anybody* regardless of nationality or affiliation can apply for the Nordic time, and proposals are ranked solely on the basis of scientific merit. Thus, over the past 5 years, over 20% of the Nordic time has been allocated by the OPC to non-Nordic PIs. About half of these projects have been part of the OPTICON Trans-National Access Programme, which supports access for European researchers to all modern European 2–4 m class telescopes. Thus, despite its name, the NOT already serves a broad European user community.

2 NOT User Services

A systematic upgrade programme has been conducted over the last five years to improve the scheduling, operation, instrumentation, and reliability of the NOT.

Scheduling. The NOT offers both visitor and service mode observations. The emphasis is increasingly on the latter because of the flexibility offered, and the fraction of service observing nights has increased steadily to ~40% in 2005–2007.

Part of the service time is allocated through a *Fast-track* procedure with a short lead time from proposal to execution, offered since 2005. Proposals for short observing programmes (max. 4 h) can be can submitted at any time (see http://www.not.iac.es/observing/service/) and are reviewed promptly by the OPC. The successful programs are then executed by the staff in queue scheduled service mode. Available instruments are ALFOSC, NOTCam, FIES and StanCam.

The pressure factor from proposals for the Nordic time has increased gradually in recent years, to about 2.5 over the last decade. Technical downtime is typically ~1%, weather down time ~10% in summer (April–September) and ~35% in winter (october–march). However, actual usable hours vary by only ~10% over the year.

Operations. Major effort has been put into making the telescope control system fast, flexible, safe and easy to use. Telescope operators are not needed, but there is always a support astronomer to provide a thorough introduction on the first night of a visitor run, and technical support is at hand throughout.

Table 1 UBVRI zero-point magnitudes (for 1e$^-$/s)

	U	*B*	*V*	*R*	*I*
ALFOSC	24.0	25.7	25.6	25.4	24.6
MOSCA	24.6	26.2	26.0	25.7	25.2
StanCam	23.8	25.7	25.4	25.3	24.5

For safety, the telescope is linked to our weather station, and the dome closes automatically if limits of humidity or wind are exceeded. Autoguiding starts fully automatically in most observing modes, and focusing is fast and reliable.

The entire observing system (detector + instrument + telescope) can be fully controlled by the new integrated data acquisition system, the *Sequencer*, implemented for the core instruments in 2006–2008. This permits in principle to automate the entire observation, although these features are currently under development.

Only one instrument at a time can be mounted at the main focus, but 45° folding mirrors in the adaptor allow the standby CCD camera (StanCam) and the high-resolution fiber spectrograph (FIES) to be available at all times, greatly increasing flexibility on time-critical programmes.

A rare feature of the NOT is its ability to go as low as 6.4° from the horizon, which is a unique capability in studies of objects in the inner Solar System, such as the planet Mercury (see, e.g. [12]).

Documentation and data flow. A comprehensive set of documentation on the telescope, instruments, detectors, and operations is available at http//www.not.iac.es. This includes on-line tools for preparing proposals and/or observations, such as an Exposure Time Calculator and Script Generators for the core instruments. Quality control data on detectors, etc., are also available.

All data from the core instruments (ALFOSC, FIES, MOSCA, NOTCam and StanCam) are stored in standard Multi-Extension Fits (MEF) format with primary WCS information in the headers. The data are delivered on DVD, or available by ftp for *Fast-Track* or time-critical Target-of-Opportunity observations. Technical details of the core instruments are summarised in the following sections. The optical zero-point magnitudes are summarized in Table 1.

3 ALFOSC

The Andalucia Faint Object Spectrograph and Camera (ALFOSC), owned by the Instituto de Astrofísica de Andalucía (IAA), has been used at the NOT by mutual agreement since 1996. ALFOSC was upgraded with a new CCD detector (e2v CCD42-40, 13.5 μm $\times$ 2,048 $\times$ 2,048) and new camera optics in 2003. The plate scale is 0.19″/pix. The new CCD provides excellent resolution at all wavelengths, but higher fringe levels than the previous CCD. Fast photometry with windowed readout and on-line reduction with comparison star subtraction is offered.

With 16 grisms, ALFOSC offers spectroscopy in the resolution range $200 < R < 10{,}000$. A VPH grism with spectral resolution $R = 10,000$ in the 6,350–6,850 Å range offers a total system efficiency of 30%. A set of vertical slits allow fast horizontal readout in a small window for time-resolved spectroscopy, and quick-look reduction tools are available for both long-slit and echelle spectra. Multi Object Spectroscopy (MOS) with pre-fabricated slit plates is available.

Dual-beam linear and circular imaging- and spectro-polarimetry is also offered with ALFOSC. A quick-look reduction tool for linear imaging polarimetry was added in 2007 (see http://www.not.iac.es/instruments/alfosc/).

4 FIES

The FIber Echelle Spectrograph (FIES) is permanently ready for use. For mechanical and thermal stability it is located in a separate building with thermal control. The detector is an e2v CCD42-40 (13.5 μm × 2,048 × 2,048), which covers the wavelength range 370–730 nm without gaps. The fiber diameters are 2.5″ in low-resolution mode ($R = 25{,}000$), 1.3″ for medium ($R = 45{,}000$) and high resolution mode ($R = 65{,}000$; slit added). The FIEStool package provides on-line quick-look spectra and is also suitable for final reductions.

A main goal for FIES is high-precision radial-velocity work. The zero-point stability is currently $<150\,\mathrm{m\,s^{-1}}$ for a stellar spectrum followed by a separate Th–Ar lamp exposure, $<15\,\mathrm{m\,s^{-1}}$ for a stellar spectrum with *simultaneous* Th–Ar calibration. As an example, the 3 M_{Jup} transiting planet WASP-10b was confirmed with radial velocities from FIES [4]. Tests with series of daytime blue-sky spectra using simultaneous Th–Ar are also giving promising results (see Fig. 2; [5]). An iodine absorption cell is being added.

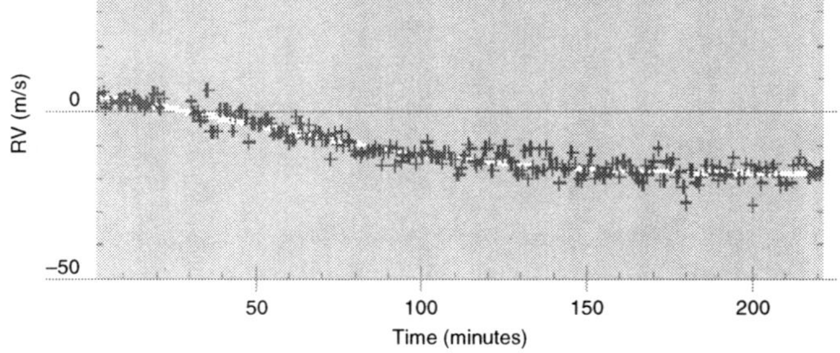

Fig. 2 A series of daytime sky spectra (reflected sunlight). The mean trend is likely due to winds in the upper atmosphere; the dispersion around the mean is $<3\,\mathrm{m\,s^{-1}}$

FIES is rapidly attracting interest and became the most-demanded instrument at the NOT among the applications for the semester 2008B. For more details, see http://www.not.iac.es/instruments/fies/.

5 NOTCam

The Nordic Optical Telescope near-infrared Camera and spectrograph (NOTCam) is a versatile instrument for the 0.8–2.5 μm wavelength range, using a recent Rockwell Hawaii-I HgCdTe detector (18 μm ×1,024 × 1,024).

The image scale can be toggled in a matter of seconds between the wide-field (WF) camera (0.234″/pix, FOV = 4′) and the high-resolution (HR) camera (0.078″/pix, FOV = 80″). The WF camera suffers from some optical distortion in the corners, but the HR camera has a high optical quality all over the FOV. With proper telescope tracking, the HR camera regularly delivers perfectly round, deep stellar images with FWHM of 0.3–0.4″.

Standard broad band filters JHK plus Y_n and 17 narrow-band filters are permanently installed. A cold shutter permits short integrations, and small cold stops are available for flux reduction of very bright targets. The zero-point magnitudes (for 1 e^-/s) are: 24.1, 24.1, and 23.5 mag, in J, H and K_S, respectively (Vega magnitudes). An example image is shown in Fig. 3.

Spectroscopy with the WF camera and a 0.6″ slit gives a resolution of $R = 2{,}500$ and covers the Y_n, J, H and K bands. The HR camera yields $R = 5{,}500$ with a 0.2″ slit and covers wavelength ranges from 1.26–1.34 μm (Paβ), 1.57–1.67 μm, to 2.07–2.20 μm (Brγ). Wollaston prisms for polarimetry, low resolution grisms [11], and broad-band ZY-filters are being considered for the future. For more detail see http://www.not.iac.es/instruments/notcam/.

Fig. 3 NOTCam image of M57 in J (*blue*), H (*green*), and a narrow-band filter centred on the H_2 line at 2.121 μm (*red*). FOV is 3.5′ × 3.2′, north up, east left. Beam-switch observations with a total on-source integration time of 360 s per filter. Reduced with the NOTCam quick-look reduction package

6 MOSCA

The MOSaic CAmera (MOSCA) is a direct imager featuring four Loral CCDs of 15 μm × 2,048 × 2,048 corresponding to 0.11″/pix and a FOV of 7.7′. The gaps between the four CCDs are 9–12″. The advantages of MOSCA are the uniformly good PSF over the entire FOV and the high throughput (see Table 1), especially in the *U* band. For more detail see http://www.not.iac.es/instruments/mosca/.

7 StanCam

The Stand-by Camera (StanCam) has a TEK 24 thinned CCD (μm ×1,024 × 1,024; 0.176″/pix, FOV = 3′). It is permanently available at a folded Cassegrain focus. Together with NOTCam it offers near-simultaneous UBVRIJHK coverage. For more detail see http://www.not.iac.es/instruments/stancam/.

8 Visitor Instruments

The Turku photo-polarimeter Turpol provides linear and circular simultaneous UBVRI polarimetry of single sources (diaphragms) with high precision through a double image chopping technique. Polarization levels below 0.01% can be detected at the NOT, and systematic errors for brighter stars are ~0.005% [10]. On-line reduction is available. See http://www.not.iac.es/instruments/turpol/.

SOFIN, a high resolution echelle spectrograph, is supported by the University of Helsinki and Astrophysikalisches Institut Potsdam. It covers the spectral range 3,500–9,000 Å with resolutions R = 27,000–170,000, and is also capable of spectro-polarimetry [6]. Observations are performed in service mode and reduction software is available. See http://www.not.iac.es/instruments/sofin/.

The PolCor *Lucky* polarimeter/coronagraph, equipped with an EMCCD (16 μm× 512×512, 0.12″/pix, FOV = 1′) with up to 33 Hz readout rate, has a rapidly turning polarizer and three coronagraphic disks. A computer controlled Lyot stop masks the M2 support vanes. The FWHM improves from initially 0.7″ to 0.4″ applying only shift and add techniques, but improves further to 0.2–0.3″ using frame selection and deconvolution [8].

Among other visitor instruments, *LuckyCam* was used at the NOT in 2002–2004 to obtain near diffraction-limited imaging in the *I* band using reference stars as faint as 16.5 mag. Keeping the 1% best frames improved the image quality from 0.7″ to 0.1″ [7]. Recently, also FastCam [9] also obtained diffraction-limited *I* band imaging using 1–5% of the frames.

9 The Educational Role of the NOT

In a world of 8-m telescopes it makes sense to spend a modest amount of time on a 2 m-class facility to train the next generation of astronomers – at least those who will be involved in designing and building the next generation of telescopes and their instrumentation. At the NOT, we are taking a systematic approach to this challenge and are developing a varied set of offers of educational services to universities, covering the whole educational *food chain*.

The traditional use of the NOT in education has been for on-site courses, where groups of typically 12 PhD or MSc students spend 1–2 weeks on La Palma learning how to observe with the NOT, reduce and analyse their data, and formulate their next observing time applications effectively. Such courses are highly educational, motivating, and popular with the students, giving them real hands-on observing experience. The NOT supports these courses by providing a small *class room* at ORM, with the control room screens projected on the wall and with 12 standard laptops available for data reduction and analysis.

These courses have become very popular, but cannot handle large volumes of students. We are therefore also developing remote observing as a means to involve hands-on observing in university courses without actually taking the students to La Palma. The observing courses at Molėtai Observatory in Lithuania have pioneered this technique, most recently in 2008.

Finally, the NOT currently hosts five PhD or MSc students, who typically spend a year at La Palma, devoting 75% of their time to their thesis projects and 25% on support duties, service observing, or various developments. Experience shows that this experience is a significant asset for their future careers, whether in or outside astronomy.

10 Future Perspectives

The world of today and tomorrow is scientifically, technically, financially, and politically very different from that of 1984 when NOTSA was founded. The great majority of our users have access to state-of-the-art 8-m telescopes; European cooperation and coordination have developed enormously; and the stand-alone paradigm for NOT has become obsolete. An International Evaluation of the NOT in 2006 supported this view (see report at http:/www.not.iac.es/news/reports/).

Given NOTSAs active participation in both the OPTICON and ASTRONET European networks (see http://www-astro.opticon.org and http://www.astronet-eu.org), it was natural to view the role of the NOT as part of a future pan-European facility of modern 2–4 m telescopes. Through a series of discussions in the user community, this vision was turned into a specific long-term strategy [1] and *Development Plan* [3].

The overarching long-term goal is to optimise the performance of the NOT by specialising its performance in specific areas, in concert with similar moves at other telescopes. This only works if an overall plan exists, and the ASTRONET Board is therefore appointing a European Telescope Strategy Review Committee with the charge to consider innovative ways to plan, equip, and operate a set of European 2–4 m telescopes so as to achieve optimum scientific returns and cost-effectiveness, including better coordination with other disciplines such as space astronomy. The NOT is committed to becoming part of such a new, common European 2–4 m facility.

Meanwhile, we continue to specialise and optimise the NOT for high-impact Nordic science. This includes a thorough review of the instrumentation, operating modes, and scheduling of projects, with emphasis on remaining competitive in studies of transient and variable sources. As part of this strategy, we are seeking to move to a single set of permanently mounted instruments, which would enable us to respond quickly and flexibly to new events and optimise scientific productivity in areas where the NOT can still be competitive if deployed intelligently. Detector and data acquisition systems are being upgraded in parallel.

Acknowledgement Thanks to Thomas Augusteijn for valuable comments.

References

1. Andersen, J., *NOT beyond 2009: Towards the Common Northern Observatory* (2006). http://www.not.iac.es/news/reports/NOT_StrategyReport_Fin.pdf
2. Ardeberg, A., *Some properties of the Nordic Optical Telescope* (1990). http://www.astro.lu.se/Resources/NOT/proptxt.html
3. Augusteijn, T., *Development plan 2008–2010. Astronomer in charge's report* (2007). http://www.not.iac.es/news/reports/Dev2008_2010.pdf
4. Christian, D.J., Gibson, N.P., Simpson, E.K., et al., MNRAS 392, 1585 (2009). arXiv:0806.1482
5. Frandsen, S., Telting, J., Buchhave, L., *FIES*, NOT Annual Report 2007 (2007). http://www.not.iac.es/news/reports/annual/NOT_AnnRep07.pdf
6. Ilyin, I., PhD thesis, Oulu 2000 University Press. ISBN 951-42-5724-3 (2000)
7. Mackay, C.D., Baldwin, J.E., Tubbs, R., Cox, G., *High-resolution "Lucky" imaging*, NOT Annual Report 2003 (2003). http://www.not.iac.es/news/reports/annual/NOT_AnnRep03.pdf
8. Olofsson, G., Florén, H.G., *The PolCor "Lucky" Polarimeter/coronagraph* (2008). http://www/not.iac.es:/instruments/polcor/PolCor-2.pdf
9. Oscoz, A., Rebolo, R., & the FastCam Team, *FastCam: Diffraction-limited optical imaging*, NOT Annual Report 2007 (2007). http://www.not.iac.es/news/reports/annual/NOT_AnnRep07.pdf
10. Piirola, V., in *Astrophysics with the NOT*, eds. H. Karttunen, V. Piirola, University of Turku, Turku, 63 (1999)
11. Telting, J., *Spectroscopy at the NOT. Current instrument status and possible efficiency/functionality upgrades* (2004). http://www.not.iac.es:/instruments/development/JHTspecUpgrades.pdf
12. Warell, J., Karlsson, O., Planet. Space Sci. 55, 2037 (2007)

The Space Telescope for Ultraviolet Astronomy WSO-UV

Ana I. Gómez de Castro, B. Shustov, M. Sachkov, N. Kappelmann, M. Huang, and K. Werner

Abstract The World Space Observatory UltraViolet (WSO-UV) project is an international space observatory designed for observations in the ultraviolet domain where some of the most important astrophysical processes can be efficiently studied with unprecedented sensitivity. WSO-UV is a multipurpose observatory, made by a 170 cm aperture telescope, capable of high-resolution spectroscopy, long-slit low-resolution spectroscopy, and deep UV and optical imaging. With a nominal mission life time of 5 years, and a planned extension to 10 years, from a geosynchronous orbit with an inclination of 51.8°, WSO-UV will provide observations of exceptional importance for the study of many astrophysical problems. WSO-UV is implemented in the framework of a collaboration between Russia (chair), China, Germany, Spain, and Ukraine. This article is an updated account of the status of the project and its implementation in Spain.

1 Introduction

The World Space Observatory Ultraviolet Project has grown out from the needs of the Astronomical community to have access to the Ultraviolet (UV) range of the spectrum. The success of the International Ultraviolet Explorer (IUE) observatory

A.I. Gómez de Castro
Fac. de CC. Matemáticas, Universidad Complutense de Madrid, Plaza de Ciencias 3, 28040 Madrid, Spain
e-mail: aig@mat.ucm.es

B. Shustov and M. Sachkov
Institute of Astronomy, Russian Academy of Sciences, Pyatnitskaya 48, 119017 Moscow, Russia

N. Kappelmann and K. Werner
Institut für Astronomie und Astrophysic, Abteilung Astronomie (IAAT), Universitüat Tübingen, Tubingen 72076, Germany

M. Huang
National Astronomical Observatories, CAS, Beijing, China

J.M. Diego et al. (eds.), *Highlights of Spanish Astrophysics V*, Astrophysics and Space Science Proceedings, DOI 10.1007/978-3-642-11250-8_22,

and successor instruments such as the GHRS, STIS spectrographs and WFPC2 and ACS imagers on-board the Hubble Space Telescope (HST), demonstrate the major impact that observations in the UV wavelength range have had on modern astronomy. Since 1978, astronomers have enjoyed continuous access to the 1,000–3,000 Å spectral range. At present, the Galaxy Evolution Explorer (GALEX) is a survey mission providing broadband imaging and low resolution grism spectroscopy between 1,300 and 2,800 Å. However, no future UV missions were planned for the post-HST era. As a result, WSO-UV has been driven by the needs of scientists from many different countries to have an UV facility in space, in the horizon of the next decade.

The WSO-UV consists of a UV telescope in orbit, incorporating a primary mirror of 1.7 m diameter feeding a set of UV low and high spectral resolution spectrographs and UV-optical imagers. With a telescope with just half the collecting area of HST, but taking advantage of the modern technology for astronomical instrumentation, and of a high altitude, high observational efficiency orbit, WSO-UV will provide UV-optical astronomical data quantitatively and qualitatively comparable to the exceptional database collected by HST. In addition to the newly observed targets, with the synergy of the HST archive, WSO-UV will allow long term photometric, spectroscopic, and astrometric monitoring of a variety of astronomical objects.

2 Technical Overview

The T-170M telescope is a Ritchey-Chretien with reflection optics and a focal length of 17 m. The diameter of the primary mirror is 1.7 m. The telescope provides a corrected field of view of 0.5° on the telescope focal surface.

Structurally, telescope T-170M consists of the optical system, structural module and service complex (see Fig. 1).

WSO-UV is equipped with multipurpose instrumentation to carry out:

- High resolution spectroscopy ($R \sim 55{,}000$) of point sources in the range 1,020–3,200 Å by means of two high resolution echelle set-ups: HIgh Resolution Double Echelle Spectrograph (HIRDES). The sensitivity of this instrument is about ten times better than that of the Space Telescope Imaging Spectrograph (STIS) on a similar configuration.
- Long-slit low resolution ($R \sim 1{,}500$–$2{,}500$) spectroscopy in the 1,020–3,200 Å range with the Long Slit Spectrograph (LSS).
- High sensitivity imaging and slit-less spectroscopy ($R \sim 300$) in the 1,200–2,000 Å range with spatial resolution 0.1 arcsec and field of view of 2.1 arcmin for weak UV sources.
- Simultaneous imaging in the 1,200–2,700 Å and 2,700–7,000 Å range of large fields of view ($> 3.4' \times 3.4'$) for surveys, with spatial resolution of 0.1 arcsec.

In the instrument compartment, see Fig. 2, the optical bench (OB), used as reference plane for all the onboard instrumentation, is aligned and maintained in the correct position with respect to the primary mirror unit (PMU) using a three rods

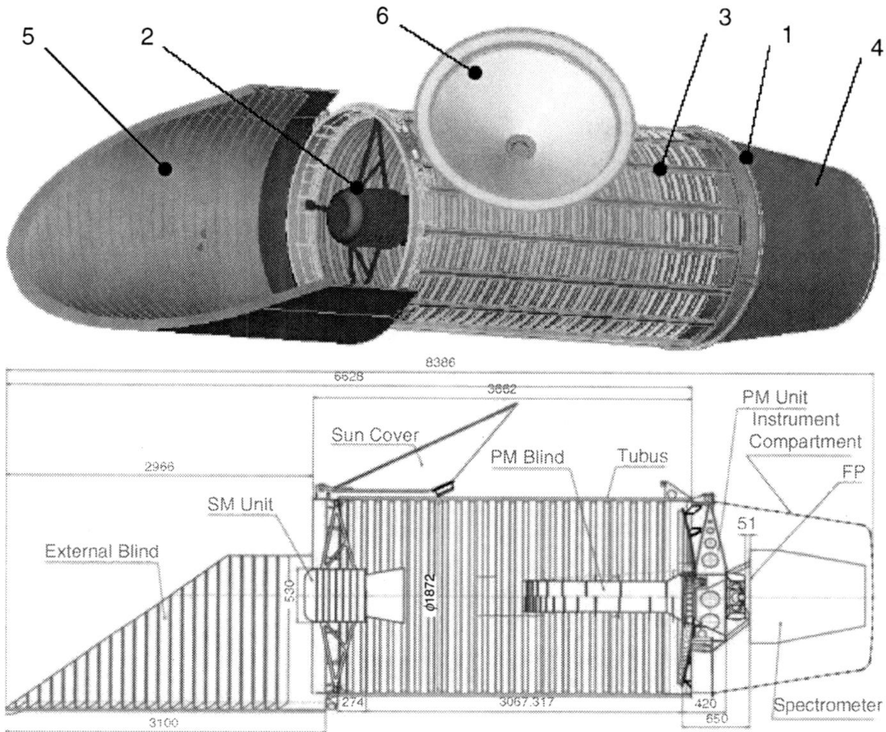

Fig. 1 The structural elements of the T-170M. Primary mirror unit (PMU) (1) is the main telescope's structural element (structural unit). The tube (3), technological (demountable) dust moisture protective cover and casing of instrumentation compartment (4) are fastened to the PMU frame. There are three attachment points of the telescope to spacecrafts service module in the bottom frames part. Optical bench with scientific instrumentations devices and primary mirrors baffle are mounted on the PMU frame

system. The Imaging and Slit less Spectroscopy Instrument for Surveys (ISSIS) is mounted on the upper basis of the optical bench, in the space available between the PMU and the OB itself, while the two spectrographs HIRDES and LSS are mounted to the OB bottom basis. The Fine Guidance System, that uses three sensors placed in the focalplane close to the spectrograph entrance slits, ensures a pointing stability better than 0.1 arcsec (3σ). Each of the focal science plane instruments has its own power supply and data handling unit in a service box mounted on the external side of the instrument compartment.

WSO-UV uses the NAVIGATOR platform designed in the Lavochkin Science & Technology Association (Russia) as a unified unit for several missions. The first two launches of Russian spacecrafts, SPECTRUM-R and ELECTRO, using this platform are scheduled for the end of this decade, just before the WSO-UV launch. The orbit is geosynchronous with an inclination of 51.8°. The launch vehicle is a Zenit-2SB.

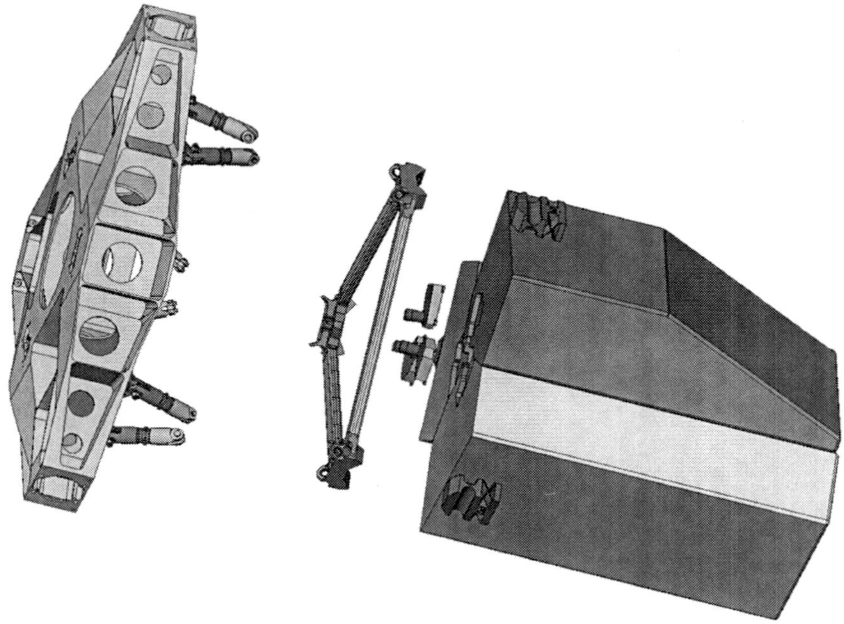

Fig. 2 Mounting of HIRDES and LSS on the optical bench

3 Science: Instruments and Operations

3.1 Imaging and Slitless Spectroscopy Instrument for Surveys: ISSIS

ISSIS design is driven by two key objectives:

- To provide a high sensitivity instrument for the characterization of the far-UV Universe. The spatial resolution has been optimized to the platform performance (0.1 arcsec). Slit-less spectroscopy ($R \sim 300$) is foreseen as the most used mode though some filters will be provided for direct imaging.
- To provide an efficient instrument for UV surveys to be carried out either in pointed mode or in parallel mode while performing long exposures with the spectroscopic instrumentation: HIRDES or LSS.

These two objectives are shown in the instrument design by the presence of two independent channels: the High Sensitivity Channel (HSC) and the Channel for Surveys (CfS). A summary of the instrument main characteristics is given in Table 1.

Table 1 Characteristics of the ISSIS imaging and slitless spectroscopy instrument

	HSC	CfS/FUV	CfS/UVO
Spectral range (Å)	1,200–2,000	1,200–6,000	1,200–6,000
Operating spec. range (Å)	1,200–2,000	1,200–2,700	2,700–6,000
Spatial resolution	0.1	0.1	0.1
Spatial sampling (arcsec)	0.05	0.05	0.05
Field of view (arcmin)	>2.1	>3.4	>3.4
Temporal resolution	40 ms	60 s	60 s
Detector type	CsI MCP	Full-frame CCD	Full-frame CCD
Detector format	> 2,048 × 2,048 pix	4,096 × 4,096 pix	4,096 × 4,096 pix
Spectral filters	10	10	10
Slitless spectroscopy	Yes, $R \sim 300$	Yes, $R \sim 300$	Yes, $R \sim 300$
Coronagraphy	TBC	No	Yes

Table 2 Properties of the HIRDES high resolution spectrographs

		UVES	VUVES
Spectral range (nm)		174.5–310.0	102.8–175.6
Dispersion		50,000	55,000
Properties at the minimum echelle order	Wavelength (nm)	310.0	175.6
	Order number	148	165
	Bandwidth/pixel (pm)	2.07	1.07
	Spectral range (nm)	2.09	1.06
	Order separation (μm)	180	565
Properties at the maximum echelle order	Wavelength (nm)	174.5	102.8
	Order number	263	282
	Bandwidth/pixel (pm)	1.16	0.63
	Spectral range (nm)	0.66	0.36
	Order separation (μm)	600	200

3.2 High Resolution Double Echelle Spectrograph: HIRDES

HIRDES is made by two echelle instruments: VUVES (1,020–1,760 Å), and UVES (1,740–3,100 Å), able to deliver high resolution spectra ($R \sim 55{,}000$). The HIRDES design uses the heritage of the ORFEUS (Orbiting and Retrievable Far and Extreme Ultraviolet Spectrometer) missions successfully flown on the Space Shuttle in 1993 and 1996 [1]. The entrance slits of the two spectrographs lie in the focal plane, on a circle with diameter 100 mm which also hosts the LSS slits. The position of the target in the slit is monitored by an (optical) sensor of an Internal Fine Guidance System, which is part of every spectrograph. The VUVES and UVES detectors are photon counting devices based on Microchannel Plates, read out by means of a Wedge&Strip Anode based on the ORFEUS detector design. Technical details of the high resolution spectrographs are given in Table 2. A comparison between the performance of the high resolution spectrographs in HIRDES and HST/STIS is provided in Fig. 3.

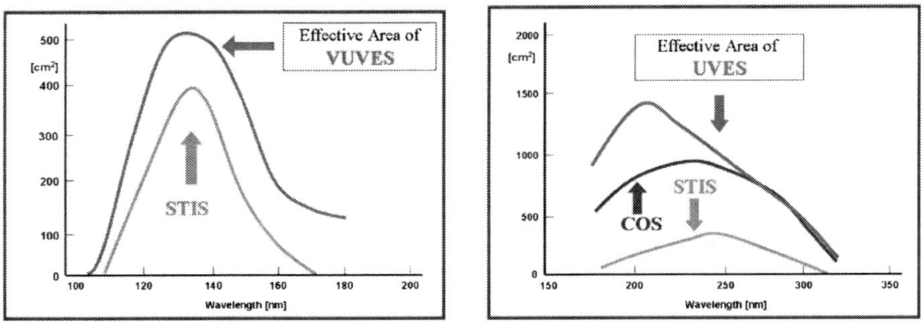

Fig. 3 Comparison of WSO/HIRDES ($\lambda/\delta\lambda$ = 50,000) and HST/STIS ($\lambda/\delta\lambda$ = 37,000) effective areas

3.3 The Long Slit Spectrograph

The Long Slit Spectrograph will provide low resolution ($R \sim$ 1,500–2,500) spectra in the 1,020–3,200 Å spectral range with a design that emphasizes sensitivity for observing faint objects. According to preliminary results of phase A study ongoing at NAOC (China), the most promising design for this instrument is to have two channels working in 1,020–1,610 Å (FUV) and 1,600–3,200 Å (NUV) ranges, respectively. Each channel has its own entrance slit. Both slits have dimensions of $1'' \times 75''$. The spatial resolution is not worse than 1 arcsec (0.4 arcsec at best). In order to maximize sensitivity, both LSS channels use holographic gratings to minimize the number of reflecting surfaces in the optical path, and to enhance overall spatial resolution along the slit. Both channels use microchannel plates working in photon-counting modes as detectors. A slit-viewer similar to that used in HIRDES is under study.

3.4 Scientific Operations

The WSO-UV Ground Segment (GS) is made by all the infrastructure and facilities involved in the preparation and execution of the WSO-UV mission operations, which typically encompass real-time monitoring and control of the spacecraft as well as reception, processing and storage of the scientific data. The ground segment is under development by Russia and Spain, which will coordinate the Mission and Scientific operations and will provide the satellite tracking stations for the project. Scientific (and mission) operations will be shared, in a 50–50% basis, between the Russian and the Spanish operations centers.

Key drivers to the WSO-UV ground segment development are a modular design, the use of existing systems, and the potentiality for distributed processing of the scientific data through a network of national Scientific Data Processing Centres.

4 Science: Core Program and Scientific Policy

The project is managed by a consortium led by the Federal Space Agency, Roscosmos, Russia, which provides the telescope, the platform, the launcher, the integration facilities and it is the main responsible of science and mission operations. The instruments for WSO-UV are provided by Spain (ISSIS), Germany (HIRDES) and China (LSS).

The project observing time is planned to be distributed as (a) Core Program; (b) Funding Bodies Programs; (c) Open Time for the international community. The Core Programme will be designed by a Core Programme Team, including scientists of the participating countries and other international scientists appointed by the WSO-UV consortium, and should be carried out during the first 2 years of the project. The time for Funding Bodies Programs is allocated by a national panel for each of the WSO-UV funding countries. The observing time granted for each country will be proportional to its contribution to the project. Finally, there will be a large fraction (up to 40%) of Open Time for the international community.

The Observatory core program will focuss on:

- Galaxy formation: determination of the diffuse baryonic content of the Universe and its chemical evolution. The Milky Way evolution is included.
- The physics of accretion and outflow: the astronomical engines.
- The Milky Way formation and evolution
- Extrasolar planetary atmospheres and astrochemistry in the presence of strong-UV radiation fields. We described some of the key science issues that WSO-UV will address during its lifetime in our recent review [2].

Acknowledgements The authors thank their colleagues in the national WSO-UV teams. The scientific participation of Spain in the WSO-UV project is funded by the Ministry of Science and Education through grant: ESP2006-27265-E and the Ministry of Science and Innovation through grant: AYA2008-06423-C03. The technical and industrial participation of Spain in the WSO-UV project is funded by the Center for the Technological and Industrial Development of Spain.

References

1. Bamstedt, J. et al., A&AS 134, 561 (1999)
2. Gómez de Castro, A.I., et al., Ap&SS, in *New quests in stellar astrophysics II: ultraviolet properties of evolved stellar populations* (2009)

Science in the Spanish Virtual Observatory

Enrique Solano

Abstract The Virtual Observatory (VO) is an international initiative that was born in 2000 with the objective of ensuring the optimum scientific exploitation of the astronomical archives and services. *Who is behind the Virtual Observatories? How can I be part of the VO? What are VO-tools? How can I use them? VO-Science? Is it already a reality?...* are some of the questions tackled in this paper.

1 The Virtual Observatory

Astronomy has traditionally been at the forefront of the development of on-line services. National and international ground- and space-based observatories produce terabytes of data which are publicly available all around the world. Results published in electronic journals are also on-line. For their daily research work, astronomers routinely use this data network, consisting of large volumes of distributed, heterogeneous data.

The already existing archival information, together with the all-sky surveys foreseen in the coming years, will cover large areas of the sky in most of the wavelength ranges. Although this fact will have a clear positive impact on multiwavelength astronomy there is, however, an important limiting factor: the lack of standardization (different access and retrieval protocols, data models, formats, policies,...) among the astronomical archives and services makes it very inefficient the data identification and retrieval from more than one resource.

Added to this lack of interoperability is the management of the volume of data presently available from astronomical archives: SDSS and 2MASS, with millions of observations, are good examples of this. The situation will be even worse in the coming years with projects like LSST which, by scanning the visible sky every few nights, will produce of the order of TBs per night, a factor of a thousand larger

E. Solano
SVO/LAEFF-CAB/INTA-CSIC, Apdo 78, 28691 Villanueva de la Cañada, Madrid, Spain
e-mail: esm@laeff.inta.es

J.M. Diego et al. (eds.), *Highlights of Spanish Astrophysics V*, Astrophysics and Space Science Proceedings, DOI 10.1007/978-3-642-11250-8_23,

than SDSS or 2MASS. In this scenario the classical approach of transferring data from the archive to the local desktop for their further analysis is not applicable. Moreover, the problem is not only constrained to the data transfer itself but also has an impact on the local management of data (storage problems, need of backup policies, databases, index-based searches, . . .) as well as on the analysis techniques to be used as most of them are not able to handle such amount of data.

The Virtual Observatory is an international initiative that was born with the objective of solving the problems that the lack of interoperability and the inefficient management of large volumes of data create to astronomical research. VO aims at providing seamless unified access to data holdings: all archives speaking the same *language*, accessed in an uniform way, and analysable by the same tools. The Virtual Observatory concept goes one step further than just giving access to distributed computational resources or to the data. It also permits operations on the data and returns results.

All the international VO initiatives are organised around IVOA (International Virtual Observatory Alliance[1]). IVOA was created in June 2002 with the goal of coordinating the assessment of the VO architecture and the development of interoperability standards. Formed by 16 members, the Spanish Virtual Observatory (SVO)[2] became an IVOA member in 2004. At national level, the SVO project provided the seed around which the SVO Thematic Network was created in 2006. With almost 100 members from more than 20 laboratories and departments, the Network is conceived as a forum to enhance the collaboration among the Spanish groups with interest in the VO. With two Schools already organised and a number of VO-Science projects on-going, the initiative is being a great success.

2 The Role of Science in the Virtual Observatories

The Virtual Observatory project was driven by science and it is becoming a science driver. This concept was clearly understood by the main VO projects which set up Science Working Groups to provide scientific advice. One of the major tasks of these groups was the identification, with a clear emphasis on the definition of the science requirements, of cases that could benefit from the use of the VO technology. The Euro–VO Science Reference Mission[3] is a good example of this.

With the rapid development of VO projects, and after several years where technology has been its major component, the Virtual Observatory concept is now mature enough to be used as a research tool for the astronomical community. The uniqueness of the Virtual Observatory as a discovery tool, based on its capability of correlating and statistically analyzing large, multi-dimension data sets (in the

[1] http://www.ivoa.net.

[2] http://svo.laeff.inta.es.

[3] http://www.euro-vo.org/pub/fc/cases.html.

wavelength and time domains), is opening the possibility for discovering new features in known phenomena as well as totally unexpected astrophysics.

The first major discovery made with the VO is described in [11]: the VO science team involved in the project discovered 31 previously undetected supermassive black holes in the GOODS fields following a VO methodology. The classification of ROSAT sources using multiwavelength information and data mining techniques [10], the discovery of massive, dust-enshrouded, carbon stars in nearby galaxies [14] or the discovery of an extremely-rare object [6] are also excellent examples of the science that can be done with the Virtual Observatory.

VO–Science reviews took place during the General Assembly of the International Astronomical Union in 2006 and during the JENAM meeting in 2007. The high number of presentations covering a wide range of topics from the Sun–Earth connection to Cosmology demonstrated the growing interest of the international astronomical community about the VO. A list of VO refereed-papers can be found at http://www.euro-vo.org/pub/fc/papers.html.

3 The Role of Science in the Spanish Virtual Observatory

One of the potential problems that may affect the growth of the Virtual Observatory is its novelty. The absence of links between the VO groups and the research community may strongly limit the VO scientific impact. So, for instance, if the services developed by the VO are not scientifically oriented, they will not be used by the scientific community.

In an attempt to avoid this situation the Spanish VO is working in establishing at national level effective liaisons with research groups who have identified scientific drivers to motivate the adoption of a Virtual Observatory methodology in their science cases. The SVO role in the collaboration is to evaluate the science case from the VO point of view, to provide information and support about the existing tools to tackle the scientific problem and, if necessary, to develop new analysis tools. In what follows I will briefly describe some of the Science Cases presently carried out in the framework of the Spanish VO.

3.1 Identification of Accreting Brown Dwarfs Using VO Tools

Brown dwarfs (BDs), substellar mass objects that do not stabilize on the hydrogen-burning main sequence, cool and fade continuously with time as they shrink to increasingly degenerate configurations. They start as relatively warm objects, spectral class M, and evolve to cooler temperatures, characterized spectroscopically as spectral classes L and T.

The formation process of brown dwarfs is still a matter of debate. Although the star-like formation scenario is widely accepted, there are other approaches that have to be taken into account. One of these alternative scenarios is the ejection theory [12], which suggests that brown dwarfs could be leftovers of a prematurely interrupted accretion process. One of the major drawbacks of this theory are the associated high spatial velocities, incompatible with the formation of circumstellar disks (an observational evidence in brown dwarfs). Moreover, these expected high velocities have not been seen in star-forming regions like Taurus or Chamaeleon. A possible solution to this problem can be found in [13] who proposed that brown dwarf would form by the fragmentation of the outer parts of the stellar disks. Once formed, the objects would be gently released into the field by interactions among themselves. *Gently* in this context means low velocity dispersion what gives the ability of retaining the disks that produce the IR excesses and accretion phenomena seen in many young brown dwarfs.

In order to shed light to this problem, we have started a study on the spatial distribution of brown dwarfs [15]. Due to mass accretion processes, many young low-mass stars and brown dwarfs show Hα emission stronger than the emission expected from chromospheric activity. Studying the Hα equivalent width and the spectral type using low-resolution spectra it can be determined whether or not a star is accreting [2]. Thus, Hα surveys constitute a valuable tool to identify very young stars and BDs that are still accreting from their disks. In this work we have made use of IPHAS (INT Photometric Hα survey of the Northern Galactic Plane), a survey covering 1,800 square degrees of the northern Milky Way that provides (Sloan) r', i' and narrowband Hα photometry down to a magnitude limit of $r' = 20$. More information on IPHAS can be found at [9].

So far the overwhelming majority of the surveys for young very low-mass objects are concentrated in the known star-forming regions and nearby young clusters (see, for instance, recent examples in [1] or [8]). In this sense our search is a pioneering work as the wide spatial coverage provided by IPHAS offers, for the first time, the possibility of searching accreting objects well outside the known star-forming regions and clusters.

The identification of ultracool objects usually requires mining the sky through an appropriate combination of attributes available from different archives (e.g. colours and/or proper motion information). Using VO tools we have cross-correlated the IPHAS point source catalogue with 2MASS selecting the potential candidates on the basis of their near-infrared colours and Hα emission. Four thousand objects were identified. A spectroscopic follow-up of some of these candidates confirmed that 33 showed strong Hα indicative of disk accretion for their spectral type. Twenty-three of them have spectral class M4 or later, 10 of which have classes in the range M5.5–M7.0 and thus could be very young brown dwarfs (Fig. 1).

3.2 Automated Determination of Physical Parameters Using VOSA: The Case of Collinder 69

One of the most interesting star-forming regions is associated to the O8III star λ Orionis, located at about 400 pc from the Sun and presenting very low extinction

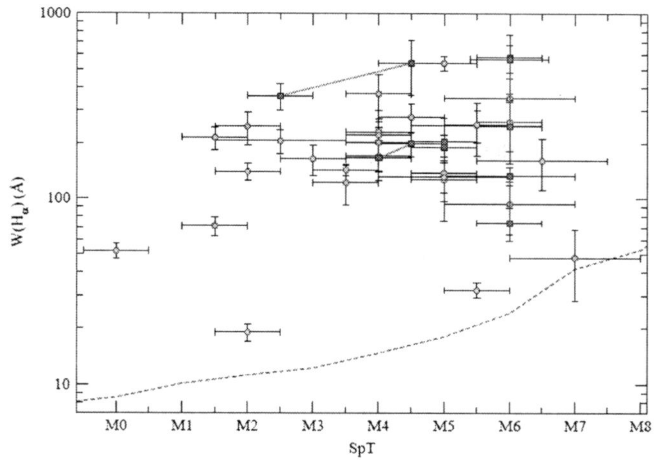

Fig. 1 Hα equivalent width against adopted spectral type for our objects. The *dashed line* denotes the dividing line between chromospheric activity and disk accretion. Our objects are clearly above the *dashed line* and hence they are likely undergoing mass accretion [15]

($A_V = 0.36$ mag) in its inner area. This star dominates the eponymous cluster (also designated as Collinder 69), with an age of about 5 Myr [3].

Our goal is to determine physical parameters of ~170 candidate members of this cluster. Studying the physical parameters of a large population of sources belonging to the same cluster is advantageous, as we can infer properties not only of the individual sources but also of the association as a whole, for example its age, assuming that all objects are coeval.

The physical parameters have been determined by comparing observed SEDs with theoretical data. This methodology requires, as a first step, gathering all the photometric/spectral information available for each of our sources. Once the observational SED has been built it has to be compared with different collections of models (which may translate into thousands of individual models). These tasks, if performed with classical methodologies, can easily become tedious and even unfeasible when applied to large amount of data. On the contrary, the Virtual Observatory represents the adequate framework where to tackle them.

In order to efficiently perform this analysis a new VO-tool was built by the Spanish Virtual Observatory. The tool was named *VOSA*[4] (Virtual Observatory SED Analyzer, [5]). In short, the tasks performed by VOSA are as follows:

- Query several photometric catalogs accessible through VO services in order to increase the wavelength coverage of the data to be analysed.

[4] http://www.laeff.inta.es/svo/theory/vosa/.

- Query VO-compliant theoretical models (spectra) for a given range of physical parameters. The models are accessible in a VO-environment from the SVO theoretical data server.[5]
- Calculate the synthetic photometry of the theoretical spectra (within the required range of physical parameters) for the set of filters previously chosen by the user.
- Perform a statistical $\chi 2$ test to decide which set of synthetic photometry reproduces best the observed data. Determine physical parameters like effective temperature, surface gravity and metallicity.
- Use the best-fit model as the source of a bolometric correction. Determine the luminosity.
- Generate a Hertzsprung–Russell diagram using the theoretical isochrones and evolutionary tracks available in the SVO theoretical data server to determine the age and mass of each individual target of the sample and the age of the association as a whole.

Using this workflow we have independently confirmed the classification from [4] of non-members for 16 of the sources of the sample, and we have added a new possible non-member to this list (a possible field L dwarf). We have also derived an upper-limit for the age of 12.3–16 Myr consistent with previous estimations in the literature (Fig. 2).

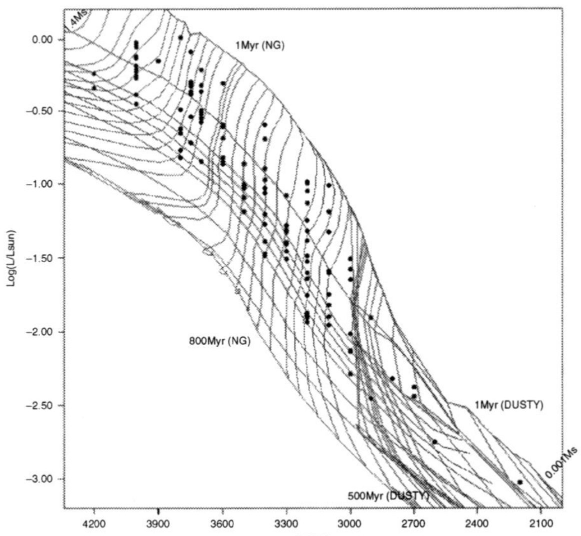

Fig. 2 HR diagram of the members of Collinder 69 for which no infrared excess was detected. Isochrones corresponding to ages of 1, 5, 10, 12.5, 16, 20, 25, 50, 100 and 800 Myr are displayed for the NextGen collection, and those corresponding to ages of 1, 5, 10, 50, 100,120 and 500 Myr for the DUSTY collection. Evolutionary tracks are also displayed for masses between $0.001 M_{\odot}$ and $1.4\ M_{\odot}$ (including both, NextGen and DUSTY collections [5])

[5] http://laeff.inta.es/svo/theory/db2vo/.

3.3 *Young Stars and Brown Dwarfs Around Alnilam and Mintaka*

Brown dwarfs are much brighter when younger. However, young star-forming regions typically have variable extinction that hinders the characterisation of the recently-born brown dwarfs. The σ Orionis cluster in the Ori OB1b association represents an exception to this rule due to its youth ($\sim$3 Myr), closeness ($d \sim 385$ pc) and low extinction ($A_V \leq 0.3$ mag).

To compare substellar mass functions, spatial distributions or disc frequencies and to look for new brown dwarfs and planetary-mass objects it is necessary, therefore, to search as many new locations as possible. Since the new hunting grounds for the search of substellar objects must resemble σ Orionis, it is natural to look for them not far away. Reference [7] have investigated the stellar populations surrounding two bright supergiants in the Orion Belt: Alnilam (ϵ Ori) and Mintaka (δ Ori).

We have performed a comprehensive, inclusive, massive Virtual Observatory analysis and bibliographic data compilation of more than 107,000 sources in the vicinity of these stars. The brightest sources were analyzed by cross-correlating Tycho-2 and 2MASS whereas DENIS and 2MASS were used for the less massive objects (Fig. 3). From this analysis we have found 136 stars displaying features of extreme youth like early spectral types, lithium in absorption, or mid-infrared flux excess. Two young brown dwarf and 289 star candidates were identified from an optical/near-infrared colour-magnitude diagram. Seventy-four objects that might belong to the association as well as foreground and background sources were also listed. The catalogue, ranging from the two massive OB-type supergiants to intermediate M-type substellar objects, provides a characterization of the mass function from 15 to 0.07 solar masses and constitutes an excellent starting point for further, more dedicated, follow-up studies of the stellar and high-mass substellar populations in the Orion Belt.

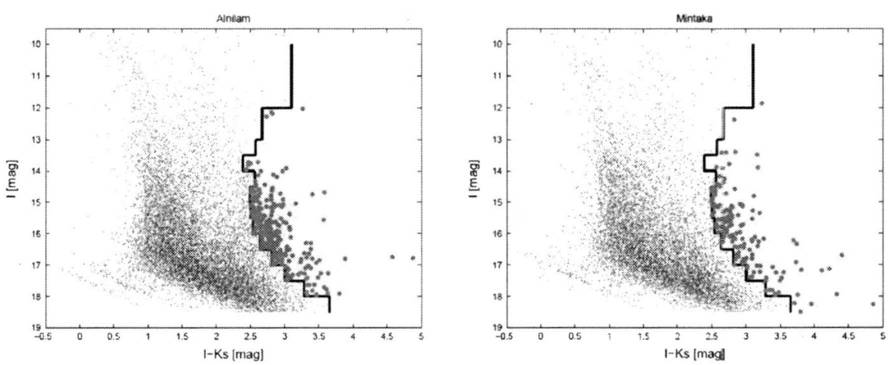

Fig. 3 DENIS / 2MASS color–magnitude diagram

References

1. Allers, K.N., Jaffe, D.T., Luhman, K.L., Liu, M.C., Wilson, J.C., Skrutskie, M.F., Nelson, M., Peterson, D.E., Smith, J.D., Cushing, M.C., ApJ 657, 511 (2007)
2. Barrado y Navascués, D., Martín, E.L., AJ 126, 2997 (2003)
3. Barrado y Navascués, D., Stauffer, J.R., Bouvier, J., Jayawardhana, R., Cuillandre, J.-C., ApJ 610, 1064 (2004)
4. Barrado y Navascués, D., Stauffer, J.R., Morales-Calderón, M., Bayo, A., Fazzio, G., Megeath, T., Allen, L., Hartmann, L.W., Calvet, N., ApJ 664, 481 (2007)
5. Bayo, A., Rodrigo, C., Barrado y Navascués, D., Solano, E., Gutiérrez, R., Morales-Calderón, M., Allard, F., A&A 492, 277 (2008)
6. Caballero, J.A., Solano E., ApJ 665, L151 (2007)
7. Caballero, J.A., Solano, E., A&A 485, 931 (2008)
8. Caballero, J.A., Béjar, V.J.S., Rebolo, R., Eislöffel, J., Zapatero Osorio, M.R., Mundt, R., Barrado y Navascués, D., Bihain, G., Bailer-Jones, C.A.L., Forveille, T., Martín, E.L., A&A 470, 903 (2007)
9. González-Solares, E.A., Walton, N.A., Greimel, R., Drew, J.E., Irwin, M.J., Sale, S.E., Andrews, K., Aungwerojwit, A., Barlow, M.J., van den Besselaar, E., et al., MNRAS 388, 89 (2008)
10. McGlynn, T.A., Suchkov, A.A., Winter, E.L., Hanisch, R.J., White, R.L., Ochsenbein, F., Derriere, S., Voges, W., Corcoran, M.F., Drake, S.A., Donahue, M., ApJ 616, 1284 (2004)
11. Padovani, R., Allen, M.G., Rosati, P., Walton, N.A., A&A 424, 545 (2004)
12. Reipurth, B., Clarke, C., AJ 122, 432 (2001)
13. Stamatellos, D., Hubber, D.A., Whitworth, A.P., MNRAS 382, L30 (2007)
14. Tsalmantza, P., Kontizas, E., Cambrésy, L., Genova, F., Dapergolas, A., Kontizas, M., A&A 447, 89 (2006)
15. Valdivielso, L., Martín, E.L., Bouy, H., Solano, E., Drew, J.E., Greimel, R., Gutiérrez, R., Unruh, Y.C., Vink, J.S., A&A 497, 973 (2009)

Part VII
Teaching and Outreach of Astronomy

Contributions of the Spanish Astronomical Society to the International Year of Astronomy 2009

B. Montesinos

Abstract The Spanish Astronomical Society, SEA in the Spanish acronym of "Sociedad Española de Astronomía", is one of the many institutions contributing to the large number of activities coordinated by the Spanish node of the International Year of Astronomy 2009 (IYA-2009). In this paper I describe the activities programmed with a large participation of members of the Society.

1 Introduction

The Spanish Astronomical Society (http://sea.am.ub.es, SEA hereafter), despite its youth (it is only about 16 years old), is an active and enthusiastic collective. About 80% of the Spanish astronomers (staff, post-doctoral researchers, PhD students, $\sim$600 people in total) belong to SEA. SEA organizes every other even year a Scientific Meeting, whose size and contents are comparable to those of an international conference. More than 200 participants have attended the last meetings in La Laguna (1998), Santiago de Compostela (2000), Toledo (2002), Granada (2004), Barcelona (2006) and Santander (2008), with a very busy programs and busy plenary and splinter sessions.

It was a pleasure to see that, in the Santander Meeting, the session devoted to *Teaching and Outreach of Astronomy* had many participants and a lively series of contributions. In my opinion, shared by many astronomers, it is extremely important to return to the society, translated into an easy language, the knowledge we accumulate and the discoveries we do. Newspapers, televisions and radios are offering on regular bases information about science in general and Astronomy in particular: the discovery of new planetary systems, facts about the expansion of the Universe, colliding galaxies, exotic objects, etc., are topics of interest for the general public.

B. Montesinos (on behalf of the Spanish Astronomical Society)
CAB/LAEFF (CSIC–INTA), POB 78, 28691 Villanueva de la Cañada, Madrid, Spain
e-mail: bmm@laeff.inta.es

J.M. Diego et al. (eds.), *Highlights of Spanish Astrophysics V*, Astrophysics and Space Science Proceedings, DOI 10.1007/978-3-642-11250-8_24,

Joining the celebration of the International Year of Astronomy 2009 (IYA-2009 hereafter), SEA decided to participate, coordinating some projects, or collaborating in several activities. The web page (http://www.astronomia2009.es/) of the Spanish node describes in detail the whole set of activities programmed by many different agents (museums and planetariums, amateur astronomers, individual institutions, high-school teachers, etc.). In the next sections I describe briefly those projects with a high participation of SEA members.

2 '12 Months, 12 Themes'

The official web portal of the IYA-2009 (*A portal to the Universe*) will be devoted each month to a particular topic. The coordinator of the web page (Emilio García, Instituto de Astrofísica de Andalucía) approached SEA in order to look for 12 teams of three to five persons that will be in charge of each of the topics. The idea is to make and shoot an interview with an astronomer who is specialist in that particular area and add an article with images, links and any useful information. The interviews will be 15–20 min long and will be edited and split in small chunks, corresponding to each question, in that way the reader will be able to choose among these small bits of information. The topics cover more or less all the areas in Astronomy: Archaeoastronomy, stellar physics, planets of our Solar System, exoplanets, extragalactic physics, cosmology, instrumentation, telescopes and observatories, how do astronomers work. A 13th interview will be make to Montserrat Villar, the person who is coordinating the Spanish node.

SEA has offered a pool of about 45 astronomers and, at the time of writing this contribution, several interviews have been already completed and the first articles, which will appear in the first months of 2009, are ready or very advanced.

3 'Astronomy Made in Spain'

In 2007, the scientific magazines *Nature* and *Science* were awarded with the Prize Príncipe de Asturias to international cooperation. These two journals are frequently quoted in the media, so the general public has heard of them and knows that the discoveries they publish are relevant. Emilio Alfaro (Instituto de Astrofísica de Andalucía), current SEA President, had the idea of collecting all the papers published in these journals with a Spanish first author in the field of Astronomy in the last 30 years.

The first step of the project consists of editing and publishing a book where each author explains what was the context where his or her research was done, what was the specific problem tackled in the paper, the impact in the area and any other aspect related with the process that led to the publication of that work.

A second step is the organization of small cycles of talks where the researches themselves tell and explain the audience about their discoveries. The peculiarity of this project is that the science will be explained directly by the person *who did it, who was there*.

Some of the authors will also participate in the project 'A University, a Universe' that I describe in the next section.

4 'A University, a Universe'

At the end of 2007, during a meeting with the Board of SEA, we had the idea of breaking the dichotomy *science and humanities* or *science and culture* that lies in the concept that the society has of the several disciplines of knowledge. It is a common place to find people who argue that since they have studied *humanities* (history, literature, philosophy, etc.), they do not feel obliged to know the very basics of the theory of gravitation, or what are – even in a very simple way – the contributions of Einstein to modern Physics.

The original idea was that, in those universities with Astronomy departments, their members could deliver talks to people in faculties whose subjects had nothing to do with science. This idea was taken and enlarged by Ana Ulla (Universidad de Vigo) in a more ambitious way: why not organize at least one talk on Astronomy in *each one* of the universities in Spain?

The project, coordinated by Ana Ulla, and with a hight participation of SEA members, took shape under the name 'A University, a Universe' ('Una Universidad, un Universo' in Spanish, and hence, its acronym is U4). The chancellors of the 75 Spanish universities (public and private) have been approached and there are contact persons almost in all of them to coordinate the talks. A pool of more than 70 astronomers is now available to give seminars and the next steps are mainly logistics, since we have to organize by geographical proximity how and when a given person delivers a talk even in the most modest university.

5 Collaboration SEA: elpais.com

One of the goals of SEA for the IYA-2009 is to reach the general public. This can be done individually, for example through collaborations in radio or television programs, or articles in newspapers or periodical magazines. However, we wanted to give a step forward, doing it in a more systematic way to reach as many people and many places as possible.

After some talks with journalists and the management of *El País*, one of the Spanish newspapers with a large number of readers, they offered us space in the digital version (http://elpais.com).

A call for volunteers done through the SEA mailing list had an enthusiastic response, and we have about 100 people willing to participate in this endeavour. There are several sections: astronomical images, articles on the history of Astronomy, articles devoted to recent discoveries and hot topics, biographies, a cumulative glossary of astronomical terms, games, etc.

This is a unique opportunity because this digital newspaper has more than one million hits per day, not only in Spain but in other Spanish-speaking countries.

6 Funding

Most of the tasks that will be carried out in these projects are completely altruistic. They will be done by SEA members during their free time. However, some of them require funding. For example, the edition and printing of the book in the project 'Astronomy Made in Spain', or the management of the organization and logistics of 'A University, a Universe'. The Spanish Foundation of Science and Technology (FECyT, Fundación Española de Ciencia y Tecnología), belonging to the former Ministry of Science and Education, (currently Ministry of Science and Innovation), granted us with 20,000 euros to cover partially these activities. Additional help from other institutions is being sought at this moment.

Acknowledgements The author is indebted to the many members of SEA who have volunteered to participate in the projects described in this contribution. SEA, and the coordinators of the projects 'Astronomy Made in Spain' and 'A University, a Universe' are grateful to FECyT for its financial support.

Confieso que Divulgo. Reflexiones y Experiencias de una Astrofísica

I. Rodríguez Hidalgo

Abstract Este artículo presenta algunas reflexiones en torno a la popularización de la Ciencia, desarrolladas a lo largo de mi trayectoria profesional, un camino inacabado desde la intuición al oficio. Tras revisar las señas de identidad de la divulgación científica, se exponen ideas, experiencias y recursos, cribados por la práctica y su posterior análisis crítico. Se destacan las actividades relacionadas con la Astronomía, que se cuentan entre las más espectaculares y gratificantes.

Confessions of a popularizer: *This paper presents some author's thoughts about scientific outreach, developed along her professional path, an unfinished way from intuition to trade. First, identity signs of outreach are revised; then, ideas, experiences and resources, sifted by practice and further critical analysis, are reviewed. Activities related to Astronomy, being one of the most spectacular and rewarding, are remarked*[1]

1 Prólogo

La *divulgación científica* es una estimulante tarea de comunicación y formación que toma mensajes del campo de la Ciencia (en nuestro caso, de la Astronomía) y los reescribe de forma creativa para su difusión en un ámbito más extenso que el de su origen, el del público no especializado. Vulgarizar la Ciencia no implica necesariamente trivializar o degradar su mensaje, sino afrontar el reto de hacerlo accesible al pueblo, contemplando las diferencias de sus integrantes en cuanto al acceso, posesión y producción del conocimiento.

[1] An important part of this work refers to communication strategies and author's experiences, only meaningful in Spanish. Since the original oral contribution was prepared and given in that language, Spanish was the natural choice for this text.

I.R. Hidalgo
Instituto de Astrofísica de Canarias, C/ Vía Láctea s/n, E-38200 La Laguna, Tenerife, España
e-mail: irh@iac.es

J.M. Diego et al. (eds.), *Highlights of Spanish Astrophysics V*, Astrophysics and Space Science Proceedings, DOI 10.1007/978-3-642-11250-8_25,

2 Reflexiones Basadas en Experiencias, Experiencias para Reflexionar

2.1 Señas de identidad de la divulgación científica

La divulgación científica "es" Ciencia porque nace y se alimenta de ella, representa la máxima expresión de su naturaleza pública y abierta, y comparte sus principales características: es sancionada sólo por la experiencia, nunca está concluida, y requiere audacia, imaginación y creatividad. Las comillas significan, naturalmente, que la divulgación *es* esencialmente Ciencia, aunque sea también algo más.

La divulgación es necesaria como herramienta imprescindible para promover hoy el desarrollo de una auténtica cultura científica accesible a las mayorías.

La divulgación es difícil... pero posible: para muchos científicos la divulgación carece de interés o es considerada inferior. Una de las causas, no siempre reconocida, es su dificultad: satisfacer la legitimidad científica y la credibilidad pública implica conocer profundamente el contenido y, al mismo tiempo, dominar las estrategias de la comunicación. Así, según un *Principio de indeterminación* aplicado a la divulgación, si la audiencia es muy amplia y heterogénea, los contenidos transmitidos serán menos profundos y completos, y a la inversa.

La divulgación es también un arte: así como la obra de arte sólo está concluida cuando es contemplada por el espectador, la Ciencia culmina y se completa cuando su mensaje llega a la sociedad. El artífice de la divulgación puede ser considerado un *artista* llamado a ser rastreador de nuevos lenguajes, significados, referencias, relaciones, escenarios y contextos; a provocar y transgredir; a salirse de la autopista de lo convencional.

2.2 De la intuición al oficio

Esta sección es resultado de combinar mi experiencia y su análisis, como agente de la divulgación, con la crítica que yo ejercería como público.

2.2.1 Palabra e imagen

En la comunicación de la Astronomía la imagen juega un papel fundamental, que se ve reforzado por una cuidadosa elección de la palabra. Sin pretensión de exhaustividad, se revisan algunos aspectos de estas dos herramientas.

Nuestra jerga favorita

Los científicos se comunican con sus colegas en un lenguaje muy especializado y suelen tener serias dificultades para abandonarlo o traducirlo al dirigirse al público en general. Algunos ejemplos ilustran que los astrónomos no somos una excepción:

Constelación: vale la pena explicar brevemente este concepto, ya que mucha gente espera encontrar, si no un dibujo en el cielo, sí al menos las líneas que unen las estrellas. Reconocer una constelación a ojo queda reservado a unos pocos, y sólo una selecta minoría sabe que las estrellas que la forman pueden no guardar relación entre sí, estar a muy diferentes distancias, y tener edades muy distintas.

Cuando decimos *telescopio de X metros*, es habitual que el interlocutor piense en un tubo de *X* metros de largo que, como ha visto tantas veces en la TV, sale por la abertura de la cúpula, en busca de las estrellas...

La expresión *reducción de datos* no tiene contrapartida en la vida normal, y lo natural es imaginar que los datos (¿qué son exactamente?) recogidos por el astrónomo (¿dónde se compran o se crían los datos?) se van haciendo, de alguna extraña manera, cada vez más pequeñitos.

Modelo: maqueta de pensamiento, imagen de la realidad que sirve para describirla y entender su funcionamiento, y que se expresa generalmente como un conjunto de enunciados y ecuaciones físico–matemáticas. Poco más de una veintena de palabras basta para explicar qué es un modelo en Ciencia. Pero, salvo que se pronuncien o escriban, el público pensará probablemente en alguna pasarela de moda con nombre de fuente o de diosa.

Es difícil escuchar a un astrofísico (o físico en general) sin que mencione repetidas veces la palabra *campo*: magnético, eléctrico, gravitatorio... Un lego imaginará un campo de fútbol, uno de golf, con suerte un campo de cereales. Aunque este último no está tan lejos del concepto físico (en cada punto del espacio, una espiga o vector) es recomendable un uso prudente del término.

Fotón es otra palabra casi imprescindible en nuestra área, que se utiliza con demasiada ligereza. Más allá de una enorme foto publicitaria, intuir su significado no es trivial, salvo que el contexto ayude mucho. Si no es prescindible, o sustituible por *luz* o *radiación*, siempre se puede apuntar el comportamiento dual de la luz, y explicar que los fotones son *algo así como* partículas de luz.

Otro de nuestros términos preferidos, *espectro*, es particularmente equívoco, al asociarse con espíritus y demás entes inmateriales esquivos a su análisis en laboratorio. Son muy útiles los ejemplos del arco iris, los pequeños espectros formados por los prismas de una lámpara de araña, o los reflejos de luz en un CD.

Para concluir, una palabra cotidiana: *equilibrio*. La gente de la calle imagina el equilibrio sobre una cuerda floja o similar; poco cuesta recordar que, en Física, nos referimos a balance o compensación entre fuerzas que actúan en sentidos opuestos.

Ejemplos, analogías y metáforas

Los *ejemplos* son eslabones imprescindibles que conectan el discurso divulgativo con la realidad más conocida. Son especialmente adecuados los más familiares, muy útiles para ilustrar conceptos abstractos, objetos inaccesibles, fenómenos sólo presentes en escenarios astrofísicos, etc. Como muestra, la magnitud de la velocidad de la luz (normalmente en km/s, una unidad poco habitual para el público en general...) se comprende mejor añadiendo *La luz puede recorrer 2500 veces en 1 segundo la distancia que separa Santa Cruz de Tenerife del aeropuerto del Sur de la isla* u otro ejemplo ad hoc.

Una *analogía* consiste en establecer una comparación o relación entre varios conceptos, objetos, experiencias... señalando las semejanzas y diferencias entre unos y otros. En divulgación astronómica son de uso común y resultan muy eficaces las analogías que ayudan a visualizar valores extremos o amplios rangos de distancia, tamaño, masa, densidad, etc. Por ejemplo: si el Sol fuese tan grande como yo, la Tierra sería aproximadamente como una uva, colocada a 150 metros (como tres piscinas olímpicas) de mí. En esta escala, la Luna sería como una letra mayúscula de un libro, situada a unos 30 cm (como un folio) de la uva; Júpiter, como un pomelo, a 15 piscinas de mí, etc. La estrella más cercana estaría 40000 kilómetros (como toda la circunferencia del Ecuador).

La *metáfora* consiste en usar una expresión con un significado distinto, o en un contexto diferente al ordinario. Se utiliza profusamente como recurso literario, es fuente de cambios semánticos en un idioma, y puede resultar muy poderosa en la divulgación científica. Así, por ejemplo, las oscilaciones solares son *el pulso del Sol*, y nosotros somos *polvo de estrellas*.

El poder de un título atractivo

Si la Ciencia resulta generalmente árdua para el público no especializado, la carta de presentación de cualquier actividad divulgativa (artículo, charla, curso, taller, etc.) es una magnífica oportunidad para despertar la curiosidad y atraer la atención.

Un título poco afortunado puede disuadir a muchos potenciales destinatarios: *Historia de los avances científicos conseguidos gracias a la observación de eclipses totales de Sol*. Un título bien elaborado debería ser correcto, preferiblemente breve, y sugerente, incluso provocativo, pero sin caer en el engaño: *Ciencia bajo la sombra de la Luna*. En muchos casos, un título insulso adquiere un inesperado valor utilizado como subtítulo: *Ciencia bajo la sombra de la Luna. Avances científicos durante eclipses totales de Sol*; o a la inversa: *Enanas blancas. Una tonelada en una cucharilla de café*. Instituciones con amplia experiencia en divulgación, como NASA, nos han regalado algunos encabezamientos memorables, como *Living with a star* que, traducido como ***Con****viviendo con una estrella*, aporta un interesante matiz emotivo.

Gráficas sí, pero...

Las gráficas son recursos imprescindibles en el quehacer astronómico. Pero sólo resultan útiles si son inteligibles, para lo que conviene cuidar algunos aspectos: caracteres del tamaño adecuado; ejes rotulados; símbolos, tipos de línea y colores bien elegidos, claros y distinguibles; etiquetas y leyendas *traducidas* para facilitar su comprensión (demasiado a menudo se utilizan figuras en inglés); representaciones simples, sin información que no será explicada... y otros detalles de sentido común, que frecuentemente se olvidan. La interpretación de las gráficas no es trivial para el público no familiarizado con ellas, así que es imprescindible guiarle en su comprensión, en la comunicación escrita y, muy especialmente, en la oral.

2.2.2 Estrategias eficaces

Para comenzar

Como sucede con el título, el inicio de un artículo o charla divulgativos es la ocasión para captar (o perder, de entrada) la atención de la audiencia. Una cita literaria o histórica, una imagen sorprendente, la primicia de un descubrimiento, el logo emulado de una productora cinematográfica, un poco de acción antes del título y los créditos o una referencia artística ajena a la Astronomía son sólo algunas propuestas para evitar un principio demasiado manido, que augura un resto poco estimulante.

Hablar de Ciencia a (y con) niños

Los niños son interlocutores siempre sorprendentes, habitualmente más curiosos, interesados y desinhibidos que cualquier otro público, muy exigentes con quien se dirige a ellos, y agudos en sus preguntas. Para hablar de Ciencia a—y con— niños, y salir bien parado del intento, ayuda un talante abierto y cercano, y hay que elegir bien la cantidad y complejidad de contenidos, según su nivel de desarrollo intelectual y formación académica. Una idea que *funciona* consiste en preguntar a los niños qué desean saber sobre un tema, y elaborar el mensaje a partir de sus respuestas. O al contrario, dirigirse a ellos con preguntas adecuadas a su curriculum, invitarles a responder y pasar a exponer la información sólo tras ese diálogo.

Tratar de comunicarse con adolescentes

El público adolescente suele mostrar poco interés (cuando no rechazar y boicotear) por todo lo que proceda de sus profesores y tutores y esté relacionado con la enseñanza... No queda más remedio que acercarse a su terreno con una oferta que no suene académica. Por ejemplo, para interrumpir los cuchicheos y captar su atención al comienzo de una charla se puede presentar a los *colaboradores*: portátil, proyector, altavoces, puntero láser, ratón IR... Otra estrategia de aproximación son las referencias a programas de TV o a sus ídolos de actualidad: por ejemplo, he impartido varias veces una conferencia para estudiantes de ESO y Bachillerato titulada *O.C.* (Operación Ciencia), en la que intento convencer a los estudiantes de que viven en una Academia y deberían aspirar a convertirse en *ciencitos*.

2.2.3 Todo vale...

La oferta de recursos para la comunicación es hoy más amplia que nunca y todos pueden ser útiles para una divulgación atractiva y eficaz: desde los tradicionales voz y texto, hasta las más espectaculares facilidades técnicas y audiovisuales, pasando por las manifestaciones artísticas de todo tipo. Todos son aplicables, además, a las experiencias interactivas como talleres didácticos y observaciones, manipulación de módulos e instrumentos, deportes, juegos, concursos... En los modernos Planetarios y Museos Científicos se ha comprobado ampliamente su eficacia para propiciar el necesario aprendizaje de actitudes positivas ante la Ciencia.

2.3 ¿Nos atrevemos?

En este apartado se exponen iniciativas de divulgación alejadas de los tradicionales artículos y charlas. El razonable éxito que todas han obtenido anima a seguir fomentando ese talante transgresor que hace de la divulgación un arte.

2.3.1 Exposiciones poco convencionales

El proyecto *másEinstein 2005* del Museo de la Ciencia y el Cosmos (MCC) de La Laguna se propuso convertir el casco antiguo de la ciudad en una sala de exposiciones. Para ello se usó un centenar de figuras 2D del científico, con algunas de sus frases, enormes lonas con llamativas ilustraciones y conceptos de relatividad, y el propio Einstein (en vinilo) viajando sentado en varias guaguas de la isla. En junio de 2008 *El Universo a tu alcance*, una iniciativa impulsada por la comunidad astrofísica internacional presente en los Observatorios de Canarias, se inspiró en esta idea para decorar un tranvía con espectaculares imágenes astronómicas obtenidas desde dichas instalaciones.

Acciones de este tipo no aspiran a transmitir profundos contenidos científicos, pero cumplen una necesaria labor de sensibilización al llegar a un público muy numeroso (impensable en otras actividades más convencionales) que, de otro modo, nunca se acercaría a la Ciencia.

2.3.2 Espacios públicos como foros de comunicación

La idea de presentar novedades científicas en tabernas y locales públicos se remonta al siglo XVII, protagonizada por destacados miembros de las recién nacidas sociedades científicas. En la estela de esa tradición, he organizado en varias ocasiones coloquios científicos en un bar de La Laguna: tres o cuatro especialistas se sientan frente a los clientes, con sólo un micro y su bebida, exponen con brevedad y lenguaje llano su aportación al tema del día, y pasan a debatir entre ellos y con la audiencia. Un moderador reparte los turnos de palabra y solicita la aclaración de términos técnicos o conceptos complicados (de esos que conforman *nuestra jerga favorita*). El ambiente relajado del local anima a los asistentes a intervenir menos inhibidos que en una sala de conferencias.

Los coloquios *No sólo de Relatividad vivió Einstein, ¿Inventó Einstein la Relatividad* y *Einstein nunca dijo eso de todo es relativo* tuvieron una excelente respuesta de público. En *Salsa Rosa Científica. Toda la verdad sobre el descubrimiento de la estructura del ADN* la tertulia abordó la faceta más humana de la Ciencia. *Tres miradas expertas. El Sol visto por... un astrofísico, un ingeniero y un médico dermatólogo* ofreció una visión multidisciplinar de un apasionante tema astronómico.

2.3.3 ¡Arriba el telón!

La vocación didáctica del teatro, tan antigua como él mismo, inspiró la idea de transmitir conceptos científicos insertados en una historia aparentemente ajena a la Ciencia. Como modelo se utilizó *El club de la comedia*, un espacio televisivo en el que un actor o actriz ofrece un monólogo sobre un cuidado guión humorístico, sin más escenografía que la luz y un taburete. Se eligió un tema tan popular como la astrología, con el propósito de introducir conceptos astronómicos básicos y evidenciar las falacias astrológicas. Para poner a prueba la afirmacin *una carcajada vale por mil silogismos*, me atreví a interpretar mi propio texto *Amores horoscopales*, en el que ironizaba sobre mis amores desafortunados por culpa de las estrellas.

2.3.4 Sinergias infrecuentes

En 2005 se estrenó en Tenerife *Harmonices Mundi, un espectáculo astrofísico–musical*, un concierto programático de tema astronómico. Una orquesta interpreta en directo un conjunto de obras musicales seleccionadas (quizá compuestas para la ocasión), según un guión divulgativo estructurado a modo de movimientos de una sinfonía. La música se combina con la narración en vivo, efectos luminosos y proyecciones de imágenes y animaciones astronómicas en una pantalla gigante.

En la misma línea el MCC presentó en abril y diciembre de 2007 *Poesía bajo las estrellas*. Se estrenaron las obras poéticas *Ruido o luz* y *Poemas del origen*, con los propios poetas como narradores, música de fondo original interpretada en directo, efectos sonoros y visuales, y el cielo del planetario en movimiento sobre los presentes.

El programa infantil de planetario *Meteorito, una roca del espacio* fue íntegramente elaborado en el MCC. Sus elementos principales son un guión planteado como una aventura participativa, lleno de contenido astronómico, narrado por un famoso y querido payaso; varias marionetas representando cuerpos celestes, grabadas en *chroma-key*, que *vuelan* por la cúpula estrellada; y una banda sonora de piezas clásicas seleccionadas.

2.3.5 Para llegar a todos los públicos

El proyecto *Astro para todos los públicos* ha sido elegido como uno de los once emblemáticos de ámbito nacional del Año Internacional de la Astronomía en España. Su objetivo es hacer irrumpir la Astronomía en la vida diaria, dando a conocer espectaculares imágenes astronómicas obtenidas en telescopios situados en territorio español. Los vehículos de difusión serán objetos cotidianos como bonos de transporte, billetes de sorteos, y contenidos personalizables para móviles y ordenadores.

Para lectores interesados

Muchas de las experiencias presentadas en esta contribución están descritas en detalle en anteriores trabajos, pero la limitación de espacio impide incluir una lista de referencias. Invito a los lectores interesados a consultar las Actas de los *Congresos de Comunicación Social de la Ciencia* celebrados en Granada (1999), Valencia (2001), Coruña (2005) y Madrid (2007), así como los Proceedings de los encuentros *Communicating Astronomy with the Public* de Tenerife (2002), Garching (2005) y Atenas (2007).

3 Epílogo

No sé si bien, pero... sí, confieso que divulgo. Y lo hago sin vergüenza ni arrepentimiento, con toda la responsabilidad, seriedad y pasión de que soy capaz.

Lo expuesto en este artículo no es un trabajo de investigación al uso, sino la crónica de un largo (e inconcluso) proceso de experimentación. Espero que su lectura estimule el interés de los astrónomos por acercarse al mundo de la divulgación, por explorarlo y vivirlo en primera persona.

Part VIII
Abstracts of the Contributions in the Online Extra Materials

Galaxies and Cosmology

VIMOS-VLT Two-Dimensional Kinematics of Local Luminous Infrared Galaxies

Julia Alfonso-Garzón, Ana Monreal-Ibero, Santiago Arribas, and Luis Colina

Abstract In this work, preliminary results of a kinematic study based on optical integral field spectroscopy with the VIMOS (Visible Multi-object Spectrograph) instrument on the VLT (Very Large Telescope) of some representative (U)LIRGs ((Ultra) Luminous Infrared Galaxies) is presented. Velocity fields and velocity dispersion distributions of the ionized gas are obtained from Hα 6,563 Å emission line. Two representative examples, an isolated galaxy (NGC 3110) and a merger (IRAS F01159-4443), are shown. The isolated galaxy presents a velocity field typical of a rotating spiral galaxy with a peak to peak velocity difference of $\sim$440 km s^{-1}. The merger shows a more perturbed kinematics although independent rotation for each individual galaxy has been found with a peak to peak velocity of $\sim$260 km s^{-1} in the northern galaxy and of $\sim$250 km s^{-1} in the southern one and a relative velocity between the two galaxies of $\sim$130 km s^{-1}.

J. Alfonso-Garzón
CAB/LAEFF (CSIC-INTA), POB 78 28691 Villanueva de la Cañada, Madrid, Spain
e-mail: julia@laeff.inta.es

A. Monreal-Ibero
ESO, Karl-Schwarzschild-Strasse 2, D-85748 Garching bei München, Germany
e-mail: amonreal@eso.org

S. Arribas and L. Colina
DAMIR-IEM-CSIC, Serrano 121, 28006 Madrid, Spain
e-mail: arribas@damir.iem.csic.es, colina@damir.iem.csic.es

J.M. Diego et al. (eds.), *Highlights of Spanish Astrophysics V*, Astrophysics and Space Science Proceedings, DOI 10.1007/978-3-642-11250-8_26,

Recovering the Real-Space Correlation Function from Photometric Redshift Surveys

Pablo Arnalte-Mur, Alberto Fernández-Soto, Vicent J. Martínez, and Enn Saar

Abstract The error on the redshift determination associated to photometric redshift surveys produces a smaller correlation and a loss of isotropy in the observed galaxy distribution. We present a method to recover the real-space correlation function, $\xi(r)$ from this kind of observations. The method is similar to that used in spectroscopic surveys to avoid the effects of peculiar velocities, and uses the fact that correlations are conserved in the plane perpendicular to the line-of-sight. We apply this method to mock photometric surveys with errors $\Delta z/(1+z) = 0.05 - 0.005$ obtained from the cosmological simulation of Heinämäki et al. (2005, arXiv:astro-ph/0507197). Our method allows to recover $\xi(r)$, within the error, for the cases with smaller Δz. For $\Delta z/(1+z) = 0.05$, the need to integrate a long range in the line-of-sight direction makes the method fail for $r > 2\,h^{-1}$ Mpc.

P. Arnalte-Mur and V.J. Martínez
Observatori Astronòmic and Departament d'Astronomia i Astrofísica, Universitat de València, Apartat de Correus 22085, E-46071 València, Spain
e-mail: pablo.arnalte@uv.es

A. Fernández-Soto
Instituto de Física de Cantabria (CSIC-UC), Avda de los Castros s/n, E-39005 Santander, Spain

E. Saar
Tartu Observatoorium, Tõravere, 61602 Estonia

J.M. Diego et al. (eds.), *Highlights of Spanish Astrophysics V*, Astrophysics and Space Science Proceedings, DOI 10.1007/978-3-642-11250-8_27,

Probing Outer Disk Stellar Populations

Judit Bakos, Ignacio Trujillo, and Michael Pohlen

Abstract We have explored radial color and stellar surface mass density profiles for a sample of 85 late-type galaxies with available deep (down to $\sim$27.0 mag/arcsec^2) SDSS g'- and r'-band surface brightness profiles. About 90% of the light profiles have been classified as broken exponentials, exhibiting either truncations (Type II galaxies) or antitruncations (Type III galaxies). Their associated color profiles show significantly different behavior. For the truncated galaxies a radial inside-out bluing reaches a minimum of $(g' - r') = 0.47 \pm 0.02$ mag at the position of the break radius, this is followed by a reddening outwards. The antitruncated galaxies reveal a more complex behavior: at the break position (calculated from the light profiles) the color profile reaches a plateau region–preceded with a reddening– with a mean color of about $(g' - r') = 0.57 \pm 0.02$ mag. Using the color to calculate the stellar surface mass density profiles reveals a surprising result. The breaks, well established in the light profiles of the Type II galaxies, are almost gone, and the mass profiles resemble now those of the pure exponential Type I galaxies. This result suggests that the origin of the break in Type II galaxies are most likely to be a radial change in stellar population, rather than being caused by an actual drop in the distribution of mass. The antitruncated galaxies on the other hand preserve their shape to some extent in the stellar surface mass density profiles. We find that the stellar surface mass density at the break for truncated (Type II) galaxies is $13.6 \pm 1.6\,M_\odot\,pc^{-2}$ and $9.9 \pm 1.3\,M_\odot\,pc^{-2}$ for the antitruncated (Type III) ones. We estimate that $\sim$15% of the total stellar mass in case of Type II galaxies and $\sim$9% in case of Type III galaxies are to be found beyond the measured break radii.

J. Bakos and I. Trujillo
Instituto de Astrofísica de Canarias, Calle Vía Lactea, 38200, La Laguna, Tenerife, Spain
e-mail: jbakos@iac.es, trujillo@iac.es

M. Pohlen
Cardiff University, School of Physics & Astronomy, Cardiff, CF24 3AA, Wales, UK
e-mail: Michael.Pohlen@astro.cf.ac.uk

J.M. Diego et al. (eds.), *Highlights of Spanish Astrophysics V*, Astrophysics and Space Science Proceedings, DOI 10.1007/978-3-642-11250-8_28,

Deconstructing the K-Band Number Counts

G. Barro, J. Gallego, P.G. Pérez-González, M.C. Eliche-Moral, M. Balcells, V. Villar, N. Cardiel, D. Cristobal-Hornillos, A. Gil de Paz, R. Guzmán, R. Pelló, M. Prieto, and J. Zamorano

Abstract We present a study that links the Number Counts (NCs) to the rest-frame luminosity functions (LFs) at the passbands probed by the observed K-band at different epochs. Making use of a large K-band selected sample in the Groth Field, HDFN and CDFS ($\sim$0.27 deg^2), we have derived highly reliable photometric redshift estimates that allow us to estimate LFs in the redshift range (0.25–1.25). We find that the larger flattening in the slope of the K-band NCs is mostly a consequence of a prominent decrease in the characteristic density (ϕ^*) around $z \sim 1$, and an almost flat evolution of M^*.

G. Barro, J. Gallego, P.G. Pérez-González, M.C. Eliche-Moral, V. Villar, N. Cardiel, A. Gil de Paz, and J. Zamorano
Universidad Complutense de Madrid (UCM), Spain
e-mail: gbc@astrax.fis.ucm.es

M. Balcells and M. Prieto
Instituto de Astrofísica de Canarias (IAC), Spain

D. Cristobal-Hornillos
Instituto de Astrofísica de Andalucía (IAA), Spain

R. Guzmán
Universidad de Florida, USA

R. Pelló
Université de Toulouse, France

J.M. Diego et al. (eds.), *Highlights of Spanish Astrophysics V*, Astrophysics and Space Science Proceedings, DOI 10.1007/978-3-642-11250-8_29,

Extremely Compact Massive Galaxies at $1.7 < z < 3$

Fernando Buitrago, Ignacio Trujillo, and Christopher J. Conselice

Abstract We measure and analyze the sizes of 82 massive ($M \geq 10^{11}\,\mathrm{M}_\odot$) galaxies at $1.7 \leq z \leq 3$ utilizing deep HST NICMOS data taken in the GOODS North and South fields. Our sample provides the first statistical study of massive galaxy sizes at $z > 2$. We split our sample into disk-like (Sérsic index $n \leq 2$) and spheroid-like (Sérsic index $n > 2$) galaxies, and find that at a given stellar mass, disk-like galaxies at $z \sim 2.3$ are a factor of 2.6 ± 0.3 smaller than present day equal mass systems, and spheroid-like galaxies at the same redshift are 4.3 ± 0.7 times smaller than comparatively massive elliptical galaxies today. We furthermore show that the stellar mass densities of very massive galaxies at $z \sim 2.5$ are similar to present-day globular clusters with values $\sim 2 \times 10^{10}\,\mathrm{M}_\odot\,\mathrm{kpc}^{-3}$.

F. Buitrago
School of Physics and Astronomy, University of Nottingham, NG7 2RD, UK
e-mail: ppxfb@nottingham.ac.uk

I. Trujillo
Instituto de Astrofísica de Canarias, Vía Láctea s/n 38200, La Laguna, Tenerife, Spain
e-mail: trujillo@iac.es

C.J. Conselice
School of Physics and Astronomy, University of Nottingham, NG7 2RD, UK
e-mail: conselice@nottingham.ac.uk

J.M. Diego et al. (eds.), *Highlights of Spanish Astrophysics V*, Astrophysics and Space Science Proceedings, DOI 10.1007/978-3-642-11250-8_30,

Cold Dark Matter Halos Based on Collisionless Boltzmann–Poisson Polytropes

J. Calvo, E. Florido, O. Sánchez, E. Battaner, J. Soler, and B. Ruiz-Granados

Abstract The aim of this work is to give some insight into the controversy between N-body simulations and observations of cold dark matter (CDM) halos by considering polytropic DM spheres associated to a collisionless gravitational Boltzmann–Poisson (BP) system. Our resulting polytrope model is used to make predictions on the behavior of the CDM halos in those regions in which the numerical models cannot produce detailed results, i.e. near the center <1 kpc (due to resolution limitations), and at the rim (as halos cannot have infinite extent). These provides a complementary information where other models present difficulties to make predictions.

J. Calvo, O. Sánchez, and J. Soler
Dpto. Matemática Aplicada, Universidad de Granada, Spain
e-mail: juancalvo@ugr.es, ossanchez@ugr.es, jsoler@ugr.es

E. Florido, E. Battaner, and B. Ruiz-Granados
Dpto. Física Teórica y del Cosmos, Universidad de Granada, Spain
e-mail: estrella@ugr.es, battaner@ugr.es, bearg@ugr.es

J.M. Diego et al. (eds.), *Highlights of Spanish Astrophysics V*, Astrophysics and Space Science Proceedings, DOI 10.1007/978-3-642-11250-8_31,

Use of Neural Networks for the Identification of New $z \geq 3.6$ QSOs from FIRST–SDSS DR5

R. Carballo, J.I. González-Serrano, C.R. Benn, and F. Jiménez-Luján

Abstract We aim to obtain a complete sample of $z \geq 3.6$ radio QSOs from FIRST sources having star-like counterparts in the SDSS DR5 photometric survey ($r_{AB} \leq 20.2$). The starting sample of FIRST–DR5 pairs includes 4,250 objects with DR5 spectra, 52 of these being $z \geq 3.6$ QSOs. Simple supervised neural networks, trained on these sources, using optical photometry and radio data, are very effective for identifying high-z QSOs, yielding 96% completeness and 62% efficiency. Applying these networks to the 4,415 FIRST–DR5 sources *without* DR5 spectra we found 58 $z \geq 3.6$ QSO candidates. We obtained spectra of 27 of them, confirming 17 as high-z QSOs. Spectra of 13 additional candidates from the literature and SDSS DR6 revealed seven more $z \geq 3.6$ QSOs, giving an overall efficiency of 60% (24/40). None of the non-candidates with spectra from NED or DR6 is a $z \geq 3.6$ QSO, consistently with a high completeness. The initial sample of high-z QSOs is increased from 52 to 76 sources (a factor 1.46). From the new identifications and candidates we estimate an incompleteness of SDSS for the spectroscopic classification of FIRST $3.6 \leq z \leq 4.6$ QSOs of 15% for $r \leq 20.2$.

R. Carballo
Dpto. de Matemática Aplicada y Ciencias de la Computación, Univ. de Cantabria. ETSI Caminos, Canales y Puertos, Avda de los Castros s/n, E-39005 Santander, Spain
e-mail: carballor@unican.es

J.I. González-Serrano
Instituto de Física de Cantabria (CSIC-Universidad de Cantabria), Avda de los Castros s/n, E-39005 Santander, Spain
e-mail: gserrano@ifca.unican.es

C.R. Benn
Isaac Newton Group, Apartado 321, E-38700 Santa Cruz de La Palma, Spain
e-mail: crb@ing.iac.es

F. Jiménez-Luján
Instituto de Física de Cantabria (CSIC-UC) and Dpto. Física Moderna, Avda de los Castros s/n, E-39005 Santander, Spain
e-mail: jimenezf@ifca.unican.es

J.M. Diego et al. (eds.), *Highlights of Spanish Astrophysics V*, Astrophysics and Space Science Proceedings, DOI 10.1007/978-3-642-11250-8_32,

Integral Field Spectroscopy of Local Luminous Compact Blue Galaxies: NGC 7673, a Case Study

A. Castillo-Morales, J. Gallego, J. Pérez-Gallego, R. Guzmán, C. Garland, D.J. Pisano, F.J. Castander, N. Gruel, and J. Zamorano

Abstract Luminous Compact Blue Galaxies (LCBGs) are high surface brightness starburst galaxies bluer than a typical SBc and with typical luminosities $\approx L^\star$, which are undergoing a major burst of star formation. LCBGs are the closest counterparts to the large population of starburst galaxies observed at high redshift, including Lyman-break galaxies at $z \approx 2$. However, because LCBGs are very rare in the nearby Universe, their properties are still largely unknown. We have selected a representative sample of LCBGs from the SDSS and UCM databases which, although small, provides an excellent reference for comparison with current and future surveys of similar starbursts at high-z. We are carrying out a 3D optical and radio spectroscopic study of this LCBG sample, including spatially resolved maps of kinematics, extinction, SFR and metallicity. This will allow us to characterize their star formation history and mass assembly, and the role of mergers and supernova galactic winds. Here we show our results of this comprehensive multiwavelength study for NGC 7673, a prototypical LCBG in the nearby Universe.

A. Castillo-Morales, J. Gallego, and J. Zamorano
Universidad Complutense de Madrid, Spain
e-mail: acm@astrax.fis.ucm.es

J. Pérez-Gallego, R. Guzmán, and N. Gruel
University of Florida, USA

C. Garland
Castleton State College, Vermont, USA

D.J. Pisano
National Radio Astronomy Observatory, Green Bank, USA

F.J. Castander
Instituto de Estudios Espaciales de Cataluña, Spain

J.M. Diego et al. (eds.), *Highlights of Spanish Astrophysics V*, Astrophysics and Space Science Proceedings, DOI 10.1007/978-3-642-11250-8_33,

Blue Massive Stars in NGC 55: A First Quantitative Study

N. Castro, A. Herrero, M. Garcia, C. Trundle, F. Bresolin, W. Gieren, G. Pietrzynski, R.-P. Kudritzki, and R. Demarco

Abstract We present the first census of blue massive stars in NGC 55, a galaxy of the Sculptor group at 1.94 Mpc. This study is based on optical low-resolution spectra of approximately 200 objects taken with VLT–FORS2. We have performed the spectral classification of these objects. The estimated stellar radial velocities show general agreement with the existing HI rotational velocity data. A first qualitative study of the stellar metallicity suggests that its global distribution over NGC 55 is close to that of the LMC, as derived from previous studies of H II regions. We have also determined the stellar parameters of one star showing that the resolution and the quality of the spectra are reliable to perform a quantitative analysis.

N. Castro, A. Herrero, and M. Garcia
Instituto de Astrofísica de Canarias, C/ Vía Láctea s/n, E-38200, Tenerife, Spain
e-mail: norberto@iac.es

C. Trundle
Astronomy Research Centre, Department of Physics & Astronomy, School of Mathematics & Physics, The Queen's University of Belfast, Belfast, UK

F. Bresolin and R.-P. Kudritzki
Institute of Astronomy, 2680 Woodlawn Drive, Honolulu, HI 96822, USA

W. Gieren and G. Pietrzynski
Departamento de Física, Astronomy Group, Universidad de Concepción, Casilla 160-C, Concepción, Chile

R. Demarco
Department of Physics and Astronomy, The Johns Hopkins University, Baltimore, USA

J.M. Diego et al. (eds.), *Highlights of Spanish Astrophysics V*, Astrophysics and Space Science Proceedings, DOI 10.1007/978-3-642-11250-8_34,

A Morphological Study of Sigma-Drop Galaxies

S. Comerón, J.H. Knapen, and J.E. Beckman

Abstract We study a sample of 20 galaxies with an observed central drop in the stellar velocity dispersion (σ-drop) matched with a control sample of galaxies without a σ-drop. We search for correlations between σ-drops and the properties, primarily morphological, of the inner zones and discs of their host galaxies. Both morphological parameters and luminosity profiles are used to classify the samples. We find a larger fraction of Hα rings and nuclear dust spirals in the σ-drop sample. We also find that the fraction of Seyfert galaxies in the σ-drop sample is bigger than that of LINERs and that the reverse is true for the control sample. Our findings indicate that a σ-drop is very probably due to inflow-induced star formation in a dynamically cool disc, or in a gas ring, shock focused by an inner Lindblad resonance above a certain critical density level. The same mechanism that feeds the nuclear ring or the nuclear disc may well be responsible for the higher rate of Seyfert galaxies among the σ-drop hosts.

S. Comerón, J.H. Knapen, and J.E. Beckman
Instituto de Astrofísica de Canarias, E-38200 La Laguna, Spain
e-mail: sebastien@iac.es, jhk@iac.es, jeb@iac.es

J.E. Beckman
Consejo Superior de Investigaciones Científicas, Spain

J.M. Diego et al. (eds.), *Highlights of Spanish Astrophysics V*, Astrophysics and Space Science Proceedings, DOI 10.1007/978-3-642-11250-8_35,

Average Iron Line Emission from Distant AGN

A. Corral, M.J. Page, F.J. Carrera, X. Barcons, S. Mateos, J. Ebrero, M. Krumpe, A. Schwope, J.A. Tedds, and M.G. Watson

Abstract We have developed a new method to construct a high SNR X-ray average spectrum for type 1 and type 2 AGN in order to measure the main properties of the Fe Kα emission line. Our method takes into account all the possible contributions to the continuum around the emission line, and an estimate of the significance of any excursion above this continuum can be obtained from it. We applied this method to the identified AGN within AXIS and XWAS, 606 type 1 and 117 type 2 AGN, and obtained average spectra for both AGN types. We found an unresolved emission line on the final average spectrum of type 1 and 2 AGN with an EW ~100 eV, corresponding to neutral or low-ionization Fe Kα emission, thus emitted far from the central source. We also found that the EW of this narrow line becomes weaker as the luminosity increases, the so-called Iwasawa–Taniguchi effect. A clear relativistic component in the Fe Kα line is not present in the average spectra and the continuum is best represented by a mixture of absorbed power laws plus a moderate reflection component for type 1 AGN. In the case of type 2 AGN, the statistics turned out to be insufficient to distinguish between a reflection component and a relativistic line. We estimated the EW of any relativistic contribution to be <400 eV and <300 eV at 3σ confidence level for type 1 and type 2 AGN, respectively. Our results are in excellent agreement with studies of local AGN, whereas we obtain a much lower value for the relativistic line EW than studies at higher redshifts.

A. Corral
Instituto de Física de Cantabria (CSIC-UC), 39005 Santander, Spain, and INAF-OAB, via Brera 28, 20121 Milan, Italy
e-mail: amalia.corral@brera.inaf.it

F.J. Carrera, X. Barcons, and J. Ebrero
Instituto de Física de Cantabria (CSIC-UC), 39005 Santander, Spain

M.J. Page
Mullard Space Science Laboratory, University College London, Holmbury St. Mary, Dorking, Surrey RH5 6NT, UK

S. Mateos, J.A. Tedds, and M.G. Watson
X-ray & Observational Astronomy Group, Department of Physics and Astronomy, University of Leicester, Leicester LE1 7RH, UK

M. Krumpe and A. Schwope
Astrophysikalisches Institut Potsdam, An der Sternwarte 16, D-14482 Postdam, Germany

J.M. Diego et al. (eds.), *Highlights of Spanish Astrophysics V*, Astrophysics and Space Science Proceedings, DOI 10.1007/978-3-642-11250-8_36,

The WMAP Cold Spot

M. Cruz, E. Martínez-González, and P. Vielva

Abstract The WMAP cold spot was found by applying spherical wavelets to the first year WMAP data. An excess of kurtosis of the wavelet coefficient was observed at angular scales of around 5°. This excess was shown to be inconsistent with Gaussian simulations with a p-value of around 1%. A cold spot centered at ($b = -57°, l = 209°$) was shown to be the main cause of this deviation. Several hypotheses were raised to explain the origin of the cold spot. After performing a Bayesian template fit, a collapsing cosmic texture was found to be the most probable hypothesis explaining the spot. Here we review the properties of the cold spot and the possible explanations.

M. Cruz, E. Martínez-González, and P. Vielva
IFCA, CSIC-Univ. de Cantabria, Avda. los Castros, s/n, 39005-Santander, Spain
e-mail: cruz@ifca.unican.es, martinez@ifca.unican.es, vielva@ifca.unican.es

J.M. Diego et al. (eds.), *Highlights of Spanish Astrophysics V*, Astrophysics and Space Science Proceedings, DOI 10.1007/978-3-642-11250-8_37,

Constraints on the Non-linear Coupling Parameter $f_{\rm nl}$ Using the CMB

A. Curto, E. Martínez-González, and R.B. Barreiro

Abstract We present a review of the constraints on the non-linear coupling parameter $f_{\rm nl}$ using the Cosmic Microwave Background (CMB) temperature anisotropies at high resolution. In particular we summarize the recent results presented in the analysis by Curto et al. (2008; arXiv:0807.0231) and compare it with other works. In Curto et al. (2008) several third order estimators based on the spherical Mexican hat wavelet are considered. These quantities are combined to perform a χ^2 analysis. The χ^2 statistic is used to test the Gaussianity of the data as well as to constrain the $f_{\rm nl}$ parameter using the WMAP data. This analysis is based on simulations that take into account the CMB and the instrumental properties. Most of the analyzes confirm that the data of these experiments are compatible with Gaussianity. The best estimate at present of the non-linear coupling parameter is $-8 < f_{\rm nl} < +111$ at 95% CL (Curto et al. 2008) using wavelets.

A. Curto
Instituto de Física de Cantabria, CSIC-Universidad de Cantabria and Dpto. de Física Moderna, Universidad de Cantabria, Avda. de los Castros s/n, 39005 Santander, Spain
e-mail: curto@ifca.unican.es

E. Martínez-González and R.B. Barreiro
Instituto de Física de Cantabria, CSIC-Universidad de Cantabria, Avda. de los Castros s/n, 39005 Santander, Spain

J.M. Diego et al. (eds.), *Highlights of Spanish Astrophysics V*, Astrophysics and Space Science Proceedings, DOI 10.1007/978-3-642-11250-8_38,

Kinematics of Inner Bars. The Stellar σ-Hollows

Adriana de Lorenzo-Cáceres, Jesús Falcón-Barroso, Alexandre Vazdekis, and Inma Martínez-Valpuesta

Abstract We present SAURON stellar kinematical analysis for four double-barred early-type galaxies: NGC 2859, NGC 3941, NGC 4725 and NGC 5850. The presence of the inner bar is not evident from the radial velocity, but it appears to have an important effect in the stellar velocity dispersion maps: we find two σ-hollows of amplitudes between 10 and 40 km s^{-1} at either sides of the center, at the ends of the inner bars. We have performed numerical simulations to explain these features. Ruling out other possibilities, we finally conclude that, although the σ-hollows might be originated by a younger stellar population component with low velocity dispersion. More likely they are an effect of the contrast between two kinematically different components: the bulge, with its high velocity dispersion, and the inner bar, characterized by its low velocity dispersion (ordered motion).

A. de Lorenzo-Cáceres, A. Vazdekis, and I. Martínez-Valpuesta
Instituto de Astrofísica de Canarias, C/ Vía Láctea S/N, E-38205 La Laguna, Spain
e-mail: adlcr@iac.es, vazdekis@iac.es, imv@iac.es

J. Falcón-Barroso
European Space Agency / ESTEC, Keplerlaan 1, Postbus 299, 2200 Noordwijk, The Netherlands
e-mail: jfalcon@rssd.esa.int

J.M. Diego et al. (eds.), *Highlights of Spanish Astrophysics V*, Astrophysics and Space Science Proceedings, DOI 10.1007/978-3-642-11250-8_39,

Gas on the Virgo Cluster from WMAP and ROSAT Observations

Jose M. Diego and Yago Ascasibar

Abstract WMAP observations at mm wavelengths are sensitive to the Sunyaev–Zel'dovich (SZ) effect in galaxy clusters. Among all the objects in the sky, the Virgo cluster is expected to provide the largest integrated signal. Based on models compatible with the X-ray emission observed in the ROSAT All Sky Survey, we predict an approximately two-sigma detection of the SZ effect from Virgo in the WMAP 3-year data. Our analysis reveals a 3-sigma signal on scales of 5°, although the frequency dependence deviates from the theoretical expectation for the SZ effect. The main sources of uncertainty are instrumental noise, and, most importantly, possible contamination from point sources and diffuse back/foregrounds.

J.M. Diego
IFCA, Universidad de Cantabria-CSIC, 39005 Santander, Spain
e-mail: jdiego@ifca.unican.es

Y. Ascasibar
Astrophysikalisches Institut Potsdam, Germany & Universidad Autonoma de Madrid, Spain
e-mail: yago.ascasibar@uam.es

J.M. Diego et al. (eds.), *Highlights of Spanish Astrophysics V*, Astrophysics and Space Science Proceedings, DOI 10.1007/978-3-642-11250-8_40,

N-body Simulations of the Rees-Sciama Effect

J.M. Diego, E. Martínez-González, and G. Yepes

Abstract Using N-body cosmological simulations, we compute the derivative of the potential along the line of sight in order to estimate the integrated Sachs–Wolfe (ISW) effect and the Rees–Sciama (R-S) effect on CMB maps. By using the same seed to generate the initial conditions and changing the spatial and mass resolution we study the impact of numerical resolution on the predicted results. As expected, the ISW is affected less than the R-S. We compute the angular power spectrum and compare it with the corresponding one for the CMB fluctuations and find that the ISW plus R-S power spectrum is well below the level of CMB fluctuations. This is however affected by the relative small size of our simulation. Future and larger simulations will be able to probe larger scales and to higher redshifts.

J.M. Diego and E. Martínez-González
IFCA, Universidad de Cantabria-CSIC, 39005 Santander, Spain
e-mail: jdiego@ifca.unican.es, martinez@ifca.unican.es

G. Yepes
Universidad Autonoma de Madrid, Spain
e-mail: gustavo.yepes@uam.es

J.M. Diego et al. (eds.), *Highlights of Spanish Astrophysics V*, Astrophysics and Space Science Proceedings, DOI 10.1007/978-3-642-11250-8_41,

Bulges of Disk Galaxies at Intermediate Redshifts: Nuclear Densities and Colours

L. Domínguez-Palmero and M. Balcells

Abstract We analyze central surface brightness, nuclear and global colors of intermediate redshift disk galaxies in WFPC2/HST Groth strip and ACS/HST GOODS–N. The aim is to obtain empirical information of relative ages of bulges and disks and to study the relation among surface densities and star formation history of galaxies at $0.1 < z < 1.3$. We find that 60% of galaxies with bulges define a passive evolution red sequence, while the remaining 40% have bluer colors that may trace star formation activity. We also find that up to $z \sim 0.8$, nuclear and global $(U - B)$ colors strongly correlate with central surface brightness, in the sense that galaxies with brighter nuclei show redder colors, as found in the local Universe. This color–density scaling breaks down at $z > 0.8$, where blue colors are found in a fraction of the high-surface brightness nuclei. The associated nuclear star formation must lead to bulge growth inside disks. We argue that blue bulges in $0.8 < z < 1.3$ may be precursors of local pseudobulges. We do not find evidence for rejuvenation of classical bulges at the sampled z.

L. Domínguez-Palmero and M. Balcells
Instituto de Astrofísica de Canarias, E-38200 La Laguna, Tenerife, Spain
e-mail: ldp@iac.es, balcells@iac.es

J.M. Diego et al. (eds.), *Highlights of Spanish Astrophysics V*, Astrophysics and Space Science Proceedings, DOI 10.1007/978-3-642-11250-8_42,

Cosmic Evolution of Active Galactic Nuclei

J. Ebrero and F.J. Carrera

Abstract The X-ray Luminosity Function (XLF) is a key tool to understand the evolution of Active Galactic Nuclei (AGN) and the variation in the accretion rate of matter onto the supermassive black hole that resides in their centres along cosmic time. We have studied the density of sources per unit luminosity in three energy bands combining the XMS survey with other shallower and deeper samples, up to redshifts of $\sim$3. The XMS survey covers a sky area of $\sim$3.3 square degrees at medium fluxes, where the bulk of the Cosmic X-ray Background is emitted. Moreover, extragalactic surveys such as XMS are essential since they contain a large amount of obscured AGN that decisively contribute to the background emission. We have found evolution in the AGN detected in soft (0.5–2 keV), hard (2–10 keV) and ultrahard (4.5–7.5 keV) X-rays, finding a maximum in the comoving density of these objects at redshifts of $\sim$1. Since we have detected spectral information of most of the AGN detected at >2 keV, we have studied their XLF along with their intrinsic absorption properties in order to obtain purely observational results and to determine the evolution of absorbed AGN at different epochs of the Universe.

J. Ebrero and F.J. Carrera
Instituto de Física de Cantabria (CSIC-UC), Avenida de Los Castros, 39005 Santander, Spain
e-mail: ecarrero@ifca.unican.es, carreraf@ifca.unican.es

J.M. Diego et al. (eds.), *Highlights of Spanish Astrophysics V*, Astrophysics and Space Science Proceedings, DOI 10.1007/978-3-642-11250-8_43,

The Buildup of E–S0 Galaxies at $z < 2$ from Pure Luminosity Evolution Models

M.C. Eliche-Moral, M. Prieto, G. Barro, M. Balcells, J. Gallego, P.G. Pérez-González, J. Zamorano, N. Cardiel, A. Gil de Paz, R. Guzmán, R. Pelló, and V. Villar

Abstract Considering that the recent history of E–S0 galaxies can be approximated by Pure Luminosity Evolution (PLE), we have examined a set of PLE models in order to delimit the epoch in which the majority of the red galaxy population moved away from this simple evolution framework. The models assume that they were assembled and formed most of their stars at a given formation redshift (z_f), and that they have evolved without merging or substantial dust obscuration since then. Comparing the model predictions with real data, we conclude that most of E–S0's at low and intermediate luminosities must have been progressively built up at $1 < z < 2$, being the bulk of formation at $z \sim 1.5$, as recently claimed by several observational studies.

M.C. Eliche-Moral, G. Barro, J. Gallego, P.G. Pérez-González, J. Zamorano, N. Cardiel, A. Gil de Paz, and V. Villar
Universidad Complutense de Madrid, Spain
e-mail: cem@astrax.fis.ucm.es

M. Prieto
Instituto de Astrofísica de Canarias, Spain
and
Universidad de La Laguna, Spain

M. Balcells
Instituto de Astrofísica de Canarias, Spain

R. Guzmán
University of Florida, USA

R. Pelló
Laboratoire d'Astrophysique de l'Observatoire Midi-Pyrénées, France

J.M. Diego et al. (eds.), *Highlights of Spanish Astrophysics V*, Astrophysics and Space Science Proceedings, DOI 10.1007/978-3-642-11250-8_44,

Evolution of the Tully–Fisher Relation

M. Fernández Lorenzo, J. Cepa, A. Bongiovanni, H. Castañeda, A.M. Pérez García, M.A. Lara López, M. Pović, and M. Sánchez Portal

Abstract The study of the evolution of the Tully–Fisher relation has raised some problems during the last years. The main difficulty is determining the required parameters of intermediate redshift galaxies. In the present work, the rotational velocities obtained from the widths of different optical lines using DEEP2 spectra are compared. The apparent magnitudes are K and extinction corrected, and the absolute magnitudes are derived using the concordance cosmological model. Finally the B, R, and I-band Tully–Fisher relation up to $z = 1.3$ is derived. Although most studies (this included) find evidences for evolution, the results are not conclusive enough, given that the possible luminosity evolution is within the scattering of the relation and the evolution in slope is difficult to determine because at high redshift only the brightest galaxies can be measured. Nevertheless, this study shows a clear tendency, that is the same for all bands, favouring a luminosity evolution where galaxies were brighter in the past for a fixed rotation velocity.

M.F. Lorenzo, J. Cepa, A. Bongiovanni, H. Castañeda, A.M.P. García,
M.A.L. López, M. Pović, and M. Sánchez Portal
IAC, c/Vía Láctea S/N, La Laguna, Spain
e-mail: mfernan@iac.es, jcn@iac.es

J.M. Diego et al. (eds.), *Highlights of Spanish Astrophysics V*, Astrophysics and Space Science Proceedings, DOI 10.1007/978-3-642-11250-8_45,

Color Dependence of the Truncation of the Stellar Disc

E. Florido, E. Battaner, A. Zurita, and A. Guijarro

Abstract We have obtained surface brightness profiles in the near infrared J, H, K_s and in the optical (V) in the edge-on galaxy NGC 6504. We find profiles which gradually decrease and eventually drop to a complete cut-off, with no signs of a double exponential. The truncation radius is lower for larger wavelengths. We further notice that the magnetic model predicts a relation between the truncation radius and the constant rotation velocity at large radii, in good agreement with the observations.

E. Florido, E. Battaner, and A. Zurita
Departamento de Física Teórica y del Cosmos, Universidad de Granada and Instituto Carlos I, Spain
e-mail: estrella@ugr.es, battaner@ugr.es, azurita@ugr.es

A. Guijarro
Departamento de Física Teórica y del Cosmos, Universidad de Granada, Instituto Carlos I and Calar Alto Observatory, Spain
e-mail: aguijarr@ugr.es

J.M. Diego et al. (eds.), *Highlights of Spanish Astrophysics V*, Astrophysics and Space Science Proceedings, DOI 10.1007/978-3-642-11250-8_46,

Cosmological Analysis of the Satellite Galaxy Distribution

M.A. Gómez-Flechoso, L. Benjouali, and R. Domínguez Tenreiro

Abstract There exists galaxy systems in the Local Universe formed by satellite dwarf galaxies orbiting a main one. The knowledge of the satellite distribution and their characteristics gives information about the formation and assembly processes of the galaxies in the Universe. In this paper, we analyze the satellite distribution around disk galaxies in cosmological hydrodynamical simulations, both in three-dimensions (3D) and in projection along random directions, mimicking observational strategies. It has been found that, at short 3D distances, the satellite orbits in rich systems (that is, systems with high number of satellites) have on average a polar distribution. The orbital distribution in projection at short distances between the satellite and its host shows a lack (excess) of minor-axis alignments for poor (rich) systems. Therefore, the alignments onto a virtual sky would appear mostly isotropic (i.e. no-preferred major or minor axis alignments), or even planar (i.e. major-axis alignments), depending the selected sample, in consistency with most observational analyses.

M.A. Gómez-Flechoso, L. Benjouali, and R.D. Tenreiro
Univ. Autónoma de Madrid, 28049 Cantoblanco, Madrid, Spain
e-mail: mangeles.gomez@uam.es, latifa.benjouali@uam.es, rosa.dominguez@uam.es

J.M. Diego et al. (eds.), *Highlights of Spanish Astrophysics V*, Astrophysics and Space Science Proceedings, DOI 10.1007/978-3-642-11250-8_47,

Extremely Red Objects in a Hierarchical Universe

V. Gonzalez-Perez, C.M. Baugh, C.G. Lacey, and C. Almeida

Abstract We analyse whether hierarchical formation models based on Λ Cold Dark Matter Cosmology can produce enough massive red galaxies to match observations. For this purpose, we compare observations with the predictions from two published models for the abundance and redshift distribution of Extremely Red Objects (EROs; red, massive, galaxies observed at $z \geq 1$). One of the models invokes *superwinds* to regulate star formation in massive haloes, whilst the other suppresses cooling through *radio-mode* AGN feedback. The first one underestimates the number counts of EROs by an order of magnitude, whereas the radio-mode AGN feedback model gives excellent agreement with the number counts of EROs and redshift distribution of K-selected galaxies. This study highlights the need to consider AGN feedback in order to understand the formation and evolution of massive galaxies at $z \geq 1$.

V. Gonzalez-Perez
Institut de Ciències de l'Espai (CSIC/IEEC), F. de Ciencies, Torre C5 Par 2a, UAB, Bellaterra, 08193 Barcelona, Spain
e-mail: gonzalez@ieec.uab.es

C. M. Baugh, C.G. Lacey, and C. Almeida
Institute for Computational Cosmology, Department of Physics, University of Durham, South Road, Durham, DH1 3LE, UK

J.M. Diego et al. (eds.), *Highlights of Spanish Astrophysics V*, Astrophysics and Space Science Proceedings, DOI 10.1007/978-3-642-11250-8_48,

Near-Infrared and Optical Observations of Galactic Warps

A. Guijarro, R.F. Peletier, E. Battaner, J. Jiménez-Vicente, R. de Grijs, and E. Florido

Abstract Warps in galactic disks have been studied extensively in HI and in the optical, but very rarely in the near-infrared (NIR) bands that trace the older stellar populations. We have obtained observations of 20 edge-on galaxies in the NIR and in several optical bands. We compare the properties of the galactic warps as a function of wavelength in our sample galaxies. We calculate a warp curve for each galaxy, from which we obtain the characteristic warp parameters. A systematic analysis of these parameters may help us to constrain the different mechanisms that have been proposed for the development and maintenance of galactic warps. Our results show that 13 of the 20 galaxies are warped, with the warp being more pronounced in the optical than at NIR wavelengths. The transition between the unperturbed inner disk and the outer warped region is rather abrupt. Our data are compatible with lenticulars showing very small or no warps.

A. Guijarro
Dpto. de Física Teórica y del Cosmos, Universidad de Granada, Spain, and Centro Astronómico Hispano Alemán, Almería, Spain
e-mail: guijarro@caha.es

R.F. Peletier
Kapteyn Astronomical Institute, University of Groningen, The Netherlands
e-mail: peletier@astro.rug.nl

E. Battaner, J. Jiménez-Vicente, and E. Florido
Dpto. de Física Teórica y del Cosmos, Universidad de Granada, Spain
e-mail: battaner@ugr.es, jjimenez@ugr.es, estrella@ugr.es

R. de Grijs
Department of Physics & Astronomy, The University of Sheffield, UK
e-mail: R.DeGrijs@sheffield.ac.uk

J.M. Diego et al. (eds.), *Highlights of Spanish Astrophysics V*, Astrophysics and Space Science Proceedings, DOI 10.1007/978-3-642-11250-8_49,

Morphological Evolution from $z \sim 2$ in the COSMOS Field from Ks-Band Imaging

M. Huertas-Company, L. Tasca, D. Rouan, J.P. Kneib, and O. Le Fèvre

Abstract Morphology is the most accessible tracer of galaxies physical structure, although its interpretation in the framework of galaxy evolution still remains a problem. Its quantification at high redshift requires deep high angular resolution imaging, the reason why space data (HST) are normally employed. At $z > 1$, the HST visible cameras probe however the UV flux, dominated by the emission of young stars, which could bias the estimated morphologies towards late-type systems. In this paper we quantify the effects of this *morphological k-correction* at $1 < z < 2$ by comparing morphologies measured in the K and I-bands in the COSMOS area. Ks-band data have indeed the advantage of probing old stellar populations in the rest-frame for $z < 2$, enabling a determination of galaxy morphological types unaffected by recent star formation. We employ a new non-parametric method based on SVM to classify ~50,000 Ks selected galaxies in the COSMOS area observed with WIRCam at CFHT. We use a 10-dimensional volume, including 5 morphological parameters, and other characteristics of galaxies such as luminosity and redshift. The classification is globally in good agreement with the one obtained using HST/ACS for $z < 1$. Above $z \sim 1$, the I-band classification tends to find less early-type galaxies than the Ks-band one by a factor ~1.5, which might be a consequence of morphological k-correction effects. We argue therefore that studies based on I-band HST/ACS classifications at $z > 1$ could be underestimating the elliptical population.

M. Huertas-Company and D. Rouan
LESIA, Observatoire de Paris, CNRS, UPMC, Universit Paris Diderot,
5 Place Jules Janssen, 92195, Meudon, France
e-mail: marc.huertas@obspm.fr

L. Tasca, J.P. Kneib, and O. Le Fèvre
LAM, CNRS-Université de Provence, 38, rue Frédéric Joliot-Curie,
13388 Marseille cedex 13, France

J.M. Diego et al. (eds.), *Highlights of Spanish Astrophysics V*, Astrophysics and Space Science Proceedings, DOI 10.1007/978-3-642-11250-8_50,

High-Resolution Optical Spectroscopy of Radio Broad Absorption Line Quasars

F. Jiménez-Luján, J.I. González-Serrano, and C.R. Benn

Abstract We present high-resolution optical spectroscopy of several high-redshift BAL (Broad Absorption Line) quasars: 0844+0503, at redshift $z = 3.3465$, which has a known radio counterpart observed by the VLA FIRST survey; 0908+0658, at redshift $z = 3.0734$, with no radio detection recorded in the NASA/IPAC Extragalactic Database (NED); and 0217–0854, at redshift $z = 2.5720$, which also has a known radio counterpart observed by the VLA FIRST survey. They show velocity structure on a scale smaller than the separations of the two components in prominent doublets (CIV, SiIV, NV). Comparison of the residual intensities in the two components has allowed us to measure the covering factor and the column densities of several atomic species in the absorbing gas. From these, the ionisation parameters have been measured, providing constraints on the distance of the gas from the nucleus. Kinetic luminosities will be determined (through the distances estimates for plausible assumed values of electron densities) in order to know their impact on the properties and evolution of these quasars and the intergalactic medium.

F. Jiménez-Luján
Dpto. de Física Moderna, Univ. de Cantabria, Avda. de los Castros s/n, E-39005 Santander, Spain
and
Instituto de Física de Cantabria (CSIC-Universidad de Cantabria), Avda. de los Castros s/n, E-39005 Santander, Spain
e-mail: jimenezf@ifca.unican.es

J.I. González-Serrano
Instituto de Física de Cantabria (CSIC-Universidad de Cantabria), Avda. de los Castros s/n, E-39005 Santander, Spain
e-mail: gserrano@ifca.unican.es

C.R. Benn
Isaac Newton Group, Apartado 321, E-38700 Santa Cruz de La Palma, Spain
e-mail: crb@ing.iac.es

J.M. Diego et al. (eds.), *Highlights of Spanish Astrophysics V*, Astrophysics and Space Science Proceedings, DOI 10.1007/978-3-642-11250-8_51,

Metallicity Estimates with SDSS–DR6

M.A. Lara-López, J. Cepa, A. Bongiovanni, H. Castañeda, A.M. Pérez García, M. Fernández Lorenzo, M. Póvic, and M. Sánchez-Portal

Abstract We present a study of the metallicity of 20268 galaxies from the Sloan Digital Sky Survey—Data Release 6 (SDSS–DR6)—using the R_{23} method, and derive analytical calibrations from several metallicity-sensitive line ratios: [N II] λ6583/Hα, [O III] λ5007/[N II] λ6583, [N II] λ6583/[O II] λ3727, [N II] λ6583/ [S II] $\lambda\lambda$6717, 6731, [S II] $\lambda\lambda$6717, 6731/Hα, and [O III] $\lambda\lambda$4959, 5007/Hβ. We have performed the study for the resdshift interval (0.04–0.1) for all the Sloan survey release. This is the first part of a more complete work which aims to study the metallicity dependences of the star-forming galaxies in the Local Universe.

M.A. Lara-López, A. Bongiovanni, H. Castañeda, A.M.P. García, M.F. Lorenzo, and M. Póvic
Instituto de Astrofísica de Canarias, 38200 La Laguna, Spain
e-mail: mall@iac.es

J. Cepa
Instituto de Astrofísica de Canarias, 38200 La Laguna, Spain, and Departamento de Astrofísica, Universidad de La Laguna, 38205 La Laguna, Spain

M. Sánchez-Portal
Herschel Science Centre, ESAC/INSA, Madrid, Spain

J.M. Diego et al. (eds.), *Highlights of Spanish Astrophysics V*, Astrophysics and Space Science Proceedings, DOI 10.1007/978-3-642-11250-8_52,

The Merger Fraction Evolution up to $z \sim 1$

C. López-Sanjuan, M. Balcells, P.G. Pérez-González, G. Barro, C.E. García-Dabó, J. Gallego, and J. Zamorano

Abstract We present results on the disk–disk major merger fraction evolution up to $z \sim 1$ in SPITZER/IRAC selected samples in the GOODS–S field. We pick as merger remnants sources with high asymmetry (A). We take into account the experimental errors in photometric redshift and index A, that tend to overestimate the merger fraction, by maximum likelihood techniques, and avoid the loss of information with redshift (degradation of spatial resolution and cosmological dimming) by artificially redshifting all sources to a representative redshift, $z_d = 1$. We define absolute B-band and mass selected samples, for which we obtain a very different merger fraction evolution: $f_m^{mph}(z, M_B \leq -20) = 0.013(1+z)^{1.8}$, while $f_m^{mph}(z, M_\star > 10^{10}\,M_\odot) = 0.001(1+z)^{5.4}$. These results implies that only ~20% (8%) of today's $M_B \leq -20$ ($M_\star > 10^{10}\,M_\odot$) galaxies have undergone a disk–disk major merger since $z = 1$. Combined with high redshift data in the literature, we expect $1.2^{+0.4}_{-0.3}$ disk–disk major mergers since $z \sim 3$ for $M_\star > 10^{10}\,M_\odot$ galaxies, with almost all the merger activity before $z = 1$.

C. López-Sanjuan and M. Balcells
Instituto de Astrofísica de Canarias, Calle Vía Láctea s/n, E-38200, La Laguna, Tenerife, Spain
e-mail: clsj@iac.es

P.G. Pérez-González, G. Barro, J. Gallego, and J. Zamorano
Departamento de Astrofísica y Ciencias de la Atmósfera, Facultad de C.C. Físicas, Universidad Complutense de Madrid, E-28040 Madrid, Spain

C.E. García-Dabó
European South Observatory, Karl-Schwarzschild-Strasse 2, D-85748 Garching, Germany

J.M. Diego et al. (eds.), *Highlights of Spanish Astrophysics V*, Astrophysics and Space Science Proceedings, DOI 10.1007/978-3-642-11250-8_53,

New Empirical Fitting Functions for the Lick/IDS Indices Using MILES

J.M. Martín-Hernández, E. Mármol-Queraltó, J. Gorgas, N. Cardiel, P. Sánchez-Blázquez, A.J. Cenarro, R.F. Peletier, A. Vazdekis, and J. Falcón-Barroso

Abstract We are presenting new empirical fitting functions for the Lick/IDS line-strength indices as measured in MILES (Medium-resolution INT Library of Empirical Spectra). Following previous work in the field, these functions describe the empirical behaviour of the line-strength indices with the atmospheric stellar parameters ($T_{\rm eff}$, $\log g$, [Fe/H]). In order to derive the fitting functions we have devised a new procedure which, being fully automatic, provides a better description of the line-strength index variations in the stellar parameter space.

J.M. Martín-Hernández, E. Mármol-Queraltó, J. Gorgas, and N. Cardiel
Dpto. Astrofísica, Universidad Complutense de Madrid, Avda. Complutense s/n, 28040 Madrid, Spain
e-mail: josemm@astrax.fis.ucm.es

P. Sánchez-Blázquez
University Of Central Lancashire, Centre for Astrophysics, Preston, PR1 2HE, UK

A.J. Cenarro and A. Vazdekis
Instituto de Astrofísica de Canarias, Vía Láctea s/n, 38200, La Laguna, Spain

R.F. Peletier
Kapteyn Astronomical Institute, University of Groningen, 9700 AV Groningen, The Netherlands

J. Falcón-Barroso
Sterrewacht Leiden, Niels Bohrweg 2, 2333 CA, Leiden, The Netherlands

J.M. Diego et al. (eds.), *Highlights of Spanish Astrophysics V*, Astrophysics and Space Science Proceedings, DOI 10.1007/978-3-642-11250-8_54,

Modelling Starburst in HII Galaxies: from Chemical to Spectro-Photometric Evolutionary Self-Consistent Models

M.L. Martín-Manjón, M. Mollá, A.I. Díaz, and R. Terlevich

Abstract We have computed a series of realistic and self-consistent models that reproduce the properties of HII galaxies. The emitted spectrum of HII galaxies is reproduced by means of the photoionization code CLOUDY, using as ionizing spectrum the spectral energy distribution of the modelled HII galaxy, calculated using new and updated stellar population synthesis model (PopStar). This, in turn, is calculated according to a star formation history and a metallicity evolution given by a chemical evolution code. Our technique reproduces observed abundances, diagnostic diagrams, colours and equivalent width–colour relations for local HII galaxies.

M. L. Martín-Manjón and A.I. Díaz
Universidad Autónoma de Madrid, Madrid, Spain
e-mail: mariluz.martin@uam.es, angeles.diaz@uam.es

M. Mollá
CIEMAT, Madrid, Spain
e-mail: mercedes.molla@ciemat.es

R. Terlevich
INAOE, Puebla, Mexico

J.M. Diego et al. (eds.), *Highlights of Spanish Astrophysics V*, Astrophysics and Space Science Proceedings, DOI 10.1007/978-3-642-11250-8_55,

Studying Barred Galaxies by Means of Numerical Simulations

Inma Martinez-Valpuesta

Abstract We describe two morphological structures of barred galaxies with the help of numerical simulations. The first one is a feature seen in face-on barred galaxies, the ansae, probably very important dynamically speaking. The second one are the Boxy/Peanut bulges in disc galaxies. They have been associated to stellar bars, and are a result of the secular evolution of barred galaxies. We analyze their properties in a large sample of N-body simulations, using different methods to measure their strength, shape and possible asymmetry, and then inter-compare the results. Some of these methods can be applied to both simulations and observations. In particular, we seek correlations between bar and peanut properties, which, when applied to real galaxies, will give information on bars in edge-on galaxies, and on peanuts in face-on galaxies.

I. Martinez-Valpuesta
Instituto de Astrofísica de Canarias, C/Via Láctea s/n, E-38200 La Laguna, S/C Tenerife, Spain
e-mail: imv@iac.es

J.M. Diego et al. (eds.), *Highlights of Spanish Astrophysics V*, Astrophysics and Space Science Proceedings, DOI 10.1007/978-3-642-11250-8_56,

Photometric and Kinematic Characterization of Tidal Dwarf Galaxy Candidates

D. Miralles-Caballero, L. Colina, and S. Arribas

Abstract Tidal Dwarf Galaxies (TDG), or self-gravitating objects created from the tidal forces in interacting galaxies, have been found in several merging systems. This work will focus on identifying TDG candidates among a sample of Luminous and Ultraluminous Infrared Galaxies (U)LIRGs, where these interactions are occurring in order to study their formation and evolution. High angular resolution imaging from Hubble Space Telescope (HST) in B, I and H band will be used to detect these sources. Photometric measurements of these regions compared to Stellar Synthesis Population models will allow us to roughly estimate the age and the mass. Using complementary optical Integral Field Spectroscopy we will be able to explore the physical, kinematical and dynamical properties in TDGs. We present preliminary photometric results for IRAS 0857+3915, as an example of the study that will be held for the entire sample of (U)LIRGs.

D. Miralles-Caballero, L. Colina, and S. Arribas
DAMIR-IEM-CSIC, Serrano 121 28006 Madrid, Spain
e-mail: dmiralles@damir.iem.csic.es, colina@damir.iem.csic.es, arribas@damir.iem.csic.es

J.M. Diego et al. (eds.), *Highlights of Spanish Astrophysics V*, Astrophysics and Space Science Proceedings, DOI 10.1007/978-3-642-11250-8_57,

Chemical Enrichment of Spiral Galaxies: Metallicity–Luminosity Relation

M. Mollá

Abstract Elemental abundances increased very rapidly at the early times of evolution of galaxies. Therefore, to interpret the high redshift observations by using synthesis models without taking into account the chemical evolution may yield erroneous conclusions. We will show how spiral and irregular galaxies evolve using a grid of realistic chemical and spectrophotometric models, able to reproduce the galaxies data of our local universe. By using these calibrated models, we may study a possible evolution of the metallicity–luminosity relation, such as other galaxy data correlations with the redshift.

M. Mollá
CIEMAT, Avda. Complutense 22, 28040, Madrid, Spain
e-mail: mercedes.molla@ciemat.es

J.M. Diego et al. (eds.), *Highlights of Spanish Astrophysics V*, Astrophysics and Space Science Proceedings, DOI 10.1007/978-3-642-11250-8_58,

Studying the Population of Radio-Loud Broad Absorption Line Quasars (BAL QSOs) from the Sloan Digital Sky Survey

F.M. Montenegro-Montes, K.-H. Mack, C.R. Benn, R. Carballo, J.I. González-Serrano, J. Holt, and F. Jiménez-Luján

Abstract Broad Absorption Lines (BALs) seem to be the most extreme manifestations of quasar (QSO) outflows. Two main scenarios have been proposed to explain the nature of BAL QSOs. They may be a physically distinct population (e.g. newborn or recently refueled QSOs) or present in all QSOs but intercepted by only a fraction of the lines of sight to the QSOs. Our previous observations of a sample of 15 radio BAL QSOs show that they have convex radio spectra typical of Giga-Hertz Peaked-Spectrum (GPS) sources. We have selected a well-defined sample of radio bright BAL QSOs from the Sloan Digital Sky Survey-Data Release 5. Here we present preliminary results on radio continuum observations in full polarization of this sample, taken with the 100 m Effelsberg radiotelescope at 2.7, 4.8, 8.4 and 10.5 GHz. The aim is to describe the radio spectra and polarization characteristics of these radio bright BAL QSOs and compare them with our previous results from the study of a radio fainter sample of BAL QSOs and with the properties of normal QSOs where the BAL phenomenon is not seen.

F.M. Montenegro-Montes
INAF - Istituto di Radioastronomia, Via P. Gobetti 101, I-40129 Bologna, Italy
Dpto. de Astrofísica, Universidad de La Laguna, La Laguna, Spain
Instituto de Astrofísica de Canarias, La Laguna, Spain
e-mail: fmm@ira.inaf.it

K.-H. Mack
INAF - Istituto di Radioastronomia, Bologna, Italy

C.R. Benn
Isaac Newton Group, Santa Cruz de La Palma, Spain

R. Carballo
Dpto. de Matemática Aplicada y Ciencias de la Computacin, Univ. de Cantabria, Santander, Spain

J.I. González-Serrano and F. Jiménez-Luján
Instituto de Física de Cantabria (CSIC- Universidad de Cantabria), Santander, Spain
Departamento de Física Moderna, Universidad de Cantabria, Santander, Spain

J. Holt
Leiden Observatory, Leiden University, Leiden, The Netherlands

J.M. Diego et al. (eds.), *Highlights of Spanish Astrophysics V*, Astrophysics and Space Science Proceedings, DOI 10.1007/978-3-642-11250-8_59,

Origin of the Near-UV Light in the Circumnuclear Regions of Seyfert Galaxies

V.M. Muñoz Marín, T. Storchi-Bergmann, R.M. González Delgado, H.R. Schmitt, and P. Spinelli

Abstract In order to better understand the nature of the near-UV light in Seyfert (Sy) galaxies, as well as the connection between the AGN and starbursts processes, we carried out a snapshot survey among the nearest Sy nuclei with HST-ACS at F330W ($\sim U$). In a previous work (Muñoz Marín et al. 2007, AJ 134, 648), we found a variety of morphologies, including star-formation dominated objects, and also other galaxies with bright filaments, biconical structure or an extended emission, which are unlikely to trace star-formation. In this work we aim to disentangle the contribution of the different processes that may contribute to the near-UV emission, focussing in the extended emission. We use a subsample of galaxies with near-UV ACS data and WFPC2 [OIII] images, as well as optical and near-IR data. From these data we create a synthetic image of the contribution of the ionized gas to be subtracted from the near-UV data. The residuals are analyzed by means of photometry in the bands F330W ($\sim U$), F547M ($\sim V$), and F160W ($\sim H$). By these means, we are able to disentangle the different contribution and their relative importance in most objects.

V.M.M. Marín and R.M.G. Delgado
Instituto de Astrofíca de Andalucía (CSIC), P.O. Box 3004, 18080, Granada, Spain
e-mail:manuel@iaa.es, rosa@iaa.es

T. Storchi-Bergmann
Instituto de Física, Universidade Federal do Rio Grande do Sul, C.P. 15001, 91501-970, Porto Alegre, Brazil

H.R. Schmitt
Remote Sensing Division, Naval Research Laboratory, Washington, DC 20375
and
Interferometrics, Inc., Herdon, VA 20171, USA

P. Spinelli
Universitäts-Sternwarte München, Scheinerstr. 1, D-81679, München, Deutschland

J.M. Diego et al. (eds.), *Highlights of Spanish Astrophysics V*, Astrophysics and Space Science Proceedings, DOI 10.1007/978-3-642-11250-8_60,

Radial Distribution of Dust Properties in Nearby Galaxies

J.C. Muñoz-Mateos, A. Gil de Paz, S. Boissier, J. Zamorano, D.A. Dale, P.G. Pérez-González, J. Gallego, B.F. Madore, G. Bendo, M. Thornley, A. Boselli, V. Buat, D. Calzetti, and J. Moustakas

Abstract We present a detailed analysis of the radial distribution of dust properties (extinction, PAH abundance and dust-to-gas ratio) in 57 galaxies in the SINGS sample, performed on a multi-wavelength set of UV, IR and radio surface brightness profiles, combined with published molecular gas profiles and metallicity gradients.

J.C. Muñoz-Mateos, A. Gil de Paz, J. Zamorano, P.G. Pérez-González, and J. Gallego
Departamento de Astrofísica y CC. de la Atmósfera, Universidad Complutense de Madrid, Spain
e-mail: jcmunoz@astrax.fis.ucm.es

S. Boissier, A. Boselli, and V. Buat
Observatoire Astronomique de Marseille-Provence, Laboratoire d'Astrophysique de Marseille, and Centre National de la Recherche Scientifique, France

D.A. Dale
Department of Physics and Astronomy, University of Wyoming, Laramie, WY, USA

B.F. Madore
Observatories of the Carnegie Institution of Washington, Pasadena, CA, USA

G. Bendo
Astrophysics Group, Imperial College, Blackett Laboratory, London, UK

M. Thornley
Department of Physics and Astronomy, Bucknell University, Lewisburg, PA, USA

D. Calzetti
Department of Astronomy, University of Massachusetts, Amherst, MA, USA

J. Moustakas
Center for Cosmology and Particle Physics,New York University, New York, NY, USA

J.M. Diego et al. (eds.), *Highlights of Spanish Astrophysics V*, Astrophysics and Space Science Proceedings, DOI 10.1007/978-3-642-11250-8_61,

Calibration of Star Formation Rate Tracers Using Evolutionary Synthesis Models

H. Otí-Floranes and J.M. Mas-Hesse

Abstract Starburst phenomena can be characterized by their Star Formation Rate, which measures the mass converted to stars per unit time within the starburst region. Many diverse expressions for this magnitude, using the emission of the bursts at different wavelengths, have been suggested in the literature: UV radiation emitted by young stars, FIR emission from dust heated by the UV field, recombination lines from the gas nebula surrounding the stars, X-ray emission from X-ray binaries, etc. Our objective is to use last generation evolutionary synthesis models to calibrate the different Star Formation Rate (for constant stellar formation bursts), or Star Formation Strength (for instantaneous bursts) tracers in a consistent way. The first performed step has been to derive the calibration of the soft X-ray luminosity as a Star Formation Rate/Strength tracer. In this contribution we present the same kind of analyzes performed on several other tracers at lower energies, such as the UV continuum, the production of ionizing photons per unit of time, the total infrared emission, etc. We also compare the expressions yielded by the models with those frequently used in the literature and discuss the ranges of usability of the latter ones. Several results are presented, among which we stress the importance of taking into account both the age of the burst considered, as well as the type of burst (extended or instantaneous) for the sources studied.

H. Otí-Floranes
Laboratorio de Astrofísica Espacial y Física Fundamental, CAB (CSIC–INTA), POB 78, 28691 Villanueva de la Cañada, Spain
and
Dpto. de Física Moderna, Facultad de Ciencias, Universidad de Cantabria, 39005 Santander, Spain
e-mail: otih@laeff.inta.es

J.M. Mas-Hesse
Laboratorio de Astrofísica Espacial y Física Fundamental, CAB (CSIC–INTA), POB 78, 28691 Villanueva de la Cañada, Spain
e-mail: mm@laeff.inta.es

J.M. Diego et al. (eds.), *Highlights of Spanish Astrophysics V*, Astrophysics and Space Science Proceedings, DOI 10.1007/978-3-642-11250-8_62,

MAGIC Observations of Active Galactic Nuclei

I. Oya, J.L. Contreras, and D. Bose

Abstract The MAGIC Imaging Atmospheric Cherenkov Telescope is located at the Roque de los Muchachos Observatory at La Palma. Currently it is the largest detector of its kind in operation. It is able to study sources of cosmic gamma-rays of energy between 50 and 60 GeV and some TeV with sensitivity down to less than 2% of the Crab nebula flux in 50 h. In this contribution we present a review of its recent results for the AGNs in flaring and quiescent states. These results can help in understanding the mechanisms of gamma-ray production in AGN jets, estimate the distribution of Extragalactic Background Light (EBL), and detecting signs of quantum gravity.

I. Oya, J.L. Contreras, and D. Bose (for the MAGIC Collaboration)
Dpto.Física Atómica, UCM, Madrid, Spain
e-mail: oya@gae.ucm.es

J.M. Diego et al. (eds.), *Highlights of Spanish Astrophysics V*, Astrophysics and Space Science Proceedings, DOI 10.1007/978-3-642-11250-8_63,

Baryonic Matter at Supercluster Scales: The Case of the Corona Borealis Supercluster

Carmen Pilar Padilla-Torres, Rafael Rebolo, Carlos M. Gutiérrez, Ricardo Génova-Santos, and José Alberto Rubiño-Martin

Abstract In a 24 deg^2 survey for baryonic matter at 33 GHz in the Corona Borealis supercluster (CrB-SC) of galaxies ($z = 0.07$), with the Very Small Array (VSA) interferometer (Génova-Santos et al. 2005, MNRAS 363, 79; 2008, arXiv: 0804.0199), we found a very strong temperature decrement in the Cosmic Microwave Background (CMB). It has an amplitude of $-230 \pm 23\,\mu$K and is located near the center of the supercluster, in a position with no known galaxy clusters, and without a significant X-ray emission in the ROSAT All-Sky Survey. Monte-Carlo simulations discard the primordial CMB Gaussian field as a possible explanation for this decrement at a level of 99.6%. We therefore concluded that this could be indicative of a Sunyaev–Zel'dovich (SZ) effect produced either by a warm/hot gas distribution in the intercluster medium or by a farther unknown galaxy cluster. Here we present an optical study of the galaxy distribution in this region, aiming at elucidating whether it traces a possible warm/hot gas filamentary distribution or a galaxy cluster. First, we have studied the galaxy population down to $r \leq 20$ magnitudes in the SDSS. This reveals an overdensity by a factor of 2 with respect to nearby control fields, but lower than in the galaxy clusters member of the CrB–SC. This indicates that the associated gas could at least be partially responsible for the observed CMB decrement. Second, we obtained spectroscopic redshifts, with the William Herschel Telescope (WHT), for a sample of galaxies in the region of the cold spot, and found evidence of a substructure with redshifts extending from 0.07 to 0.10. This suggests the existence of a dense filamentary structure with a length of several tens of Mpc. Finally, we investigated the presence of at least one farther cluster in the same line-of-sight, at $z \approx 0.11$.

C.P. Padilla-Torres, C.M. Gutiérrez, R. Génova-Santos, and J.A. Rubiño-Martin
Instituto de Astrofísica de Canarias, Spain
e-mail: cppt@iac.es

R. Rebolo
Instituto de Astrofísica de Canarias and Consejo Superior de Investigaciones Científicas, Spain
e-mail: rrl@iac.es

J.M. Diego et al. (eds.), *Highlights of Spanish Astrophysics V*, Astrophysics and Space Science Proceedings, DOI 10.1007/978-3-642-11250-8_64,

Spitzer/IRS Mapping of Local Luminous Infrared Galaxies

M. Pereira-Santaella, A. Alonso-Herrero, G.H. Rieke, and L. Colina

Abstract We present results of our program Spitzer/IRS Mapping of local Luminous Infrared Galaxies (LIRGs). The maps cover the central $20'' \times 20''$ or $30'' \times 30''$ regions of the galaxies, and use all four IRS modules to cover the full $\sim 5-38\,\mu$m spectral range. We have built spectral maps of the main mid-IR emission lines, continuum and PAH features, and extracted 1D spectra for regions of interest in each galaxy. The final goal is to fully characterize the mid-IR properties of local LIRGs as a first step to understanding their more distant counterparts.

M. Pereira-Santaella, A. Alonso-Herrero, G. H. Rieke, and L. Colina
Instituto de Estructura de la Materia, CSIC, 28006 Madrid, Spain
Steward Observatory, University of Arizona, Tucson AZ85721, USA

J.M. Diego et al. (eds.), *Highlights of Spanish Astrophysics V*, Astrophysics and Space Science Proceedings, DOI 10.1007/978-3-642-11250-8_65,

Observational Evidence of Different Evolutionary Stages in Galactic Bars

I. Pérez, P. Sánchez-Blázquez, and A. Zurita

Abstract We analyze the stellar line-strength index distribution along the bar of a sample of 20 early-type galaxies derived from optical long-slit observations along the bar major axis. The aim is to study the formation and evolution of bars in galaxies. We obtain age and metallicity distributions using stellar population models. We find that the mean bar values of age, metallicity and [E/Fe] correlate with central velocity dispersion in a similar way to that of bulges, pointing to a intimate evolution of both components. Galaxies with high stellar velocity dispersions ($>170\,\mathrm{km\,s^{-1}}$) host bars with old stars while galaxies with lower central velocity dispersion show stars with a large dispersion in their ages. We find, for the first time, gradients in both age and metallicity. We find three different types of bars according to their metallicity and age distribution along the radius: (1) bars with negative metallicity gradients. They show mean young/intermediate population (<2 Gyr), and have amongst the lowest stellar velocity dispersion of the sample. (2) Bars with null metallicity gradients. The galaxies that do not show any gradient in their metallicity distribution along the bar and have positive age gradients. (3) Bars with a mean older population and positive metallicity gradients, i.e. more metal rich at the bar ends. The results indicate that most bars are long-lasting structure. These derived gradients place strong constrains on models of bar evolution. All the galaxies show disk-like central components, implying a strong role played by bars in the bulge secular evolution.

I. Pérez
Kapteyn Astronomical Institute, University of Groningen, Postbus 800, Groningen 9700 AV, the Netherlands, and Departamento de Física Teórica y del Cosmos, Campus de Fuentenueva Universidad de Granada, 18071 Granada, Spain
e-mail: isa@astro.rug.nl

P. Sánchez-Blázquez
Centre For Astrophysics, Univeristy of Central Lancashire, PR1 2HE Preston, UK
e-mail: psanchez-blazquez@uclan.ac.uk

A. Zurita
Departamento de Física Teórica y del Cosmos, Campus de Fuentenueva, Universidad de Granada 18071 Granada, Spain
e-mail: azurita@ugr.es

J.M. Diego et al. (eds.), *Highlights of Spanish Astrophysics V*, Astrophysics and Space Science Proceedings, DOI 10.1007/978-3-642-11250-8_66,

OTELO Survey: Deep BVRI Broad-Band Photometry of the Groth Strip: Number Counts and Two-Point Correlation Functions

A.M.P. García, J. Cepa, A. Bongiovanni, E.J. Alfaro, H. Castañeda, J. Gallego, J.I. González Serrano, J.J. González, and M. Sánchez-Portal

Abstract We describe the OTELO survey and present deep BVRI imaging data of the Groth field. Galaxy number counts and galaxy clustering are analyzed. We find an excellent agreement between observed and mock data number counts. We also find a evidences about a galaxy clustering evolution and a strong dependence of the angular correlation function with the observed $V - I$ color. Our data favor a flattening of the clustering amplitude with median apparent magnitude. The good general agreement between our clustering analysis and the estimates from the mock data is remarkable.

A.M. Pérez García, A. Bongiovanni, and H. Castañeda
IAC, La Laguna, Spain
e-mail: apg@iac.es, bongio@iac.es, hcastane@iac.es

J. Cepa
Departamento de Astrofísica, U. de La Laguna & IAC, La Laguna, Spain
e-mail: jcn@iac.es

E.J. Alfaro
IAA, Granada, Spain
e-mail: emilio@iaa.es

J. Gallego
Departamento de Astrofísica y CC. de la Atmósfera, UCM, Madrid, Spain
e-mail: jgm@astrax.fis.ucm.es

J.I. González-Serrano
Instituto de Física de Cantabria, Santander, Spain
e-mail: gserrano@ifca.unican.es

M. Sánchez-Portal
Herschel Science Center, INSA/ESAC, Madrid, Spain
e-mail: Miguel.Sanchez@sciops.esa.int

J.J. González
Instituto de Astronomía UNAM, México D.F, México
e-mail: jesus@astroscu.unam.mx

J.M. Diego et al. (eds.), *Highlights of Spanish Astrophysics V*, Astrophysics and Space Science Proceedings, DOI 10.1007/978-3-642-11250-8_67,

Spitzer View on the Downsizing Scenario of Galaxy Formation and the Role of AGN

P.G. Pérez-González, A. Alonso-Herrero, J. Donley, G. Rieke, G. Barro, J. Gallego, and J. Zamorano

Abstract We present the latest results of the *Spitzer* Cosmological Surveys concerning the characterization of the evolution of galaxies in the last 12 Gyr (from $z = 4$). We have analyzed the stellar mass function up to $z = 4$ using a sample of more than 28,000 galaxies selected in the rest-frame near-infrared with *Spitzer*/IRAC. Our results confirm and quantify the *downsizing* scenario of galaxy formation. Based on the study of the specific Star Formation Rates (SFRs) of X-ray emitters, we discuss the role of AGN in the evolution of galaxies, arguing against the link between nuclear activity and the quenching of the star formation in massive galaxies at $z < 1.4$.

P.G. Pérez-González, G. Barro, J. Gallego, and J. Zamorano
Universidad Complutense de Madrid, Spain

A. Alonso-Herrero
IEM/CSIC, Spain

J. Donley and G. Rieke
University of Arizona, USA

J.M. Diego et al. (eds.), *Highlights of Spanish Astrophysics V*, Astrophysics and Space Science Proceedings, DOI 10.1007/978-3-642-11250-8_68,

Search for H_α Emitters in Galaxy Clusters with Tunable Filters

Ricardo Pérez Martínez, Miguel Sánchez Portal, Jordi Cepa, Ángel Bongiovanni, and Ana Pérez García

Abstract The studies of the evolution of galaxies in Galaxy Clusters have as a traditional complication the difficulty in establishing cluster membership of those sources detected in the field of view. The determination of spectroscopic redshifts involves long exposure times when it is needed to reach the cluster peripheral regions of/or clusters at moderately large redshifts, while photometric redshifts often present uncertainties too large to offer significant conclusions. The mapping of the cluster of galaxies with narrow band tunable filters makes it possible to reach large z intervals with an accuracy high enough to establish the source membership of those presenting emission/absorption lines easily identifiable, as H_α. Moreover, the wavelength scan can include other lines as NII, OIII or H_β, allowing to distinguish those sources with strong stellar formation activity and those with an active galactic nucleus. All this makes it possible to estimate the stellar formation rate of the galaxies observed. This, together with ancillary data in other wavelengths, may lead to a good estimation of the stellar formation histories. It will shed new light over the galaxy evolution in clusters and will improve our understanding of galaxy evolution, especially in the outer cluster regions, usually less studied and with significant unexploited data that can not be correctly interpreted without redshift determination.

R.P. Martínez and M.S. Portal
ESAC/INSA
e-mail: ricardo.perez@sciops.esa.int

J. Cepa
Universidad de La Laguna and IAC, Spain

Á. Bongiovanni and A.P. García
IAC, Spain

J.M. Diego et al. (eds.), *Highlights of Spanish Astrophysics V*, Astrophysics and Space Science Proceedings, DOI 10.1007/978-3-642-11250-8_69,

Exploring Mergers of Galaxy Clusters in a Cosmological Context

Susana Planelles and Vicent Quilis

Abstract We present results of an Eulerian Adaptive Mesh Refinement (AMR) hydrodynamical and N-body simulation in a ΛCDM cosmology. The simulation incorporates common cooling and heating processes, a phenomenological description of the star formation and supernovae feedback. A specific halo finder has been designed and applied in order to extract a sample of galaxy clusters directly obtained from the simulation without considering any resimulating scheme. We have studied the evolutionary history of the cluster halos, and classified them in three categories depending on the merger events they have undergone. We pay special attention to discuss the role of merger events as a source of feedback and reheating.

S. Planelles and V. Quilis
Departament d'Astronomia i Astrofísica, Universitat de València,
46100 - Burjassot (Valencia) Spain
e-mail: susana.planelles@uv.es, vicent.quilis@uv.es

J.M. Diego et al. (eds.), *Highlights of Spanish Astrophysics V*, Astrophysics and Space Science Proceedings, DOI 10.1007/978-3-642-11250-8_70,

OTELO Survey: Properties of X-ray Emitters in the Groth Field – I. Optical Counterparts and Morphological Classification

M. Pović, M. Sánchez-Portal, A.M. Pérez García, A. Bongiovanni, J. Cepa, J.A. Acosta-Pulido, E.J. Alfaro, H. Castañeda, M. Fernández Lorenzo, J. Gallego, J.I. González-Serrano, J.J. González, and M.A. Lara-López

Abstract We present results from the study of optical broadband and X-ray properties of a large sample of active galactic nuclei (AGN) in the Groth–Westphal Strip (GWS) field. In order to determine the morphology of all objects, we obtained different structural parameters. Combining these parameters with other optical/X-ray properties, we were searching for possible correlations between them, which could point out some of the AGN characteristics (see the contribution *OTELO Survey: X-ray Emitters in the Groth Field – II. Properties of the AGN Population* by Sánchez-Portal et al., in this volume).

M. Pović, A.M.P. García, A. Bongiovanni, J.A. Acosta-Pulido, H. Castañeda, M.F. Lorenzo, and M.A. Lara-López
Instituto de Astrofísica de Canarias, 38205 La Laguna, Spain
e-mail: mpovic@iac.es

M. Sánchez-Portal
Herschel Science Centre, ESAC/INSA, Madrid, Spain
e-mail: miguel.sanchez@sciops.esa.int

J. Cepa
Instituto de Astrofísica de Canarias, 38205 La Laguna, Spain
and
Departamento de Astrofísica, Universidad de La Laguna, 38205 La Laguna, Spain

E.J. Alfaro
Instituto de Astrofísica de Andalucía-CSIC, Granada, Spain

J. Gallego
Departamento de Astrofísica y CC. de la Atmósfera, Universidad Complutense de Madrid, Madrid, Spain

J.I. González-Serrano
Instituto de Física de Cantabria, CSIC-Universidad de Cantabria, Santander, Spain

J.J. González
Instituto de Astronomía UNAM, México D.F, México

J.M. Diego et al. (eds.), *Highlights of Spanish Astrophysics V*, Astrophysics and Space Science Proceedings, DOI 10.1007/978-3-642-11250-8_71,

Red Galaxies in the GOYA Photometric Survey: Passive and Dusty Star-Forming Galaxies

Mercedes Prieto, Carlos López San Juan, and Marc Balcells

Abstract We have undertaken a study of the nature of red galaxies at low and intermediate redshifts in the GOYA photometric survey. A new photometric index based on apparent magnitudes has been obtained to separate red galaxies dominated by old populations from those dominated by star formation. A two-component mass code has been used to estimate the stellar mass and age of the galaxies. The main results are: the two types of galaxy have differences in their masses, age and number densities; the red galaxies at $z < 1$ with masses greater than 10^{11} $M_{\odot}$ are passive; the bulk of the passive galaxies at $0.3 < z < 1.5$ had their last starburst between $1 < z < 2$; the number density of the passive galaxies decreases while that of the dusty star-forming galaxies increases in the redshift range $0.3 < z < 1.5$. All these results are compatible with a scenario in which the dusty star-forming galaxies are transitional galaxies between the blue and passive galaxies.

M. Prieto, C.L.S. Juan, and M. Balcells
Instituto de Astrofísica de Canarias, C. Vía Láctea s/n, and Universidad de La Laguna
Tenerife, Spain
e-mail: mpm@iac.es, clsj@iac.es, balcells@iac.es

J.M. Diego et al. (eds.), *Highlights of Spanish Astrophysics V*, Astrophysics and Space Science Proceedings, DOI 10.1007/978-3-642-11250-8_72,

Spectral Energy Distribution of Hyper-Luminous Infrared Galaxies

A. Ruiz, F.J. Carrera, F. Panessa, and G. Miniutti

Abstract The relationship between star formation and super-massive black hole growth is central to our understanding of galaxy formation and evolution. Hyper-Luminous Infrared Galaxies (HLIRGs) are unique laboratories to investigate the starburst-AGN connection, since they exhibit extreme star formation rates, and most of them show evidence of harboring powerful AGN. We have performed an X-ray spectral study of a sample of HLIRGs observed with XMM–*Newton*, finding that the X-ray emission of most of these sources is dominated by AGN activity. To improve our estimate of the relative contribution of the AGN and starburst (SB) emission to its total bolometric output, we have built multi-wavelength (from radio to X-rays) spectral energy distributions (SED) for these HLIRGs, and we have fitted standard AGN and SB templates to these SEDs. We have found that most of our HLIRGs need an AGN template to model its SED, and this component dominates the bolometric output. We also have found that our sources classified as type 1 AGN are better modeled using a luminosity-dependent template. Extending the SED to the X-ray bands places better constraints on the relative contribution of the AGN and SB with respect to using only IR/sub-mm data.

A. Ruiz and F.J. Carrera
Instituto de Física de Cantabria (CSIC-UC), Santander 39005, Spain
e-mail: ruizca@ifca.unican.es

F. Panessa
INAF-Rome

G. Miniutti
Lab. APC-Paris

J.M. Diego et al. (eds.), *Highlights of Spanish Astrophysics V*, Astrophysics and Space Science Proceedings, DOI 10.1007/978-3-642-11250-8_73,

AMIGA Project. Radio Continuum and Nuclear Activity in a Complete Sample of Isolated Galaxies

J. Sabater, S. Leon, L. Verdes-Montenegro, U. Lisenfeld, J. Sulentic, S. Verley, D. Espada, A. Ballu, G. Bergond, and E. García

Abstract The aim of the AMIGA project (Analysis of the Interstellar Medium in Isolated GAlaxies) is building a reference sample of isolated galaxies to study the role of the environment in galactic evolution. AMIGA began in 2003 and nowadays involves more than 30 participants from 15 international institutions. Radio continuum emission in isolated spiral galaxies is coming from disk-dominated emission in spiral galaxies, in contrast to the results found in high-density environments where nuclear activity is more frequent. The radio continuum power is lower on average in our sample than in interacting galaxies or galaxies without an environment selection criterion. This confirms the relevance of our sample as a baseline to study the effects of the environment. Finally, we have studied the nuclear activity in isolated galaxies. We used different selection methods of isolated galaxies with active nucleus: (1) the far infrared colors give us a fraction of 7–20% of AGN candidates and (2) the rate of radio excess galaxies in the correlation of far infrared with radio continuum is less than 1%, which is the lowest rate found comparing with samples in other environments. This confirms the role of the environment as fundamental in the triggering of the radio nuclear activity.

J. Sabater, L. Verdes-Montenegro, S. Verley, G. Bergond, and E. García
Instituto de Astrofísica de Andalucía, CSIC,Apdo. 3004, 18080 Granada, Spain
e-mail: jsm@iaa.es, lourdes@iaa.es, simon@iaa.es, gilles@iaa.es, garcia@iaa.es

S. Leon
Instituto de RadioAstronomía Milimétrica (IRAM), Granada, Spain
e-mail: leon@iram.es

U. Lisenfeld
Dept. Física Teórica y del Cosmos, Universidad de Granada, Spain
e-mail: ute@ugr.es

J. Sulentic
Department of Astronomy, Univ. of Alabama, Tuscaloosa, USA
e-mail: jsulenti@bama.ua.edu

D. Espada
Harvard-Smithsonian Center for Astrophysics, 160 Concord Ave. M-223
Cambridge, MA 02138, USA
e-mail: despada@cfa.harvard.edu

A. Ballu
École Nationale Supérieure de Physique, Université Louis Pasteur, Strasbourg, France

J.M. Diego et al. (eds.), *Highlights of Spanish Astrophysics V*, Astrophysics and Space Science Proceedings, DOI 10.1007/978-3-642-11250-8_74,

Cosmological Vector Perturbations and CMB Anomalies

Diego Sáez and Juan Antonio Morales

Abstract Recently, it has been proved that large scale vector modes could explain most of the CMB anomalies in the first temperature multipoles. Some divergenceless (vortical) velocity fields–which are superimpositions of vector modes–can explain both the alignment of the second and third multipoles and the planar character of the octopole. In this paper we comment: (a) some papers trying to account for the mentioned anomalies, (b) our explanation based on vector modes, and (c) some current ideas about the possible origin of these modes.

D. Sáez and J.A. Morales
Departamento de Astronomía y Astrofísica, Universidad de Valencia
46100-Burjassot (Valencia) Spain
e-mail: diego.saez@uv.es, antonio.morales@uv.es

J.M. Diego et al. (eds.), *Highlights of Spanish Astrophysics V*, Astrophysics and Space Science Proceedings, DOI 10.1007/978-3-642-11250-8_75,

On the Fractal Distribution of HII Regions in Disk Galaxies

Néstor Sánchez and Emilio J. Alfaro

Abstract In this work we quantify the degree to which star-forming events are clumped. We apply a precise and accurate technique to calculate the correlation dimension D_c of the distribution of HII regions in a sample of disk galaxies. Our reliable results are distributed in the range $1.5 \lesssim D_c \lesssim 2.0$. We get significant variations in the fractal dimension among galaxies, contrary to a universal picture sometimes claimed in literature. The faintest galaxies tend to distribute their HII regions in more clustered (less uniform) patterns. Moreover, the fractal dimension for the brightest HII regions within the same galaxy seems to be smaller than for the faintest ones suggesting some kind of evolutionary effect.

N. Sánchez and E.J. Alfaro
Instituto de Astrofísica de Andalucía, CSIC, Granada, Spain
e-mail: nestor@iaa.es, emilio@iaa.es

J.M. Diego et al. (eds.), *Highlights of Spanish Astrophysics V*, Astrophysics and Space Science Proceedings, DOI 10.1007/978-3-642-11250-8_76,

OTELO Survey: X-ray Emitters in the Groth Field – II. Properties of the AGN Population

M. Sánchez-Portal, M. Pović, A.M. Pérez García, A. Bongiovanni, J. Cepa, J.A. Acosta-Pulido, E.J. Alfaro, H. Castañeda, M. Fernández Lorenzo, J. Gallego, J.I. González-Serrano, J.J. González, and M.A. Lara-López

Abstract We present the results from the analysis of the optical broadband and X-ray properties of a large sample of active galactic nuclei (AGN) in the Groth–Westphal Strip (GWS) field. The description of the observational material and data processing is described in a separate paper (see Pović et al. 2009, in this volume). Here the main findings are presented.

M. Sánchez-Portal
Herschel Science Centre, ESAC/INSA, Madrid, Spain
e-mail: miguel.sanchez@sciops.esa.int

M. Pović, A.M.P. García, A. Bongiovanni, J.A. Acosta-Pulido, H. Castañeda, M.F. Lorenzo, and M.A. Lara-López
Instituto de Astrofísica de Canarias, 38205 La Laguna, Spain

J. Cepa
Instituto de Astrofísica de Canarias, 38205 La Laguna, Spain and Departamento de Astrofísica Universidad de La Laguna, 38205 La Laguna, Spain

E.J. Alfaro
Instituto de Astrofísica de Andalucía-CSIC, Granada, Spain

J. Gallego
Departamento de Astrofísica y CC. de la Atmósfera, Universidad Complutense de Madrid Madrid, Spain

J.I. González-Serrano
Instituto de Física de Cantabria, CSIC-Universidad de Cantabria, Santander, Spain

J.J. González
Instituto de Astronomía UNAM, México D.F, México

J.M. Diego et al. (eds.), *Highlights of Spanish Astrophysics V*, Astrophysics and Space Science Proceedings, DOI 10.1007/978-3-642-11250-8_77,

A New Veto Strategy for Continuous Gravitational Wave Signals

L. Sancho de la Jordana and A.M. Sintes

Abstract In this paper we present a χ^2 veto adapted to the Hough transform searches for continuous gravitational wave signals. We characterize the χ^2–significance plane for different frequency bands and discuss the expected performance of this veto in LIGO analysis.

L. Sancho de la Jordana and A.M. Sintes
Universitat de les Illes Balears, Cra. Valldemossa km 7.5 E-07122, Palma de Mallorca, Spain
e-mail: llucia.sancho@uib.es, sintes@aei.mpg.de

J.M. Diego et al. (eds.), *Highlights of Spanish Astrophysics V*, Astrophysics and Space Science Proceedings, DOI 10.1007/978-3-642-11250-8_78,

Observing Supermassive Black Hole Binary Systems with LISA

Miquel Trias and Alicia M. Sintes

Abstract We study parameter estimation of supermassive black holes by LISA using the inspiral full post-Newtonian gravitational waveforms up to 2PN. The analysis shows that LISA will observe the coalescence of a system with total mass $(10^4 - 10^8)\,M_\odot$ with signal-to-noise ratios of hundreds and errors in masses and luminosity distance smaller than 10%. For sources at $z = 1$ we find that at least 20% could be localized within a $(1^\circ \times 1^\circ)$ patch.

M. Trias
Departament de Física, Universitat de les Illes Balears, Cra. Valldemossa Km. 7.5
E-07122 Palma de Mallorca, Spain
e-mail: miquel.trias@uib.es

A.M. Sintes
Departament de Física, Universitat de les Illes Balears, Cra. Valldemossa Km. 7.5, E-07122 Palma de Mallorca, Spain, and Max-Planck-Institut für Gravitationsphysik (Albert-Einstein-Institut)
Am Mühlenberg 1, 14476 Golm, Germany
e-mail: sintes@aei.mpg.de

J.M. Diego et al. (eds.), *Highlights of Spanish Astrophysics V*, Astrophysics and Space Science Proceedings, DOI 10.1007/978-3-642-11250-8_79,

Cosmic Evolution of Stellar Disk Truncations

I. Trujillo, R. Azzollini, J. Bakos, J.E. Beckman, and M. Pohlen

Abstract We present results on the cosmic evolution of the outskirts of disk galaxies. In particular we focus on galaxies with stellar disk truncations. Using UDF, GOODS and SDSS data we show how the position of the break evolves with time since $z \sim 1$. Our findings agree with an evolution on the radial position of the break by a factor of 1.3 ± 0.14 in the last 8 Gyr for galaxies with similar stellar masses. We also present radial color gradients and how they evolve with time. At all redshift we find a radial inside-out bluing reaching a minimum at the position of the break radius, this minimum is followed by a reddening outwards. Our results favor a scenario where stars are formed inside the break radius and are relocated in the outskirts of galaxies through secular processes.

I. Trujillo, R. Azzollini, J. Bakos, and J.E. Beckman
Instituto de Astrofísica de Canarias, C/Vía Láctea s/n, 38205 La Laguna, S/C de Tenerife, Spain
e-mail: trujillo@iac.es, ruyman@iac.es, jbakos@iac.es, jeb@iac.es

M. Pohlen
Cardiff University, School of Physics & Astronomy, Cardiff, CF24 3AA, Wales, UK
e-mail: Michael.Pohlen@astro.cf.ac.uk

J.M. Diego et al. (eds.), *Highlights of Spanish Astrophysics V*, Astrophysics and Space Science Proceedings, DOI 10.1007/978-3-642-11250-8_80,

Robotic Optical Monitoring of a Compact Lens System: FBQ 0951+2635 in the i Sloan Filter

Aurora Ullán, Vyacheslav N. Shalyapin, Luis J. Goicoechea, and Rodrigo Gil-Merino

Abstract The quasar FBQ 0951+2635 is gravitationally lensed by an early-type galaxy. The presence of two close quasar images (A and B) is the most notorious consequence of this phenomenon. In this contribution, we describe an experiment on the variability of FBQ 0951+2635 using the Liverpool Robotic Telescope (LQLM programme). Our 100 day monitoring campaign in the i Sloan passband allow us to resolve both images and to produce individual light curves. There is no evidence of short-timescale extrinsic (e.g. microlensing) variability in the new records. Moreover, we measure an accurate i-band flux ratio (corrected by the time delay, but uncorrected for the galaxy contamination on B): $A/B = 2.74 \pm 0.02$ (1σ), which agrees with the i-band ratio of 2.8 ± 0.1 (1σ) from the SDSS frame of the system. At the red end of the optical continuum, the SDSS and LQLM experiments suggest the absence of a substantial extrinsic gradient on a timescale of a few years.

A. Ullán
Robotic Telescopes Group, Centro de Astrobiología (CSIC-INTA), Ctra de Ajalvir, km 4, 28850 Torrejón de Ardoz, Madrid, Spain
e-mail: ullanna@inta.es

V. N. Shalyapin
Institute for Radiophysics and Electronics, National Academy of Sciences of Ukraine, 12 Proskura St., Kharkov 61085, Ukraine
e-mail: vshal@ire.kharkov.ua

L. J. Goicoechea
Departamento de Física Moderna, Universidad de Cantabria, Avda. de Los Castros s/n, 39005 Santander, Spain
e-mail: goicol@unican.es

R. Gil-Merino
Instituto de Física de Cantabria (CSIC-UC), Avda. de Los Castros s/n, 39005 Santander, Spain
e-mail: gilmerino@ifca.unican.es

GLENDAMA project (http://grupos.unican.es/glendama/index.htm)

J.M. Diego et al. (eds.), *Highlights of Spanish Astrophysics V*, Astrophysics and Space Science Proceedings, DOI 10.1007/978-3-642-11250-8_81,

Distance Determination to the Andromeda Galaxy Using Variable Stars

F. Vilardell, I. Ribas, C. Jordi, E.L. Fitzpatrick, and E.F. Guinan

Abstract Distance determinations to Local Group galaxies conform the basic rungs of the cosmological distance scale. The chief goal of our project is obtaining direct and accurate distance determinations to M 31 from two important stellar populations: eclipsing binaries and Cepheids. A variability survey in the North–Eastern quadrant of M 31 was performed (in B and V passbands) with the Isaac Newton Telescope (La Palma) to identify suitable targets. The resulting catalog of 3,964 variable stars contains 437 eclipsing binaries and 416 Cepheids with ~250 epochs per filter. The selection of the 68 Cepheids less affected by blending were selected to determine a distance to M 31 of $(m - M)_0 = 24.32 \pm 0.12$ mag. At the same time, the analysis of the eclipsing binary sample (with the acquisition of Gemini/GMOS spectroscopy) has provided two direct distance determinations to M 31: $(m - M)_0 = 24.44 \pm 0.12$ mag and $(m - M)_0 = 24.46 \pm 0.19$ mag. All the obtained distances are in complete agreement and provide a direct and accurate distance determination to M 31. The combination of these results with additional data could well reduce the distance uncertainty to M 31 to better than 4%.

F. Vilardell and C. Jordi
Departament d'Astronomia i Meteorologia, Universitat de Barcelona, c/ Martí i Franquès, 1, E-08028 Barcelona, Spain
e-mail: francesc.vilardell@am.ub.es

I. Ribas
Institut de Ciències de l'Espai (CSIC-IEEC), Campus UAB, E-08193 Bellaterra, Spain

E.L. Fitzpatrick and E.F. Guinan
Department of Astronomy and Astrophysics, Villanova University, 800 Lancaster Avenue, Villanova, PA 19085, USA

J.M. Diego et al. (eds.), *Highlights of Spanish Astrophysics V*, Astrophysics and Space Science Proceedings, DOI 10.1007/978-3-642-11250-8_82,

Star Formation in Bars: Where and Why?

Almudena Zurita and Isabel Pérez

Abstract We present first results in a project devoted to understanding the conditions which inhibit/favor star formation in bars and their spatial relation with the main bar dynamical features. We have carried out a detailed analysis of the strong bar of NGC 1530, our pilot target. The analysis comprises N–body/SPH modeling and $BVRKs$ and Hα photometry. The simulations reproduce remarkably well the observed general morphology and kinematics, but fail to predict the loci of massive star formation. We find differences in the Hα equivalent width of the H II regions of the bar, which are related to their position with respect to the bar dust-lanes: the Hα equivalent width of the H II regions located downstream the bar dust-lane are lower than the rest. The possible factors producing this difference have been carefully analyzed and age has been confirmed to be the most plausible explanation. This result implies that the H II regions located further away from the bar dust-lane, in its leading side (downstream), are older than the rest. The possible scenarios explaining this result are discussed.

A. Zurita
Dpto. de Física Teórica y del Cosmos, Campus de Fuentenueva, Universidad de Granada, 18071–Granada, and Instituto Carlos I, Spain
e-mail: azurita@ugr.es

I. Pérez
Kapteyn Institute, U. of Groningen, Groningen 9700 AV (the Netherlands), Dpto. de Física Teórica y del Cosmos, Campus de Fuentenueva, Universidad de Granada, 18071–Granada, and Instituto Carlos I, Spain
e-mail: isa@astro.rug.nl, isa@ugr.es

J.M. Diego et al. (eds.), *Highlights of Spanish Astrophysics V*, Astrophysics and Space Science Proceedings, DOI 10.1007/978-3-642-11250-8_83,

The Galaxy and Its Components

Origin of the Moving Groups and Their Contribution to the Determination of the Large-scale Galactic Potential

T. Antoja, D. Fernández, F. Figueras, E. Moreno, B. Pichardo, J. Torra, and O.Valenzuela

Abstract We evaluate the hypothesis that relates the formation of moving groups to the resonances of the non-axisymmetric component of the Galactic potential (spiral arms and central bar). We apply multiscale techniques to an extensive sample of more than 24,000 stars of the solar neighborhood to characterize the kinematic structures in the U–V – age – [Fe/H] space. The observational data is compared with the distribution in the kinematic plane obtained through the computation of test particle orbits under analytic models for the potential. Results from two different models for the spiral structure and/or the Galactic bar are presented here. We show that the simulations near the position of the ILR $-1{:}4$ of the spiral structure and the OLR 1:2 of the Galactic bar are able to reproduce kinematic structures similar to the observed ones. We discuss the fundamental role of the age and metallicity distributions in the confirmation of the hypothesis.

T. Antoja, D. Fernández, F. Figueras, and J. Torra
Departament d'Astronomia i Meteorologia and IEEC-UB, Institut de Ciències del Cosmos de la Universitat de Barcelona, Martí i Franquès, 1, E-08028 Barcelona, Spain
e-mail: tantoja@am.ub.es

E. Moreno, B. Pichardo, and O. Valenzuela
Instituto de Astronomía, Universidad Nacional Autónoma de México, A.P. 70-264
04510 México D.F., México
e-mail: barbara@astroscu.unam.mx

J.M. Diego et al. (eds.), *Highlights of Spanish Astrophysics V*, Astrophysics and Space Science Proceedings, DOI 10.1007/978-3-642-11250-8_84,

Spectroscopy of Pre-CV Candidates in the Open Cluster M 67

L. Balaguer-Núñez, D. Galadí-Enríquez, C. Jordi, and S. Sánchez

Abstract The systematic study of selected open clusters by our team has lead to the production of the best set of Strömgren photometry ever obtained of the old open cluster M 67. Its analysis has shown a previously unknown clump of more than 50 stars in the HR diagram, located below the cluster MS. The spatial distribution of these stars indicates that most of them could be cluster members. Two alternative hypothesis would explain their photometry: (1) if members, they would be binary systems composed by a white dwarf and a red dwarf, i.e. pre-cataclysmic variable systems; (2) if non-members, they would constitute a stream of G-type stars placed behind the cluster. Medium dispersion spectra taken by our team using the PMAS/PPAK at 3.5 m telescope in Calar Alto[1] will show the composite or single nature of the objects, and will allow to deblend the spectral contributions from the white and the red dwarfs, if the pre-CV hypothesis would turn out to be true. Also good spectrophotometric calibration will allow to determine precise spectral types, luminosities, surface temperatures and gravities, thus providing a preliminary astrophysical characterization of these systems. The same spectrophotometric calibration will yield separate and accurate synthetic photometry for both components, if present, in different broad and intermediate band systems.

[1] Based on observations collected at the German–Spanish Astronomical Center, Calar Alto, jointly operated by the Max-Planck-Institut für Astronomie Heidelberg and the Instituto de Astrofísica de Andalucía (CSIC).

L. Balaguer-Núñez
Dpt. d'Astronomia i Meteorologia, Universitat de Barcelona, ICC,
Avda. Diagonal 647. E-08028 Barcelona, Spain
e-mail: lola.balaguer@am.ub.es

D. Galadí-Enríquez and S. Sánchez
Centro Astronómico Hispano Alemán CAHA. E-18008 Granada, Spain
e-mail: dgaladi@caha.es, sanchez@caha.es

C. Jordi
Dpt. d'Astronomia i Meteorologia, Universitat de Barcelona, ICC-IEEC,
Avda. Diagonal 647. E-08028 Barcelona, Spain
e-mail: carme@am.ub.es

J.M. Diego et al. (eds.), *Highlights of Spanish Astrophysics V*, Astrophysics and Space Science Proceedings, DOI 10.1007/978-3-642-11250-8_85,

Spitzer/IRAC Young Stellar Objects Candidates in 30 Doradus

A. Bayo, J.R. Stauffer, and D. Barrado y Navascués

Abstract We have analyzed several sets of images of the 30 Doradus (a massive Star Forming Region located in the Large Magellanic Cloud) obtained with the IRAC and MIPS cameras of the Spitzer Space Telescope. Our goal is to identify the youngest stars and even protostars in this region, either by their very red colors or by their photometric variability. In order to do this, we have cross-correlated the Spitzer photometry with data obtained at other wavelengths. We identify a set of ~30 new candidate protostars, provide new data for some of the previous candidates, and attempt to place the new candidates in context. We have used these data to empirically place constraints on the physical processes that cause the photometric variability of high mass protostars.

A. Bayo
LAEFF-CAB, INTA-CSIC, P.O. Box 78, E-28691 Villanueva de la Cañada, Madrid, Spain
e-mail: abayo@laeff.inta.es

J.R. Stauffer
Spitzer Science Center, Caltech, Pasadena, CA 91125, USA
e-mail: stauffer@ipac.caltech.edu

D.Barrado y Navascués
LAEFF-CAB, INTA-CSIC, Spain
e-mail: barrado@laeff.inta.es

J.M. Diego et al. (eds.), *Highlights of Spanish Astrophysics V*, Astrophysics and Space Science Proceedings, DOI 10.1007/978-3-642-11250-8_86,

An Application of the *Mayrit* Catalogue: Very Wide Binaries in the σ Orionis Cluster

José A. Caballero

Abstract The young σ Orionis cluster in the Orion Belt is an incomparable site for studying the formation and evolution of high-mass, solar-like, and low-mass stars, brown dwarfs, and substellar objects below the deuterium burning mass limit. The first version of the Mayrit catalogue was a thorough data compilation of cluster members and candidates, which is regularly used by many authors of different disciplines. I show a new application of the catalogue and advance preliminary results on very wide binarity in σ Orionis. The making-up of a new version of the Mayrit catalogue with additional useful data is in progress.

J.A. Caballero
Departamento de Astrofísica y Ciencias de la Atmósfera, Facultad de Física,
Universidad Complutense de Madrid, E-28040 Madrid, Spain
e-mail: caballero@astrax.fis.ucm.es

J.M. Diego et al. (eds.), *Highlights of Spanish Astrophysics V*, Astrophysics and Space Science Proceedings, DOI 10.1007/978-3-642-11250-8_87,

Preliminary Results on a Virtual Observatory Search for Companions to Luyten stars

J.A. Caballero, F.X. Miret, J. Genebriera, T. Tobal, J. Cairol, and D. Montes

Abstract The Aladin sky atlas of the Virtual Observatory has shown to be a powerful and easy-handling tool for the discovery, confirmation, and characterization of high proper-motion, multiple stellar systems of large separation in the solar vicinity. Some of these systems have very low mass components (at the star/brown dwarf boundary) and are amongst the least bound systems found to date. With projected physical separations of up to tens of thousands astronomical units, these systems represent a challenge for theoretical scenarios of formation of very low-mass stars and brown dwarfs. Here we show preliminary results of a novel *virtual* search of binary systems and companions to Luyten stars with proper motions between 0.5 and 1.0 arcsec a^{-1}.

J.A. Caballero and D. Montes
Departamento de Astrofísica y Ciencias de la Atmósfera, Facultad de Física, Universidad Complutense de Madrid, E-28040 Madrid, Spain
e-mail: caballero@astrax.fis.ucm.es

F.X. Miret, T. Tobal, and J. Cairol
Observatori Astronòmic del Garraf, Barcelona, Catalunya, Spain

J. Genebriera
Observatorio de Tacande, El Paso, La Palma, Islas Canarias, Spain

J.M. Diego et al. (eds.), *Highlights of Spanish Astrophysics V*, Astrophysics and Space Science Proceedings, DOI 10.1007/978-3-642-11250-8_88,

High-Energy Emission from Microquasars (with BH)

M.D. Caballero-García, J.M. Miller, and A.C. Fabian

Abstract The source of the high-energy emission from black holes, in the energy range observed by INTEGRAL and XMM–Newton, is a key point in the understanding of several processes occurring in black holes (i.e. as QPOs, relativistic Fe line, measurements in radio). In this paper we review the different models proposed for this high-energy emission and how these could be related with other phenomena observed in black holes. We give some insights into the results obtained from the spectral analysis of INTEGRAL and XMM–Newton data of GX 339–4.

M.D. Caballero-García and A. C. Fabian (on behalf of a large collaboration team)
Institute of Astronomy, Madingley road, Cambridge CB3 0HA, USA
e-mail: mcaballe@ast.cam.ac.uk

J.M. Miller
University of Michigan, Department of Astronomy, 500 Church Street,
Dennison 814, Ann Arbor, MI 48105, USA

J.M. Diego et al. (eds.), *Highlights of Spanish Astrophysics V*, Astrophysics and Space Science Proceedings, DOI 10.1007/978-3-642-11250-8_89,

Tidal Remnants Around the Galactic Globular Clusters NGC 1851 and NGC 1904

J.A. Carballo-Bello and D. Martínez-Delgado

Abstract Deep photometry around the Galactic globular clusters NGC 1851 and NGC 1904 obtained with the Wide Field Imager in the 2.2 m ESO telescope reveals a distinct main-sequence of a metal-poor stellar population, consistent with the presence of a very low surface brightness stellar system. The unveiled population might belong to an unknown tidal stream in the Milky Way but other possibilities are here discussed.

J.A. Carballo-Bello
Instituto de Astrofísica de Canarias (IAC), Spain
e-mail: jacb@iac.es

D. Martínez-Delgado
Instituto de Astrofísica de Canarias (IAC), C/ Vía Láctea s/n - 38205 La Laguna, Spain
e-mail: ddelgado@iac.es

J.M. Diego et al. (eds.), *Highlights of Spanish Astrophysics V*, Astrophysics and Space Science Proceedings, DOI 10.1007/978-3-642-11250-8_90,

Calibration Model for Gaia Photometry and Spectrophotometry

J.M. Carrasco, H. Voss, C. Jordi, C. Fabricius, and F. Figueras

Abstract From the astrometric measurements of unfiltered (white) light, Gaia will produce broad band G magnitudes, while the spectral energy distribution of each source will be sampled by a dedicated spectrophotometric instrument providing low resolution spectra in the blue (BP, 330–680 nm) and the red (RP, 650–1,050 nm). We present the data reduction scheme for this data. It is foreseen as an iterative process updating the mean spectra and the calibration parameters. Uncertainties from a prototype of the data reduction chain are also evaluated.

J.M. Carrasco, H. Voss, C. Jordi, C. Fabricius, and F. Figueras
University of Barcelona, ICC-IEEC, Martí i Franquès 1, E-08028 Barcelona, Spain
e-mail: carrasco@am.ub.es

J.M. Diego et al. (eds.), *Highlights of Spanish Astrophysics V*, Astrophysics and Space Science Proceedings, DOI 10.1007/978-3-642-11250-8_91,

Testing the Initial–Final Mass Relationship of White Dwarfs

S. Catalán, J. Isern, E. García–Berro, and I. Ribas

Abstract The initial–final mass relationship connects the mass of a white dwarf with the mass of its progenitor in the main-sequence. Although this function is of fundamental importance to several fields in modern astrophysics, it is not well constrained either from the theoretical or the observational points of view. In this contribution we revise the present semi-empirical initial–final mass relationship by re-evaluating the available data. The distribution obtained from grouping all our results presents a considerable dispersion, which is larger than the uncertainties. We have carried out a weighted least-squares linear fit of these data and a careful analysis to give some clues on the dependence of this relationship on metallicity. Finally, we have also performed a test of the initial–final mass relationship by studying its effect on the luminosity function and on the mass distribution of white dwarfs.

S. Catalán, J. Isern, E. García–Berro, and I. Ribas
Institut d'Estudis Espacials de Catalunya, c/ Gran Capità 2–4, 08034 Barcelona, Spain
e-mail: catalan@ieec.uab.es

S. Catalán, J. Isern, and I. Ribas
Institut de Ciències de l'Espai, CSIC, Facultat de Ciències, UAB, 08193 Bellaterra, Spain

E. García–Berro
Departament de Física Aplicada, Escola Politècnica Superior de Castelldefels, Universitat Politècnica de Catalunya, Avda. del Canal Olímpic s/n, 08860 Castelldefels, Spain

J.M. Diego et al. (eds.), *Highlights of Spanish Astrophysics V*, Astrophysics and Space Science Proceedings, DOI 10.1007/978-3-642-11250-8_92,

Unveiling New Quiescent Black Holes with IPHAS

J.M. Corral-Santana, J. Casares, and I.G. Martínez-Pais

Abstract Soft X-ray transients (SXTs) provide the best evidence for the existence of stellar-mass black holes (BHs) with almost 20 confirmed systems based on dynamical studies. However, there is an estimated population of a few thousand dormant BH binaries, which slowly reveal themselves through secular X-ray outbursts. Therefore, new strategies aimed at unveiling this dormant population are clearly needed. We propose to use the IPHAS catalogue, together with several diagnostic diagrams, as a shortcut to unveil the brightest members of the Galactic population of BH binaries. Here we present some of the diagrams we are using to distinguish between SXTs and other types of object found by the IPHAS survey.

J.M. Corral-Santana and J. Casares
Instituto de Astrofísica de Canarias (IAC), Spain
e-mail: jcorral@iac.es, jcv@iac.es

I.G. Martínez-Pais
Instituto de Astrofísica de Canarias (IAC), Vía Láctea, s/n - 38205 La Laguna, Spain
e-mail: igm@iac.es

J.M. Diego et al. (eds.), *Highlights of Spanish Astrophysics V*, Astrophysics and Space Science Proceedings, DOI 10.1007/978-3-642-11250-8_93,

Results of the Analysis of Several Galactic Sources Observed by MAGIC

M.T. Costado, C. Delgado, and R.J. García López

Abstract One of the tasks carried out by the Astroparticle Physics Group of the IAC is the analysis of objects observed by the MAGIC telescope looking for very high energy gamma-ray emission. The main goal of these observations is to search for acceleration of cosmic rays sites and the identification of accelerated species. We have analyzed MAGIC data of five observed objects, and the results are presented here. These objects are: three Supernova Remnants (IC 443, W44, W66), the globular cluster M 13 and the region of star formation M 42. High energy γ-ray emission has been found in one of these objects.

M.T. Costado and R.J.G. López (on behalf of the MAGIC Collaboration)
Instituto de Astrofísica de Canarias, E-38205 La Laguna, Tenerife, Spain
and
Departamento de Astrofísica de la Universidad de La Laguna, Universidad, E-38206 La Laguna, Tenerife, Spain
e-mail: mcostado@iac.es, rgl@iac.es

C. Delgado
Centro de Investigaciones Energéticas, Medioambientales y Tecnológicas, Madrid, Spain
e-mail: carlos.delgado@ciemat.es

J.M. Diego et al. (eds.), *Highlights of Spanish Astrophysics V*, Astrophysics and Space Science Proceedings, DOI 10.1007/978-3-642-11250-8_94,

Weak Flares on M-Dwarfs

I. Crespo-Chacón, J. López-Santiago, D. Montes, M.J. Fernández-Figueroa, G. Micela, F. Reale, D. García-Álvarez, M. Caramazza, and I. Pillitteri

Abstract We have investigated the physics of flares in M-dwarfs by means of optical/X-ray observations and modeling. The great efficiency of current optical spectrographs and detectors has allowed us to detect and analyze a great number of non white-light flares with intermediate spectral resolution and high temporal resolution. Although this kind of flares is the most typical on the Sun, few such events have been so far recorded on stars. We have obtained the physical parameters of the chromospheric flaring plasma (electron temperature, electron density, optical depth and temperature of the underlying source) by using a model that minimizes the difference between the observed Balmer decrements and the calculated ones, which result from solving the radiative transfer equation. On the other hand, the great sensitivity, wide energy range, high energy resolution and continuous time coverage of the EPIC detectors (on-board the XMM–Newton satellite) have enabled us to study both the effect of weak flares on the corona of active stars and the spatial properties of coronal flaring loops. The results are consistent with interpreting stellar flares as scaled-up versions of solar flares and show multiple evidence for flares being an important heating agent of the outer atmospheric stellar layers.

I. Crespo-Chacón, J. López-Santiago, D. Montes, and M.J. Fernández-Figueroa
Departamento de Astrofísica, Facultad de Ciencias Físicas, Universidad Complutense de Madrid, E-28040 Madrid, Spain
e-mail: icc@astrax.fis.ucm.es, jls@astrax.fis.ucm.es, dmg@astrax.fis.ucm.es, mjf@astrax.fis.ucm.es

G. Micela, F. Reale, and M. Caramazza
INAF - Osservatorio Astronomico di Palermo, Piazza del Parlamento 1, I-90134 Palermo, Italy
e-mail: giusi@astropa.unipa.it, reale@astropa.unipa.it, mcarama@astropa.unipa.it

F. Reale, M. Caramazza, and I. Pillitteri
Dip. di Scienze Fisiche e Astronomiche - Sez. di Astronomia - Università di Palermo, Piazza del Parlamento 1, I-90134 Palermo, Italy
e-mail: pilli@astropa.unipa.it

D. García-Álvarez
Imperial College London, Blackett Laboratory, Prince Consort Road, London SW7 2AZ, UK
and
Instituto de Astrofísica de Canarias and GTC Project Office, E-38205 La Laguna, Tenerife, Spain
e-mail: d.garcia-alvarez@imperial.ac.uk

J.M. Diego et al. (eds.), *Highlights of Spanish Astrophysics V*, Astrophysics and Space Science Proceedings, DOI 10.1007/978-3-642-11250-8_95,

Herbig–Haro Objects and Protoplanetary Discs

Luis Cuesta Crespo

Abstract Herbig–Haro objects are associated with T Tau stars, being the structures that were formed from their winds. Also, the discs that form in the planetary nebulae are very similar to the protoplanetary ones and the dynamics they generate is very similar, so the models used to explain them are also applicable here. Information on the characteristics of the disc as density, and its spatial variation, and temperature are obtained from these models. This study will serve to independently verify the results on protoplanetary discs.

L.C. Crespo
Centro de Astrobiología, INTA, Carretera de Ajalvir, km 4,
28850 Torrejón de Ardoz, Madrid, Spain
e-mail: cuestacl@inta.es

J.M. Diego et al. (eds.), *Highlights of Spanish Astrophysics V*, Astrophysics and Space Science Proceedings, DOI 10.1007/978-3-642-11250-8_96,

Diffusion of Cosmic-Rays and Gamma-Ray Sources

E. de Cea del Pozo, D.F. Torres, and A.Y. Rodríguez Marrero

Abstract It is commonly accepted that supernova remnants (SNR) are one of the most probable scenarios of leptonic and hadronic cosmic-ray (CR) acceleration. Such energetic CR can interact with interstellar gas to produce high-energy gamma rays, which can be detected through ground-based air Cherenkov detectors and space telescopes. Here we present a theoretical model that explains the high energy phenomenology of the neighborhood SNR IC 443, as observed with the Major Atmospheric Gamma Imaging Cherenkov (MAGIC) telescope and the Energetic Gamma-ray Experiment Telescope (EGRET). We interpret MAGIC J0616+225 as delayed TeV emission of CR diffusing from IC 443, what naturally explains the displacement between EGRET and MAGIC sources.

E. de Cea del Pozo, D.F. Torres, and A.Y.R. Marrero
Institut de Ciències de l'Espai (IEEC-CSIC), Campus UAB, Torre C5, 2a planta, 08193 Barcelona, Spain
e-mail: decea@ieec.uab.es, arodrig@ieec.uab.es, dtorres@ieec.uab.es

D.F. Torres
[2]Institució Catalana de Recerca i Estudis Avançats (ICREA), Spain

J.M. Diego et al. (eds.), *Highlights of Spanish Astrophysics V*, Astrophysics and Space Science Proceedings, DOI 10.1007/978-3-642-11250-8_97,

SWIFT J195509+261406: Dramatic Flaring Activity from a New Galactic Magnetar

A. de Ugarte Postigo, A.J. Castro-Tirado, J. Gorosabel, T.A. Fatkhullin, V.V. Sokolov, M. Jelínek, D. Sluse, P. Ferrero, D.A. Kann, S. Klose, M. Bremer, J.M. Winters, D. Nurenberger, D. Pérez-Ramírez, M.A. Guerrero, J. French, G. Melady, L. Hanlon, B. McBreen, F.J. Aceituno, R. Cunniffe, P. Kubánek, S. Vitek, S. Schulze, A.C. Wilson, R. Hudec, J.M. González-Pérez, T. Shahbaz, S. Guziy, L. Pavlenko, E. Sonbas, S. Trushki, N. Bursov, N.A. Nizhelskij, and L. Sabau-Graziati

A. de Ugarte Postigo and D. Nurenberger
European Southern Observatory (ESO), Casilla 19001, Santiago 19, Chile
e-mail: adeugart@eso.org

A.J. Castro-Tirado, J. Gorosabel, M. Jelínek, M.A. Guerrero, F.J. Aceituno, R. Cunniffe, P. Kubánek, and S. Vitek
Instituto de Astrofísica de Andalucía (IAA-CSIC), C. Bajo de Huétor 50, E-18080 Granada, Spain

T.A. Fatkhullin, V.V. Sokolov, E. Sonbas, S. Trushki, N. Bursov, and N.A. Nizhelskij
Special Astroph. Obs. (SAO-RAS), Nizhnij Arkhyz, Karachai-Cirkassian Rep., 369167 Russia

D. Sluse
Lab. d´Astroph., Ecole Pol. Fédérale de Lausanne (EPFL) Obs., 1290 Sauverny, Switzerland

P. Ferrero, D.A. Kann, S. Klose, and S. Schulze
Thüringer Landessternwarte Tautenburg, Sternwarte 5, D-07778 Tautenburg, Germany

M. Bremer and J.M. Winters
Inst. de Radioast. Milim. (IRAM), 300 rue de la Piscine, 38406 Saint Martin d´Héres, France

D. Pérez-Ramírez
Facultad de Ciencias Experimentales, U. de Jaén, Campus Las Lagunillas, E-23071 Jaén, Spain

J. French, G. Melady, L. Hanlon, and B. McBreen
University College, Belfield, Dublin 4, Ireland

A.C. Wilson
Department of Astronomy, University of Texas, Austin, TX 78712, USA

R. Hudec
Astronomical Institute of the Czech Academy of Sciences, Ondrejov. Czech Republic

J.M. González-Pérez and T. Shahbaz
Instituto de Astrofísica de Canarias (IAC), Via Láctea s/n, La Laguna, Tenerife, Spain

S. Guziy
Nikolaev State University, Nikolskaya 24, 54030 Nikolaev, Ukraine

L. Pavlenko
Crimean Astrophysical Observatory, Nauchny, Ukraine

L. Sabau-Graziati
Instituto Nacional de Técnica Aerospacial (INTA), Ctra. de Ajalvir km. 4, 28750 Torrejón de Ardoz (Madrid), Spain

J.M. Diego et al. (eds.), *Highlights of Spanish Astrophysics V*, Astrophysics and Space Science Proceedings, DOI 10.1007/978-3-642-11250-8_98,

Abstract Most of the transient sources that are detected in the gamma-ray sky are produced by extragalactic gamma-ray bursts (GRBs). However, it is known that there are some other astronomical objects that can produce high-energy bursts within the Milky Way. SWIFT J195509+261406, just one degree off the Galactic plane, is one of them. It was discovered on the 10th July 2007 by the *Swift* satellite and was since then observable for a period of a fortnight. During this time SWIFT J195509+261406 experimented dramatic flaring activity that could be observed in near infrared, optical and X-rays. We gathered multi-wavelength observations of SWIFT J195509+261406 including optical, near infrared, millimeter and radio observations. Our dataset covers the time from 1 min after the burst onset to more than 4 months later. Following the initial burst in the gamma-ray band, we recorded more than 40 flaring episodes in the optical bands (reaching up to $I_c \sim 15$) over a time span of 3 days, plus a faint infrared flare that was observed at late times. After this time, the source slowly faded away until it became undetectable. Using the observations compiled in this work we propose that this source is part of the magnetar family, linking soft gamma-ray repeaters and anomalous X-ray pulsars to dim isolated neutron stars.

Pulsating B and Be Stars in the Magellanic Clouds

P.D. Diago, J. Gutiérrez-Soto, J. Fabregat, C. Martayan, and J. Suso

Abstract Stellar pulsations in main-sequence B-type stars are driven by the κ-mechanism due to the Fe-group opacity bump. The current models do not predict the presence of instability strips in the B spectral domain at very low metallicities. As the metallicity of the Magellanic Clouds (MC) has been measured to be around $Z = 0.002$ for the Small Magellanic Cloud (SMC) and $Z = 0.007$ for the Large Magellanic Cloud (LMC), they constitute a very suitable objects to test these predictions. The aim of this work is to investigate the existence of B-type pulsators at low metallicities, searching for short-term periodic variability in a large sample of B and Be stars from the MC with accurately determined fundamental astrophysical parameters.

P.D. Diago, J. Gutiérrez-Soto, J. Fabregat, and J. Suso
Observatori Astronòmic de la Universitat de València, Ed. Instituts d'Investigació, Polígon La Coma, 46980 Paterna, València, Spain
e-mail: Pascual.Diago@uv.es

J. Gutiérrez-Soto, J. Fabregat, and C. Martayan
GEPI, Observatoire de Paris, CNRS, Université Paris Diderot, Place Jules Janssen 92195 Meudon Cedex, France

C. Martayan
Royal Observatory of Belgium, 3 Avenue Circulaire, B-1180 Brussels, Belgium

J.M. Diego et al. (eds.), *Highlights of Spanish Astrophysics V*, Astrophysics and Space Science Proceedings, DOI 10.1007/978-3-642-11250-8_99,

Constraints to the Proposed Close-in Perturber to GJ 436 b

A. Font-Ribera, I. Ribas, J.P. Beaulieu, J.C. Morales, and E. García-Melendo

Abstract The planet-hosting star GJ 436 has been extensively observed over the last months, including both spectroscopic and photometric data. Here we review the last studies and comment the validity of the scenario of a close-in perturber in GJ 436, concluding that it is not only plausible, but, in our opinion, is also the most likely to explain the observations.

A. Font-Ribera
Institut de Ciències de l'Espai (IEEC-CSIC), Campus UAB, Bellaterra 08193, Spain
e-mail: font@ieec.uab.es

I. Ribas and J.C. Morales
Institut de Ciències de l'Espai (IEEC-CSIC), Spain
e-mail: iribas@ieec.uab.es, jcmorales@am.ub.es

J.P. Beaulieu
Institut d'Astrophysique de Paris (IAP), CNRS (UMR 7095), Paris, France
e-mail: beaulieu@iap.fr

E. García-Melendo
Esteve Duran Observatory Foundation, Montseny 46, 08553 Seva, Spain
e-mail: duranobs@astrogea.org

J.M. Diego et al. (eds.), *Highlights of Spanish Astrophysics V*, Astrophysics and Space Science Proceedings, DOI 10.1007/978-3-642-11250-8_100,

FR Cnc Nature Revisited

M.C. Gálvez, A. Golovin, M. Hernán-Obispo, E. Pavlenko, M. Andreev, D. Montes, J.C. Pandey, A. Sergeev, Yu. Kuznyetsova, and V. Krushevska

Abstract The results of photometric and spectroscopic monitoring of FR Cnc reported a tricky nature. We carried out several observations at different observatories in India, Russia, Ukraine and Spain during several years to characterize and discover the source of its radial velocity (RV) variations. After discarding a binary nature in first instance due to its high level of activity, further detailed and complete study lead as to still take into account the presence of a stellar companion possibility. We present here the study of this star and preliminary conclusions about its real nature.

M.C. Gálvez
Centre for Astrophysic Research, University of Hertfordshire, AL109AB, U.K
and
Universidad Complutense de Madrid, E-28040 Madrid, SPAIN
e-mail: M.Galvez-Ortiz@herts.ac.uk

A. Golovin, Yu. Kuznyetsova, and V. Krushevska
Main Astronomical Observatory of National Academy of Science of Ukraine, Kyiv, UKRAINE

M. Hernán-Obispo and D. Montes
Departmento de Astrofísica, Universidad Complutense de Madrid, E-28040 Madrid, SPAIN

E. Pavlenko
Crimean Astrophysical Observatory, Nauchny, UKRAINE

M. Andreev and A. Sergeev
Terskol Branch of Institute of Astronomy RAS, RUSSIA

J.C. Pandey
The Indian Astronomical Observatory, Mt. Saraswati, Hanle, INDIA

J.M. Diego et al. (eds.), *Highlights of Spanish Astrophysics V*, Astrophysics and Space Science Proceedings, DOI 10.1007/978-3-642-11250-8_101,

Massive Stars with Weak Winds

Miriam García and Artemio Herrero

Abstract The theory of radiatively-driven winds successfully explains the key points of the stellar winds of hot massive stars. However, it has been recently found that some O- and B-type stars do not exhibit as strong wind as predicted by theory. In this work we explain the main issues concerning this topic. We also describe our analysis of the UV spectra of a set of massive stars in a Galactic star-forming region (the Orion Complex) plus two reference stars, to determine their wind properties. The wind momentum of the very young OB-stars located in the region and also one of the reference stars are smaller than expected. Our results hint that stellar weak winds may be related to luminosity, rather than age or magnetic fields.

M. García and A. Herrero
Instituto de Astrofísica de Canarias, Vía Láctea S/N, E-38200 La Laguna (Tenerife), Spain
e-mail: mgg@iac.es, ahd@iac.es

J.M. Diego et al. (eds.), *Highlights of Spanish Astrophysics V*, Astrophysics and Space Science Proceedings, DOI 10.1007/978-3-642-11250-8_102,

Numerical Modeling of Type Ia Supernovae Explosions

D. García-Senz and E. Bravo

Abstract A better knowledge of the mechanism behind the explosion of Type Ia supernovae (SNIa) is necessary to use these events in cosmological applications such as to study the large scale geometry of the universe or to find its equation of state. We review the present status of the subject with special emphasis in the so-called pulsating models which reproduce the gross features of the explosions without using free parameters.

D. García-Senz and E. Bravo
Departament de Fisica i Enginyeria Nuclear (UPC). Barcelona, Spain
e-mail: domingo.garcia@upc.edu, eduardo.bravo@upc.edu

J.M. Diego et al. (eds.), *Highlights of Spanish Astrophysics V*, Astrophysics and Space Science Proceedings, DOI 10.1007/978-3-642-11250-8_103,

GALEP, Spectral Mapping of the Inner Galaxy with EMIR

F. Garzón, P.L. Hammersley, C. González, and A. Cabrera

Abstract We propose to use EMIR to obtain near IR spectroscopy of many thousands Galaxy sources, mainly located in the inner regions. These will be selected from their position on IR colour–magnitude (CM) diagrams and will include disc, bar bulge and ring sources. The principal aim is to accurately classify the sources to provide a better understanding of, particularly, the redder parts of a infrared CM diagram. Without this information there will remain ambiguities in the interpretation of structures in the inner Galaxy.

F. Garzón
Instituto de Astrofísica de Canarias (IAC), 38200-La Laguna, S.C. Tenerife, Spain
e-mail: fgl@iac.es

P.L. Hammersley, C. González, and A. Cabrera
IAC, Spain
e-mail: plh@iac.es, cgonzal@iac.es, acabrera@iac.es

J.M. Diego et al. (eds.), *Highlights of Spanish Astrophysics V*, Astrophysics and Space Science Proceedings, DOI 10.1007/978-3-642-11250-8_104,

And the Oscar Goes to: BD+20 1790 for "The Mystery of the Unseen Companion"

M. Hernán-Obispo, M.C. Gálvez, G. Anglada-Escudé, S.R. Kane, E. de Castro, and M.Cornide

Abstract BD+20 1790 is a young, rich metal and very active late–type K5Ve star. Our group has been developing a study of stellar activity and kinematics for this star over the past years. Previous results show a high level of stellar activity, with the presence of prominence-like structures, spots on surface and strong flare events. Radial velocity (RV) variations with a semi-amplitude of up to 1 km s^{-1} were detected. When the nature of these variations was investigated it was found that they are not due to stellar activity. Based upon the analysis of bisector velocity span, as well as Ca II H & K emission, we report that the best explanation for RV variations is the presence of a sub–stellar companion. The Keplerian fit of the RV data yields an orbital solution for a close-in massive planet with an orbital period of 7.783 days. Also, the presence of this close–in massive planet could be an interpretation for the high level of stellar activity detected.

M. Hernán-Obispo, M.C. Gálvez, E. de Castro, and M. Cornide
Dpto. de Astrofísica y CC. de la Atmósfera, Facultad de Física, Universidad Complutense de Madrid, Avda. Complutense s/n, E-28040, Madrid, Spain
e-mail: mho@astrax.fis.ucm.es

M.C. Gálvez
Centre for Astrophysics Research, Science & Technology Research Institute, University of Hertfordshire, College Lane, Hatfield, Hertfordshire AL10 9AB, UK

G. Anglada-Escudé
Departament d ′Astronomia i Meteorologia. Universitat de Barcelona, Martí i Franqués 1, Barcelona, 08028, Spain
and
Department of Terrestrial Magnetism, Carnegie Institution of Washington, 5241 Broad Branch Road, NW, Washington, DC 20015-1305, USA

S.R. Kane
NASA Exoplanet Science Institute, Caltech, MS 100-22, 770 South Wilson Avenue, Pasadena, CA 91125, USA

J.M. Diego et al. (eds.), *Highlights of Spanish Astrophysics V*, Astrophysics and Space Science Proceedings, DOI 10.1007/978-3-642-11250-8_105,

GUMS & GOG: Simulating the Universe for Gaia

Y. Isasi, F. Figueras, X. Luri, and A.C. Robin

Abstract A key component of the Gaia mission development is its simulator. This system can generate a realistic simulation of the data stream that will be produced by Gaia, a critical milestone to verify the data reduction programs. A core piece of the simulator is its universe model, a module that can generate synthetic catalogues of objects to obtain simulated observations. GUMS is the application that can generate the object catalogue up to a particular magnitude limit. As a example of application that uses the model, we describe GOG, a simulator of Gaia catalogue intermediate and final data. It combines the source observation of the universe model with the satellite instrumental specifications and the available error models.

Y. Isasi, F. Figueras, and X. Luri
Departament d'Astronomia i Meteorologia and IEEC-UB, Institut de Ciències del Cosmos de la Universitat de Barcelona, Martí i Franquès, 1, E-08028 Barcelona, Spain
e-mail: isasi@am.ub.es

A.C. Robin
Institut UTINAM, CNRS-UMR6213, Observatoire de Besanon, France
e-mail: annie@obs-besancon.fr

J.M. Diego et al. (eds.), *Highlights of Spanish Astrophysics V*, Astrophysics and Space Science Proceedings, DOI 10.1007/978-3-642-11250-8_106,

HD 64315: A Very Massive Spectroscopic Binary

Javier Lorenzo, Ignacio Negueruela, and Sergio Simón

Abstract We present the results of a spectroscopic campaign on the early-type double-lined binary HD 64315 conducted between October 2006 and February 2007. Alter the reduction of 100 echelle spectra, we fit model atmospheres to the two components and find $T_{\rm eff1} \sim 40{,}000$ K and $T_{\rm eff2} \sim 38{,}000$ K. We calculate the radial velocity curve for both components and we obtain a period of 2.71 ± 0.01 days and mass ratio $q = 1.05$. We derive minimum masses for both components of the binary system $7.5 \pm 0.2\,M_{\odot}$ and $7.1 \pm 0.2\,M_{\odot}$. The projected semimajor axes are $9.75 \pm 0.05\,R_{\odot}$ and $10.23 \pm 0.05\,R_{\odot}$. We also compare these results with those obtained in a previous study (Solivella & Niemela 1986).

J. Lorenzo and I. Negueruela
Alicante University, P.O. Box 99, E-03080, Spain
e-mail: javihd64315@gmail.com, ignacio@dfists.ua.es

S. Simón
Observatoire Astronomique, Université de Genève, Switzerland
e-mail: Sergio.Simon-Diaz@obs.unige.ch

J.M. Diego et al. (eds.), *Highlights of Spanish Astrophysics V*, Astrophysics and Space Science Proceedings, DOI 10.1007/978-3-642-11250-8_107,

Spectroscopic Studies of Nearby Cool Stars: The DUNES Sample

J. Maldonado, R.M. Martinez-Arnáiz, C. Eiroa, and D. Montes

Abstract At the universities of Madrid we are carrying out a systematic analysis of the spectroscopic properties of the nearby ($d < 25$ pc), late-type stellar population with the aim of contributing to the knowledge of the stellar formation history in the solar neighbourhood. Part of our sample will be observed by DUNES, a Herschel OTKP aiming at detecting and studying cold, faint dust disks around nearby stars. In this contribution we present some preliminary results on the kinematics of the DUNES sample.

J. Maldonado and C. Eiroa
Universidad Autónoma de Madrid, Dpto. Física Teórica C-XI, Facultad de Ciencias, Madrid, Spain
e-mail: jesus.maldonado@uam.es

R.M. Martínez-Arnáiz and D. Montes
Universidad Complutense de Madrid, Dpto. Astrofísica, Facultad de Físicas, Madrid, Spain

J.M. Diego et al. (eds.), *Highlights of Spanish Astrophysics V*, Astrophysics and Space Science Proceedings, DOI 10.1007/978-3-642-11250-8_108,

Simultaneous Modelling of the Complete SN1993J Expansion and Radio Light Curves

I. Martí-Vidal, J.M. Marcaide, and A. Alberdi

Abstract We report on our modelling of all the available VLBI data and radio light curves of supernova SN1993J. We have used the most complete expansion curve of a supernova ever, which spans more than a decade at several frequencies. For the data modelling, we have developed a new software capable of simulating the evolution of the radio emission of a supernova. We find that for explaining both the radio light curves and the expansion curve simultaneously, a radial structure of the magnetic field inside the radiating region and opacity effects from the ejected material have to be considered, together with a constant pre-supernova mass-loss rate (contrary to some results found by other authors).

I. Martí-Vidal and J.M. Marcaide
Univ. Valencia, Dr. Moliner 50, 46100 Burjassot (Valencia), Spain
e-mail: I.Marti-Vidal@uv.es, J.M.Marcaide@v.es

A. Alberdi
IAA (CSIC), Camino bajo de Huétor 50, 18008 Granada, Spain
e-mail: Antxon@iaa.es

J.M. Diego et al. (eds.), *Highlights of Spanish Astrophysics V*, Astrophysics and Space Science Proceedings, DOI 10.1007/978-3-642-11250-8_109,

High Resolution Spectroscopic Characterization of the FGK Stars in the Solar Neighbourhood

R. Martínez-Arnáiz, J. Maldonado, D. Montes, C. Eiroa, B. Montesinos, I. Ribas, and E. Solano

Abstract We present the most recent results of our ongoing long-term high resolution spectroscopic study of nearby ($d \leq 25$ pc) FGK stars which aim is to characterize the local properties of the Galaxy, in particular, the star-formation history. A thorough analysis has been carried out for 253 cool stars in the solar neighborhood. This includes radial and rotational velocities determinations, chromospheric activity levels inference, kinematic analysis, and age estimates. This study does not only shed new light on the issue of stellar formation history but also contributes to any present or future mission aiming to detect extra-solar planets. Exo-planets are likely to be found orbiting around nearby cool stars and their detection and characterization is highly dependent on the precise determination of fundamental stellar parameters such as age, and activity levels. Therefore, our study is of paramount importance to ensure the success of any such mission.

R. Martínez-Arnáiz and D. Montes
Universidad Complutense de Madrid, Spain

J. Maldonado and C. Eiroa
Universidad Autónoma de Madrid, Spain

B. Montesinos and E. Solano
Laboratorio de Astrofísica Espacial y Física Fundamental LAEFF, INTA, Spain

I. Ribas
Institut de Ciénces de l'Espai, CSIC, Spain

J.M. Diego et al. (eds.), *Highlights of Spanish Astrophysics V*, Astrophysics and Space Science Proceedings, DOI 10.1007/978-3-642-11250-8_110,

Optical Spectroscopic Monitoring of Pre-Main Sequence Stars: The UXOr Sub-Sample

Ignacio Mendigutía, Benjamín Montesinos, Carlos Eiroa, and Alcione Mora

Abstract The EXPORT consortium carried out in 1998–1999 photo-polarimetric and spectroscopic monitoring over 74 Pre-Main Sequence (PMS) and Main Sequence (MS) stars with infrared excesses, 43 of them showing the Hα line in emission. We have measured the equivalent widths (EWs) of the Hα, [OI]6300, HeI5876 and NaID lines in 370 intermediate resolution optical spectra, obtained along four different observing campaigns, of those 43 objects. In this contribution we focus on the simultaneous spectroscopic and photometric behaviour of 12 UX-Ori type variable stars included in the sample. The main results indicate that the $EW(\mathrm{H}\alpha) - V$ correlation is a main feature of UXOrs. This correlation is also found in the [OI]6300 line in most of the objects. The variability in EW(HeI5876)–but not in the flux (F) in several cases–does not depend on the V-behaviour, as well as the complex patterns displayed in the NaI D lines. We interpret these results according to the dusty obscuration effect generated by the disks surrounding these stars, and propose that different accretion mechanisms could operate in different objects, depending on their $F(\mathrm{HeI}5876) - V$ behaviour.

I. Mendigutía
Laboratorio de Astrofísica Espacial y Física Fundamental (CAB/LAEFF/INTA), European Space Astronomy Centre, PO Box 78, E-28691 Villanueva de la Cañada, Madrid, Spain
e-mail: Ignacio.Mendigutia@laeff.inta.es

B. Montesinos
Laboratorio de Astrofísica Espacial y Física Fundamental (CAB/LAEFF/INTA), Spain

C. Eiroa
Universidad Autónoma de Madrid (UAM)

A. Mora
Universidad Autónoma de Madrid (UAM), Spain

J.M. Diego et al. (eds.), *Highlights of Spanish Astrophysics V*, Astrophysics and Space Science Proceedings, DOI 10.1007/978-3-642-11250-8_111,

Identification of Isolated Post T Tauri Stars in the Solar Neighborhood

D. Montes, J. López-Santiago, and R.M. Martínez-Arnáiz

Abstract Post T Tauri stars (PTTS) are late-type stars in the age range between 10 and 100 Myr filling the gap between T Tauri (TTS) and zero-age main sequence phases. This period of evolution remains ambiguous and until now different studies of young stars have failed to find the numbers of PTTS that are expected. In the last years, some PTTS have been identified among the X-ray detected pre-main sequence stars in some star-forming regions. More recently, additional PTTS have been identified in young associations and moving groups (β Pic, TW Hya, Tucana/Horologium, and the AB Dor). However, many isolated PTTS still remain undiscovered. In this contribution, we compile the PTTS previously identified in the literature, and identify new candidates using the information provided by the high resolution spectra obtained during our surveys of late-type stars, possible members to young moving groups, FGK stars in the solar neighborhood. To identify PTTS we applied an age-oriented definition using relative age indicators (Li abundance, chromospheric and coronal emission and the kinematics) as well as color–magnitude diagrams and pre-main sequence isochrones.

D. Montes, J. López-Santiago, and R.M. Martínez-Arnáiz
Universidad Complutense de Madrid, Dpt. de Astrofísica, Facultad de C.C. Físicas, 28040, Madrid, Spain
e-mail: dmg@astrax.fis.ucm.es

J.M. Diego et al. (eds.), *Highlights of Spanish Astrophysics V*, Astrophysics and Space Science Proceedings, DOI 10.1007/978-3-642-11250-8_112,

Interstellar Reddening Determination Trough the 2200 Å Dust Absorption Band

Carmen Morales, Angelo Cassatella, and Gisela Bañó

Abstract A comparison is carried out between two methods to evaluate the correction for interstellar reddening: the three ultraviolet points method, and the traditional model fitting method. The two methods have been applied to a large sample of well known stars of spectral types O, B and A to test their reliability and to asses their general applicability.

C.M. Durán
LAEFF-INTA, ESAC, E-28691 Villanueva de la Cañada, Madrid, Spain

A. Cassatella
INAF-IFSI, Via del Fosso del Cavaliere 100, 00133 Roma, Italy

G.B. Esplugues
LAEFF-INTA, ESAC, E-28691 Villanueva de la Cañada, Madrid, Spain

J.M. Diego et al. (eds.), *Highlights of Spanish Astrophysics V*, Astrophysics and Space Science Proceedings, DOI 10.1007/978-3-642-11250-8_113,

Low-Mass Stars as Tests for Stellar Models

Juan Carlos Morales, José Gallardo, Ignasi Ribas, Carme Jordi, Isabelle Baraffe, and Gilles Chabrier

Abstract Low-mass eclipsing binaries have unveiled that stellar models do not reproduce the radii and effective temperatures of its components while luminosities are correctly predicted. The magnetic activity due to the rapid rotation of these stars has been proposed as the cause of these differences between models and observations and corrections to the models have already been suggested. One of the theoretical scenarios is considering that stars could be more spotted than it is deduced from light curves. Therefore, tests to the amount of spot coverage needed in the models to reproduce observations are implemented on well-known eclipsing binary systems and consistency checks between these corrections and observations are presented.

J.C. Morales
Institut d'Estudis Espacials de Catalunya (IEEC), Edif. Nexus, C/ Gran Capità 2-4, 08034 Barcelona, Spain
e-mail: morales@ieec.uab.es

J. Gallardo
Departamento de Astronomía, Universidad de Chile, Casilla 36-D, Santiago, Chile

I. Ribas
Institut de Ciències de l'Espai (CSIC), Campus UAB, Facultat de Ciències, Torre C-5 - parell - 2a planta, 08193 Bellaterra, Spain

C. Jordi
Dept. d'Astronomia i Meteorologia, Institut de Ciències del Cosmos, Universitat de Barcelona, C/ Martí i Franquès 1, 08028 Barcelona, Spain

I. Baraffe and G. Chabrier
École Normale Supérieure de Lyon, CRAL (UMR CNRS 5574), Université de Lyon, France

J.M. Diego et al. (eds.), *Highlights of Spanish Astrophysics V*, Astrophysics and Space Science Proceedings, DOI 10.1007/978-3-642-11250-8_114,

A Kinematical Study of the Galactic Disk Using Red Clump Stars

Salvador J. Ribas, Francesca Figueras, and Jordi Torra

Abstract We determine kinematical properties of the Galactic disk using a large sample of Red Clump stars extracted from the UCAC2 catalogue using photometric data from 2MASS catalogue. The compiled large sample, more than 2 million stars with accurate enough distances and proper motions, allow us to trace the disk kinematics up to 2 Kpc from the Sun. First and second order galactic rotation parameters are determined and the Dehnen and Binney (1998) method is used to characterize the velocity distribution function in a large area of the galactic plane. Extensive simulations have permitted to evaluate the capabilities of the method as well as the selection and completeness effects.

S.J. Ribas
Consorci del Montsec, Pl. Major 1, 25691 Ager (Lleida) and Departament d'Astronomia i Meteorologia, Universitat de Barcelona, Mart i Franques 1, 08028 Barcelona, Spain
e-mail: sribas@am.ub.es

F. Figueras and J. Torra
Departament d'Astronomia i Meteorologia & IEEC-UB, Institut de Ciencies del Cosmos de la Universitat de Barcelona, Marti i Franques 1, 08028 Barcelona, Spain
e-mail: cesca@am.ub.es, jordi@am.ub.es

J.M. Diego et al. (eds.), *Highlights of Spanish Astrophysics V*, Astrophysics and Space Science Proceedings, DOI 10.1007/978-3-642-11250-8_115,

High Energy Sources Monitored with OMC

D. Risquez, A. Domingo, M.D. Caballero-García, J. Alfonso-Garzón, and J.M. Mas-Hesse

Abstract The Optical Monitoring Camera on-board INTEGRAL (OMC) provides Johnson V band photometry of any potentially variable source within its field of view. Taking advantage of the INTEGRAL capabilities allowing the simultaneous observation of different kind of objects in the optical, X and gamma rays bands, we have performed a study of the optical counterparts of different high-energy sources. Up to now, OMC has detected the optical counterpart for more than 100 sources from the High Energy Catalog (Ebisawa et al., 2003, A&A 411, L59). The photometrically calibrated light curves produced by OMC can be accessed through our web portal at http://sdc.laeff.inta.es/omc.

D. Risquez, A. Domingo, M.D. Caballero-García, J. Alfonso-Garzón, and J.M. Mas-Hesse
Centro de Astrobiología – LAEFF (CSIC–INTA), Apartado 78, E-28691 Villanueva de la Cañada, Madrid, Spain
e-mail: omc-support@laeff.inta.es

J.M. Diego et al. (eds.), *Highlights of Spanish Astrophysics V*, Astrophysics and Space Science Proceedings, DOI 10.1007/978-3-642-11250-8_116,

Photoelectric Absorption in the Stellar Wind of the Binary System 4U 1538–52/QV Nor

J.J. Rodes, J.M. Torrejón, and G. Bernabéu

Abstract We have used the Rossi X-ray Timing Explorer (RXTE) observatory to analyze the persistent X-ray source 4U 1538–52/QV Nor. The RXTE satellite observed close to one complete binary orbit on January 1997 and 2001 of this X-ray binary system. We have obtained orbital phase resolved spectra to investigate changes in the absorption column density with orbital phase. The variation of the column density over the binary orbit is caused by the movement of the X-ray source through the stellar wind of the companion star. A simple model of absorption in a stellar wind from the companion star described the orbital phase dependence reasonably. From our analysis, we have inferred a wind mass-loss rate from the companion star of $(1.3 - 2.5) \times 10^{-6}\,M_{\odot}/yr$. Such rate is in agreement with those obtained by the Ginga observations of this X-ray binary pulsar.

J.J. Rodes
Universitat d'Alacant, Apartat de correus 99, E03080 Alacant, Spain
e-mail: rodes@dfists.ua.es

J.M. Torrejón and G. Bernabéu
Universitat d'Alacant, Spain
e-mail: jmt@dfists.ua.es, bernabeu@ua.es

J.M. Diego et al. (eds.), *Highlights of Spanish Astrophysics V*, Astrophysics and Space Science Proceedings, DOI 10.1007/978-3-642-11250-8_117,

Distribution for the Regular Component of the Galactic Magnetic Field Using 5-Year WMAP Data

B. Ruiz-Granados, J.A. Rubiño-Martín, and E. Battaner

Abstract We have studied the spatial structure of the three dimensional large-scale pattern of the Galactic Magnetic Field using the polarization maps obtained by the WMAP satellite at 22 GHz. By using different models of the large-scale magnetic field of the Milky Way and a certain model for the cosmic rays electron distribution, we predict the expected polarized synchrotron emission. Using a Maximum Likelihood method, we explore the parameter values which best reproduce the WMAP data and obtain the marginal probability distribution for each of them. The model that better reproduces the observed map of polarization is an axisymmetric model with radial dependence of the strength of the magnetic field.

B. Ruiz-Granados and E. Battaner
Dpto. Física Teórica y del Cosmos. Edif. Mecenas, planta baja, Campus Fuentenueva, E-18071. Universidad de Granada, Granada, Spain
and
Instituto de Física Teórica y Computacional Carlos I, Granada, Spain
e-mail: bearg@ugr.es, battaner@ugr.es

J.A. Rubiño-Martín
Instituto de Astrofísica de Canarias (IAC), C/Vía Láctea, s/n, E-38200, La Laguna, Tenerife, Spain
e-mail: jalberto@iac.es

J.M. Diego et al. (eds.), *Highlights of Spanish Astrophysics V*, Astrophysics and Space Science Proceedings, DOI 10.1007/978-3-642-11250-8_118,

The Origin of the Galactic ^{26}Al and ^{60}Fe

Moisès Suades, Margarita Hernanz, and Nicolas de Séréville

Abstract The radioactive nucleus ^{26}Al (1 Myr lifetime) was the first cosmic radioactivity ever detected, through its gamma ray emission line at 1.809 MeV, with the HEAO-3 satellite in the 80's. More recently, COMPEL instrument onboard CGRO made the first all-sky map of its diffuse emission in the Galaxy, which revealed that 1.8 MeV photons trace the massive star population, but with room to other potential important producers like AGB stars and novae. The SPI instrument of the current ESA mission INTEGRAL has corroborated the detection of the ^{26}Al line with excellent spectroscopic resolution, and has also detected the two lines at 1.173 and 1.333 MeV of ^{60}Fe (2 Myr lifetime), yielding an observed ^{60}Fe/^{26}Al gamma ray flux ratio which can not be reproduced with current theoretical determinations based solely on massive stars. We will discuss the contribution of the different stellar scenarios to the global ^{26}Al and ^{60}Fe content of the Milky Way and give an interpretation of the recent INTEGRAL observations.

M. Suades and M. Hernanz
Institut de Ciències de l'Espai (CSIC-IEEC), Campus UAB, Facultat Ciències, Torre C5 par., 2a planta, 08193 Bellaterra, Spain
e-mail: suades@ieec.uab.es, hernanz@ieec.uab.es

N. de Séréville
Institut de Physique Nuclaire (IPN), Orsay, France
e-mail: deserevi@ipno.in2p3.fr

J.M. Diego et al. (eds.), *Highlights of Spanish Astrophysics V*, Astrophysics and Space Science Proceedings, DOI 10.1007/978-3-642-11250-8_119,

CdC-SF Catalogue.II: Application of its Proper Motions to Open Clusters

B. Vicente and F. Garzón

Abstract We present an astrometric catalogue of positions and proper motions derived from the *Carte du Ciel* plates of the San Fernando zone, photographic material with a mean epoch 1901.4 with a limiting magnitude $V \sim 15$. Digitization has been made using a conventional flatbed scanner. Special techniques have been developed to handle the combination of plate material and the large distortion introduced by the scanner. A variety of post-scan corrections are shown to be necessary. The equatorial coordinates are on the ICRS system defined by Tycho-2. Comparison with the reference catalog indicates external errors of $0\farcs2$. The UCAC2 Catalogue was used as second-epoch positions to derive proper motions with a mean accuracy of 1.2 mas/year for the proper motions for well-measure stars. The usefulness of the resulting catalogue of proper motions is demonstrated by means of a proper-motion analysis of seven open clusters ASCC 30, BOCHUM 3, NGC 2215, NGC 2302, NGC 2311, NGC 2323 and NGC 2548, determining individual membership probabilities and characterizing the gross properties of each cluster.

B. Vicente
Instituto de Astrofísica de Canarias (IAC), 38200 La Laguna, Tenerife, Spain
e-mail: bvicente@iac.es

F. Garzón
Instituto de Astrofísica de Canarias (IAC), 38200 La Laguna, Tenerife, Spain
and
Departamento de Astrofísica, Universidad de La Laguna, Tenerife, Spain
e-mail: fgl@iac.es

J.M. Diego et al. (eds.), *Highlights of Spanish Astrophysics V*, Astrophysics and Space Science Proceedings, DOI 10.1007/978-3-642-11250-8_120,

The Sun and the Solar System

Damping of Fast Magnetohydrodynamic Oscillations in Quiescent Filament Threads

I. Arregui, J. Terradas, R. Oliver, and J.L. Ballester

Abstract High-resolution observations provide evidence of the existence of small-amplitude transverse oscillations in solar filament fine structures. These oscillations are believed to represent fast magnetohydrodynamic (MHD) waves and the disturbances are seen to be damped in short timescales of the order of 1–4 periods. We propose that, due to the highly inhomogeneous nature of the filament plasma at the fine-structure spatial scale, the phenomenon of resonant absorption is likely to operate in the temporal attenuation of fast MHD oscillations. By considering transverse inhomogeneity in a straight flux tube model we find that, for density inhomogeneities typical of filament threads, the decay times are of a few oscillatory periods only.

I. Arregui, R. Oliver, and J.L. Ballester
Departament de Física, Universitat de les Illes Balears, E-07122 Palma de Mallorca, Spain
e-mail: inigo.arregui@uib.es

J. Terradas
Centre for Plasma Astrophysics, K.U. Leuven, B-3001 Heverlee, Belgium

J.M. Diego et al. (eds.), *Highlights of Spanish Astrophysics V*, Astrophysics and Space Science Proceedings, DOI 10.1007/978-3-642-11250-8_121,

Dynamics and Clouds in Jupiter Equatorial Zone

J. Arregi, J.F. Rojas, R. Hueso, and A. Sánchez-Lavega

Abstract In this work we show a study of the dynamics and clouds in the equatorial zone of Jupiter. The studied area is wider than the pure Equatorial Zone ranging from the southern limit of the South Equatorial Belt (SEB) to the northern limit of the North Equatorial Belt (NEB). We have used images from the Cassini flyby in December 2000 (wavelengths of 752 and 939 nm) and from the Galileo orbiter taken in 1999 and 2001 (wavelengths of 559 and 756 nm). When needed we have used images from the International Outer Planet Watch database to complete the time coverage of the dataset. In visible wavelengths the study of the dark-bluish regions in the northern limit of the NEB that corresponds to the infrared *hot spots* show that they have the characteristics of a Rossby wave and can be explained as some Rossby wave induced effect. Nevertheless trying to explain the smaller and more abundant dark marks situated on the southern limit of the SEB in the same way has proven to be much more difficult. We will also describe our measurements of an anticyclonic white oval situated near the SEB dark marks. Finally we will present three train gravity waves that we have found in Galileo maps near the Equator.

J. Arregi, J.F. Rojas, R. Hueso, and A. Sánchez-Lavega
Grupo de Ciencias Planetarias, Dpto Fisica Aplicada I, Escuela Universitaria Ingenieria, Universidad del Pais Vasco/Euskal Herriko Unibertsitatea, Vitoria-Gasteiz, Bilbao, Spain
e-mail: jesus.arregi@ehu.es, josef.rojas@ehu.es, ricardo.hueso@ehu.es, agustin.sanchez@ehu.es

J.M. Diego et al. (eds.), *Highlights of Spanish Astrophysics V*, Astrophysics and Space Science Proceedings, DOI 10.1007/978-3-642-11250-8_122,

Turbulence in Jupiter's Clouds

N. Barrado-Izagirre, S. Pérez-Hoyos, and A. Sánchez-Lavega

Abstract We have studied the spatial distribution of Jupiter's higher clouds in order to characterize the turbulent regime and the presence of waves in this planet's atmosphere. We have used images from the Hubble Space Telescope (HST) 1995's archive and from Cassini's ISS camera during its Jupiter fly-by in its way to Saturn in 2000 in three wavelengths: near infrared (∼940 nm), blue (∼430 nm) and ultraviolet (∼260 nm). These images were cylindrically projected and composed to obtain complete planispheres of Jupiter that cover the latitudinal range from 60°N to 60°S. When applying the Fast Fourier Transform (FFT) to each latitude reflectivity scan, we obtain brightness power spectra and periodograms that show the presence of wavy phenomena. From the spectra we study the decay of the slopes and their possible correlation with the underlying turbulent regime. We compare the turbulent structure of Jupiter's clouds with the dynamic structure characterized by the latitudinally alternating zonal wind regime (East–West) and with the meridional wind shear.

N. Barrado-Izagirre, S. Pérez-Hoyos, and A. Sánchez-Lavega
Universidad del Pais Vasco, Alda. Urkijo s/n, 48013 Bilbao, Spain
e-mail: naiara.barrado@ehu.es

J.M. Diego et al. (eds.), *Highlights of Spanish Astrophysics V*, Astrophysics and Space Science Proceedings, DOI 10.1007/978-3-642-11250-8_123,

The Importance of the Nucleus Rotation on the Size of the Dust Particle Ejected from Comets

A. Molina, F. Moreno, and F.J. Jiménez-Fernández

Abstract Dust particles attached on the nucleus surface can be dragged by cometary gas and, sometimes, these particles may remain around the comet in pseudostable orbits. In a previous work (Molina et al., 2008, EM & P, 102, 521) we show the equation of motion, including rotational terms, and we estimate the maximum diameter of a cometary dust particle that could be lifted from the nucleus surface. The purpose of this work is to analyze the importance of those rotational terms in order to obtain the values of the size of the largest grains ejected from the nucleus. We consider a strong sunward anisotropy emission as reported by Fulle (1997, A & A, 325, 1237). We discuss the obtained values and we make a comparison with those obtained using radar measurements.

A. Molina
Departamento de Fsica Aplicada, Universidad de Granada, Facultad de Ciencias, Avenida Severo Ochoa s/n, 18071, Granada, Spain

A. Molina, F. Moreno, and F.J. Jiménez-Fernández
Instituto de Astrofsica de Andaluca, CSIC, Camino Bajo de Hutor 50, 18008, Granada, Spain

J.M. Diego et al. (eds.), *Highlights of Spanish Astrophysics V*, Astrophysics and Space Science Proceedings, DOI 10.1007/978-3-642-11250-8_124,

Venus Spectrophotometry During the MESSENGER Mission Fly-By

S. Pérez-Hoyos, A. Sánchez-Lavega, R. Hueso, J. Peralta, G. Holsclaw, and W. McClintock

Abstract The NASA mission MESSENGER fly-byed planet Venus on June 2007 on its route to Mercury. This chance was took to produce coordinated observations between Messenger and ESA Venus Express spacecrafts. This work shows spectra in the wavelength range between 320 and 1450 nm retrieved with the instrument MASCS (Mercury Atmospheric and Surface Composition Spectrometer). Spectra are calibrated in absolute reflectivity (diffuse reflection by Venus clouds) and wavelength, and they are navigated in order to retrieve their position in the planet's disk. Comparing synthetic spectra with these ones for each viewing geometry we will obtain information on the vertical distribution of cloud particulates between 60 and 75 km height, approximately, as well as the SO_2 abundance, among others. This will be combined with almost simultaneous data gathered by the visible and infrared spectrograph VIRTIS onboard Venus Express spacecraft. The results of the atmospheric modeling will be presented elsewhere.

S. Pérez-Hoyos, A. Sánchez-Lavega, R. Hueso, and J. Peralta
Grupo de Ciencias Planetarias UPV/EHU, Dpto Física Aplicada I, E.T.S. Ingeniería, Bilbao, Spain
e-mail: santiago.perez@ehu.es

G. Holsclaw and W. McClintock
University of Colorado, Laboratory for Atmospheric and Space Physics, Boulder, USA

J.M. Diego et al. (eds.), *Highlights of Spanish Astrophysics V*, Astrophysics and Space Science Proceedings, DOI 10.1007/978-3-642-11250-8_125,

Three Dimensional Structure of Penumbral Filaments from Hinode Observations

K.G. Puschmann, B. Ruiz Cobo, and V. Martínez Pillet

Abstract We analyze spectropolarimetric observations of the penumbra of the NOAA AR 10953 at high spatial resolution (0.3″). The full Stokes profiles of the Fe I lines at 630.1 nm and 630.2 nm have been obtained with the Solar Optical Telescope (SOT) on board the Hinode satellite. The data have been inverted by means of the SIR code, deriving the stratifications of temperature, line-of-sight velocity, and the components of the magnetic field vector in optical depth. In order to evaluate the gas pressure and to obtain an adequate geometrical height scale, the motion equation has been integrated for each pixel taking into account the terms of the Lorentz force. To establish the boundary condition, a genetic algorithm has been applied. The final resulting magnetic field has a divergence compatible with 0 inside its uncertainties. First analyzes of the correlation of the Wilson depression with velocity, temperature, magnetic field strength, and field inclination strongly support the uncombed penumbral model proposed by Solanki & Montavon (1993, A&A, 275, 283).

K.G. Puschmann, B.R. Cobo, and V.M. Pillet
Instituto de Astrofísica de Canarias, 38200 La Laguna, Tenerife, Spain
e-mail: kgp@iac.es, brc@iac.es, vmp@iac.es

J.M. Diego et al. (eds.), *Highlights of Spanish Astrophysics V*, Astrophysics and Space Science Proceedings, DOI 10.1007/978-3-642-11250-8_126,

The Temporal Evolution of Linear Fast and Alfvén MHD Waves in Solar Coronal Arcades

S. Rial, I. Arregui, J. Terradas, R. Oliver, and J.L. Ballester

Abstract The excitation and temporal evolution of fast and Alfvén magnetohydrodynamic oscillations in a two-dimensional coronal arcade are investigated. The approach is to consider an equilibrium magnetic and plasma structure and then to introduce a perturbation trying to mimic a nearby disturbance, such as a flare or filament eruption. By numerically solving the time-dependent linearized MHD wave equations, the properties of the solutions have been studied. First, the properties of uncoupled fast and Alfvén waves are described. Then, longitudinal propagation of perturbations is included, and the properties of coupled waves are determined.

S. Rial, I. Arregui, R. Oliver, and J.L. Ballester
Departament de Física, Universitat de les Illes Balears, E-07122 Palma de Mallorca, Spain
e-mail: samuel.rial@uib.es

J. Terradas
Centre for Plasma Astrophysics, K.U. Leuven, B-3001 Heverlee, Belgium

J.M. Diego et al. (eds.), *Highlights of Spanish Astrophysics V*, Astrophysics and Space Science Proceedings, DOI 10.1007/978-3-642-11250-8_127,

Structure and Dynamics of Penumbral Filaments

B. Ruiz Cobo and L.R. Bellot Rubio

Abstract High-resolution observations of sunspots have revealed the existence of dark cores inside the bright filaments of the penumbra. Here we present the stationary solution of the heat transfer equation in a stratified penumbra consisting of nearly horizontal magnetic flux tubes embedded in a stronger and more vertical field. The tubes and the external medium are in horizontal mechanical equilibrium. This model produces bright filaments with dark cores as a consequence of the higher density of the plasma inside the flux tube, which shifts the surface of optical depth unity toward higher (cooler) layers. Our results suggest that the surplus brightness of the penumbra is a natural consequence of the Evershed flow, and that magnetic flux tubes about 250 km in diameter can explain the morphology of sunspot penumbra.

B.R. Cobo
Instituto de Astrofísica de Canarias, Tenerife, Spain
e-mail: brc@iac.es

L.R.B. Rubio
Instituto de Astrofísica de Andalucía (CSIC), Granada, Spain
e-mail: lbellot@iaa.es

J.M. Diego et al. (eds.), *Highlights of Spanish Astrophysics V*, Astrophysics and Space Science Proceedings, DOI 10.1007/978-3-642-11250-8_128,

Observatories and Instrumentation

A VO Archive for the DSS-63 Antenna at Robledo

J. Manuel Alacid, José Enrique Ruiz, Raúl Gutiérrez, Ricardo Rizzo, Lourdes Verdes-Montenegro, Enrique Solano, and Juan de Dios Santander-Vela

Abstract In this contribution we describe the development of a Virtual Observatory (VO) Archive for the DSS-63 antenna of the NASA Deep Space Communication Complex in Robledo de Chavela (Madrid). In an initial step the archive includes observations in the K-band (18–26 GHz) of the mentioned antenna with a future extension to other ranges, in particular the Q-band (40–50 GHz) and the Ka-band (32 GHz). A first version of the archive will be available from the LAEFF Scientific Data Centre[1] by the end of 2008. This work is the result of a collaboration between the AMIGA group of the IAA–CSIC and the SVO and Radioastronomy groups at LAEFF and forms part of a pioneering initiative to integrate radio astronomical data and services in the Virtual Observatory.

[1] http://sdc.laeff.inta.es

J.M. Alacid, R. Gutiérrez, and E. Solano
SVO/LAEFF-CAB/INTA-CSIC, Apdo 78, 28691 Villanueva de la Cañada, Madrid, Spain
e-mail: jmalacid@laeff.inta.es, raul@laeff.inta.es, esm@laeff.inta.es

J.E. Ruiz, L. Verdes-Montenegro, and J. de Dios Santander-Vela
Instituto de Astrofísica de Andalucia, CSIC. Apdo 3004. 18080 Granada, Spain
and
Red temática Observatorio Virtual Español
e-mail: jer@iaa.es, lourdes@iaa.es, jdsant@iaa.es

Ricardo Rizzo at LAEFF-CAB/INTA-CSIC, Apdo 78, 28691 Villanueva de la Cañada, Madrid, Spain
e-mail: ricardo@laeff.inta.es

J.M. Diego et al. (eds.), *Highlights of Spanish Astrophysics V*, Astrophysics and Space Science Proceedings, DOI 10.1007/978-3-642-11250-8_129,

BLAST: Study of the Earliest Stages of Galactic Star Formation

D. Anglés, P.A.R. Ade, J.J. Bock, C. Brunt, E.L. Chapin, M.J. Devlin, S. Dicker, M. Griffin, J.O. Gundersen, M. Halpern, P.C. Hargrave, D.H. Hughes, J. Klein, G. Marsden, P.G. Martin, P. Mauskopf, C.B. Netterfield, L. Olmi, E. Pascale, G. Patanchon, M. Rex, D. Scott, C. Semisch, M.D.P. Truch, C. Tucker, G.S. Tucker, M.P. Viero, and D.V. Wiebe

D. Anglés and L. Olmi
University of Puerto Rico, Rio Piedras Campus, Department of Physics, Box 23343, Puerto Rico
e-mail: daniel.angles@gmail.com

P.A.R. Ade, M. Griffin, P.C. Hargrave, P. Mauskopf, E. Pascale, and C. Tucker
Department of Physics & Astronomy, Cardiff University, 5 The Parade, Cardiff, CF24 3AA, UK

J.J. Bock
Jet Propulsion Laboratory, Pasadena, CA 91109-8099, USA

C. Brunt
School of Physics, University of Exeter, Stocker Road, Exeter, EX4 4QL, UK

E.L. Chapin, M. Halpern, G. Marsden, and D. Scott
Department of Physics & Astronomy, University of British Columbia, 6224 Agricultural Road, Vancouver, BC, Canada V6T 1Z1

M.J. Devlin, S. Dicker, J. Klein, M. Rex, C. Semisch, and M.D.P. Truch
Department of Physics & Astronomy, University of Pennsylvania, 209 South 33rd Street, Philadelphia, PA, 19104, USA

J.O. Gundersen
Department of Physics, University of Miami, 1320 Campo Sano Drive, Coral Gables, FL 33146, USA

D.H. Hughes
Instituto Nacional de Astrofísica Óptica y Electrónica, Aptdo. Postal 51 y 72000 Puebla, Mexico

P.G. Martin
Canadian Institute for Theoretical Astrophysics, University of Toronto, 60 St. George Street, Toronto, ON, Canada M5S 3H8

C.B. Netterfield and D.V. Wiebe
Department of Physics, University of Toronto, 60 St. George Street, Toronto, ON, Canada M5S 1A7

P.G. Martin, C.B. Netterfield, and M.P. Viero
Department of Astronomy & Astrophysics, University of Toronto, 50 St. George Street Toronto, ON, Canada M5S 3H4

L. Olmi
Instituto di Radioastronomia, Largo E. Fermi 5, I-50125, Firence, Italy

J.M. Diego et al. (eds.), *Highlights of Spanish Astrophysics V*, Astrophysics and Space Science Proceedings, DOI 10.1007/978-3-642-11250-8_130,

Abstract The Balloon-borne Large-Aperture Submillimeter Telescope (BLAST) is, until the regular operation of Herschel satellite, the most powerful submillimeter mapping telescope in the world. By operating above most of the atmosphere, BLAST provides a sensitivity (and therefore mapping speed) approximately an order–of–magnitude faster than any other existing submillimeter facilities in terms of detecting compact cores and even a greater improvement in terms of measuring diffuse structures in the interstellar medium (ISM). Using its three-band photometry at 250, 350, and 500 μm, BLAST samples the peak of the spectral energy distribution of the coldest starless cores, providing the critical coverage needed to constrain masses, luminosities, and temperatures. In this contribution we present a general description of the telescope and summarize the observations performed during the 2005 and 2006 Long Duration Balloon flights. In addition, we describe the Vela Molecular Ridge, a region extensively observed by BLAST, and discuss some of our preliminary results.

G. Patanchon
Laboratoire APC, 10, rue Alice Domon et Léonie Duquet 75205 Paris, France

G.S. Tucker
Department of Physics, Brown University, 182 Hope Street, Providence, RI 02912, USA

The EUCLID–NIS Mission

M. Balcells

Abstract Obtaining a measure of the baryonic acoustic oscillations (BAO) is among the top goals of cosmology in the beginning of the twenty-first century. Detection of BAOs offers the possibility to test the standard cosmological model, by providing strong constraints on the equation of state of dark energy. The EUCLID–NIS mission proposes to measure BAOs by mapping the 3D distribution of baryons at intermediate redshift, $1 < z < 3$. Such map will be derived from spectroscopic redshifts in the J and H bands, for the order of 10^8 galaxies, from space. In the current design, the instrument utilizes digital micromirror devices (DMD) from Texas Instruments.

M. Balcells
Instituto de Astrofísica de Canarias, C/ Vía Láctea, SN, 38200 La Laguna, Tenerife, Spain
e-mail: balcells@iac.es

J.M. Diego et al. (eds.), *Highlights of Spanish Astrophysics V*, Astrophysics and Space Science Proceedings, DOI 10.1007/978-3-642-11250-8_131,

Final Optical Design of PANIC, a Wide-Field Infrared Camera for CAHA

M.C. Cárdenas, J. Rodríguez Gómez, R. Lenzen, E. Sánchez-Blanco, and the International PANIC team

Abstract We present the Final Optical Design of PANIC (PAnoramic Near Infrared camera for Calar Alto), a wide-field infrared imager for the Ritchey–Chrtien focus of the Calar Alto 2.2 m telescope. This will be the first instrument built under the German–Spanish consortium that manages the Calar Alto observatory. The camera optical design is a folded single optical train that images the sky onto the focal plane with a plate scale of 0.45 arcsec per 18 μm pixel. The optical design produces a well defined internal pupil available to reducing the thermal background by a cryogenic pupil stop. A mosaic of four detectors Hawaii 2RG of 2 k×2 k, made by Teledyne, will give a field of view of 31.9 arcmin×31.9 arcmin.

M.C. Cárdenas, J. Rodríguez Gómez, E. Sánchez-Blanco, and the International PANIC team
Instituto de Astrofísica de Andalucía (IAA-CSIC), P.O. Box 3004, E-18080 Granada, Spain
e-mail: conchi@iaa.es

R. Lenzen and the International PANIC team
Max Plank Institut für Astronomie (MPIA), Königstuhl 17, 69117 Heidelberg, Germany

The International PANIC team:
http://www.iaa.es/PANIC/index.php/gb/panic_team

J.M. Diego et al. (eds.), *Highlights of Spanish Astrophysics V*, Astrophysics and Space Science Proceedings, DOI 10.1007/978-3-642-11250-8_132,

Searching for Good Blank Regions in the Sky for Flatfielding

Nicolás Cardiel and Miriam Aberasturi

Abstract The most important advantage of widefield cameras is, precisely, the *widefield*, since this offers the observers the possibility of obtaining vast amounts of data in a much shorter observing time. However, for a reliable data interpretation it is necessary a proper data calibration. Concerning the flatfielding of images, many times it is required to obtain several integrations in blank regions (sky patches without bright sources) nearby to the science target areas. In this work we present a systematic approach to obtain a catalogue of useful blank regions, based on the application of the Delaunay triangulation of the sky.

N. Cardiel
Departamento de Astrofísica, Facultad de Ciencias Físicas, Universidad Complutense de Madrid, Spain
e-mail: ncl@astrax.fis.ucm.es

M. Aberasturi
Laboratorio de Astrofísica Espacial y Física Fundamental, INSA, Apdo. 78, 28691 Villanueva de la Cañada, Madrid, Spain
e-mail: miriam@laeff.inta.es

J.M. Diego et al. (eds.), *Highlights of Spanish Astrophysics V*, Astrophysics and Space Science Proceedings, DOI 10.1007/978-3-642-11250-8_133,

EUREKA: The X-ray Universe in High Spectral Resolution

Francisco J. Carrera and Xavier Barcons

Abstract EURECA is a multinational project (European and Japanese) with the goal of building a prototype for a X-ray spectrometer based on Transition Edge Sensors (TES), with a view to new space missions, in particular XEUS/IXO. In this presentation we will describe the technological and scientific objectives of this project, and the Spanish participation in it, including development of software for scientific and calibration data analysis, development and building of superconducting bilayers, and characterization of LC filters for TES.

F.J. Carrera and X. Barcons
Instituto de Física de Cantabria (CSIC-UC), Avenida de los Castros, 39005 Santander (Spain)
e-mail: carreraf@ifca.unican.es, barcons@ifca.unican.es

J.M. Diego et al. (eds.), *Highlights of Spanish Astrophysics V*, Astrophysics and Space Science Proceedings, DOI 10.1007/978-3-642-11250-8_134,

AXIS–SVO Data Centre Creation

M. Teresa Ceballos

Abstract We present the process followed to create the AXIS–SVO Data Centre at the Instituto de Física de Cantabria under the standards of the Virtual Observatory using the publication tools elaborated by the ESA–VO team at the European Space Astronomy Centre (ESAC). The current content of this Data Centre is a sample of optical spectra which are part of the AXIS–XMS sample, based on observations of the XMM–Newton X-ray observatory.

M.T. Ceballos
Instituto de Física de Cantabria (CSIC-UC), Santander (Spain)
e-mail: ceballos@ifca.unican.es

J.M. Diego et al. (eds.), *Highlights of Spanish Astrophysics V*, Astrophysics and Space Science Proceedings, DOI 10.1007/978-3-642-11250-8_135,

XMM–Newton Data Analysis with SAS Software Over GRID Technology

M.T. Ceballos, I. Campos, P. Orviz, D. Tapiador, R. Álvarez, A. Ibarra, C. Gabriel, and J.R. Rodón

Abstract The processing of data obtained by the XMM–Newtonobservatory of the European Space Agency (ESA) is done using the SAS (Software Analysis System) tools provided by the ESA XMM Science Operations Centre (SOC). In order to be operative, these tools must be downloaded from the official SOC web pages and then installed (eventually upgraded) and run locally in the computer used by the data analyser. This process can be in some cases cumbersome for some users and local resources. In this presentation, we summarise the initiative developed from the SOC at the European Space Astronomy Centre (ESAC) in collaboration with the Instituto de Física de Cantabria (IFCA CSIC-UC), to run these tools in a GRID environment, and thus taking advantage of the resources distribution and the possibility of remote use of the analysis tools.

M.T. Ceballos, I. Campos, P. Orviz, and J.R. Rodón
IFCA(CSIC-UC), Avda Los Castros, Santander
e-mail: ceballos@ifca.unican.es, iscampos@ifca.unican.es, orviz@ifca.unican.es, rodon@ifca.unican.es

D. Tapiador, R. Álvarez, A. Ibarra, and C. Gabriel
ESAC(ESA), Madrid
e-mail: Daniel.Tapiador@sciops.esa.int, Ruben.Alvarez@sciops.esa.int, Aitor.Ibarra@sciops.esa.int, Carlos.Gabriel@sciops.esa.int

J.M. Diego et al. (eds.), *Highlights of Spanish Astrophysics V*, Astrophysics and Space Science Proceedings, DOI 10.1007/978-3-642-11250-8_136,

CIRCE: The Canarias InfraRed Camera Experiment

M.V. Charcos-Llorens, A.J. Cenarro, M.L. Edwards, S.S. Eikenberry, K.T. Hanna, J. Julian, N.M. Lasso Cabrera, A. Marín-Franch, C. Packham, S.N. Raines, M. Rodgers, and F. Varosi

Abstract The Canarias InfraRed Camera Experiment (CIRCE) is a near-infrared (1–2.5 μm) instrument for the Gran Telescopio Canarias (GTC) 10.4 m telescope. CIRCE has an off-axis aspheric all-reflective optical system that offers both excellent throughput and image quality. Observational modes include broad/narrow band imaging and low-resolution spectroscopy. High time-resolution data acquisition and polarimetry are available in both cases. We present an overview of the instrument and the current status of design and fabrication.

M.V. Charcos-Llorens, M.L. Edwards, S.S. Eikenberry, K.T. Hanna, J. Julian, N.M.L. Cabrera, C. Packham, S.N. Raines, and F. Varosi
University of Florida, 211 Bryant Space Science Center, Gainesville FL 32611, USA
e-mail: mcharcos@astro.ufl.edu, edwards@astro.ufl.edu, eiken@astro.ufl.edu, hanna@astro.ufl.edu, julian@astro.ufl.edu, nestor@astro.ufl.edu, packham@astro.ufl.edu, raines@astro.ufl.edu, varosi@astro.ufl.edu

A.J. Cenarro and A. Marín-Franch
Instituto de Astrofísica de Canarias, C/ Vía Láctea, s/n, E38205 La Laguna, Tenerife, Spain
e-mail: cenarro@iac.es, amarin@iac.es

M. Rodgers
Optical Research Associates, 3280 East Foothill Boulevard Suite 300, Pasadena CA 91107, USA

J.M. Diego et al. (eds.), *Highlights of Spanish Astrophysics V*, Astrophysics and Space Science Proceedings, DOI 10.1007/978-3-642-11250-8_137,

Design of Wide Band Bow-Tie Slot Antennas for Multi-Frequency Operation in CMB Experiments

Angel Colin

Abstract This report presents two proposals of antenna designs suitable to be included in arrays for multi-frequency operation in the ranges from ~10 to ~60 GHz and from ~130 to 300 GHz, both aimed to be applied in Cosmic Microwave Background (CMB) experiments. The antennas exhibit small sizes and wide bandwidth approaching 131 and 79% respectively. The radiation characteristics for a single element and the total directivity for configurations of 5×5 and 15×15 elements array were simulated by means of HFSS (Ansoft) software.

A. Colin
Instituto de Fsica de Cantabria (CSIC-UC), Av. Los Castros s/n, 39005, Santander, Spain
e-mail: acolin@ifca.unican.es

J.M. Diego et al. (eds.), *Highlights of Spanish Astrophysics V*, Astrophysics and Space Science Proceedings, DOI 10.1007/978-3-642-11250-8_138,

OAdM Observatory: Towards Fully Unattended Control

J. Colomé, I. Ribas, D. Fernández, X. Francisco, J. Isern, X. Palau, and J. Torra

Abstract The Montsec Astronomical Observatory (OAdM) is a small-class observatory working on a completely unattended control, due to the isolation of the site. Robotic operation is, then, mandatory for its routine use. The level of robotization of an observatory is given by the confidence reached to respond to environment changes and by the required human interaction due to possible alarms. These two points establish a level of human attendance to ensure low risk at any time. There are key problems to solve when a robotic control is envisaged. Learned lessons and solutions to these issues at the OAdM are discussed here. We present a description of the general control software (SW) and several SW packages developed. They specially protect the system at the identified single points of failure and constitute a distributed control of any subsystem, which is able to respond independently when an alarm is triggered thanks to a top-down control flow. All together this composes a SW suite designed to reach the complete robotization of an observatory.

J. Colomé, I. Ribas, D. Fernández, X. Francisco, J. Isern, and J. Torra
Institut d'Estudis Espacials de Catalunya (IEEC), Gran Capità 2-4 (Edifici Nexus), E-08034 Barcelona, Spain

J. Colomé, I. Ribas, and J. Isern
Institut de Ciències de l'Espai (CSIC-IEEC), Torre C-5 Parell, 2ona planta, facultat Ciències, Universitat Autònoma de Barcelona, E-08193 Bellaterra, Spain

D. Fernández and J. Torra
Institut de Ciències del Cosmos (UB-IEEC), Departament d'Astronomia i Meteorologia, Universitat de Barcelona, Av.Diagonal 647, E-08028 Barcelona, Spain

X. Palau
Fundació Joan Oró, Travessera de les Corts 272, E-08014 Barcelona, Spain

J.M. Diego et al. (eds.), *Highlights of Spanish Astrophysics V*, Astrophysics and Space Science Proceedings, DOI 10.1007/978-3-642-11250-8_139,

Status of MAGIC and the CTA Project

J.L. Contreras

Abstract The MAGIC telescope collaboration has built the world's largest single dish Imaging Atmospheric Cherenkov Telescope, which is operative in the ORM observatory at the Island of La Palma since 2004. It covers the energy range from around 50–60 GeV up to several TeV. A second MAGIC telescope is presently under construction at the same site, its completion is foreseen for September 2008. We show the present and expected performance of MAGIC and the status of the project. At the same time MAGIC, with most of the European Very High Energy (VHE) Astrophysics Community, is preparing the new Cherenkov Telescope Array (CTA) project. It will be the first global VHE open observatory, with vastly improved sensitivity over existing telescopes. We sketch its design and outline the physics goals pursued.

J.L. Contreras (for the MAGIC Collaboration), Dpto. Física Atómica, Universidad Complutense de Madrid
e-mail: contrera@gae.ucm.es

J.M. Diego et al. (eds.), *Highlights of Spanish Astrophysics V*, Astrophysics and Space Science Proceedings, DOI 10.1007/978-3-642-11250-8_140,

In the Search of Exoplanets

Luis Cuesta Crespo, and the Group of Robotic Telescopes at CAB

Abstract The Spanish *Instituto Nacional de Técnica Aeroespacial* has a network of three telescopes located in some of the best places for Astronomy in Spain: the Observatory of Calar Alto, in Almería, near Calatayud, in Zaragoza, at the summit of a 1,400 m high mountain, and at the campus of INTA, in Madrid. The three telescopes have diameters between 40 and 50 cm, and are equipped with instrumentation very adequate to identify exoplanets.

L.C. Crespo
Centro de Astrobiología, INTA, Carretera de Ajalvir, km 4, 28850 Torrejón de Ardoz, Madrid
e-mail: cuestacl@inta.es

J.M. Diego et al. (eds.), *Highlights of Spanish Astrophysics V*, Astrophysics and Space Science Proceedings, DOI 10.1007/978-3-642-11250-8_141,

First Results of the Optical Speckle Interferometry with the 3.5 m Telescope at Calar Alto (Spain): Measurements and Orbits of Visual Binaries

J.A. Docobo, V.S. Tamazian, M. Andrade, J.F. Ling, Y.Y. Balega, J.F. Lahulla, and A.F. Maximov

Abstract We present the first results of optical speckle interferometry of visual binaries obtained with the 3.5m telescope at Calar Alto (Almería, Spain) in July 2005. During this campaign fifty stars were observed with angular separations between 0.″058 and 2.″1. New orbits for binaries COU 490 and A 2257, and improved orbits for COU 606, A 225, A 2189, COU 1785, COU 2416, and COU 327 AB, along with their systemic masses, were calculated. The first estimations of distance for the COU 490 (145 pc) and the A 2257 (210 pc) were accomplished thanks to the previously obtained dynamical parallaxes and to the known spectral and photometric data. Total masses for systems with both new and improved orbits are given, those being concordant with their known spectral types and photometric data. We conclude that with the 3.5 m telescope we can routinely obtain good optical speckle data for binary systems with angular separations close to its diffraction limit.

J.A. Docobo, V.S. Tamazian, M. Andrade, and J.F. Ling
Astronomical Observatory R.M. Aller (USC), Santiago de Compostela
e-mail: joseangel.docobo@usc.es, vakhtang.tamazian@usc.es, manuel.andrade@usc.es, josefinaf.ling@usc.es

Y.Y. Balega and A.F. Maximov
Special Astrophysical Observatory (Russia), Karachaevo-Cherkesia
e-mail: balega@sao.ru, max@sao.ru

J.F. Lahulla
National Astronomical Observatory, Madrid
e-mail: lahulla@oan.es

J.M. Diego et al. (eds.), *Highlights of Spanish Astrophysics V*, Astrophysics and Space Science Proceedings, DOI 10.1007/978-3-642-11250-8_142,

The INTEGRAL–OMC Scientific Archive

A. Domingo, R. Gutiérrez-Sánchez, D. Rísquez, M.D. Caballero-García, J.M. Mas-Hesse, and E. Solano

Abstract The Optical Monitoring Camera (OMC) on-board the INTEGRAL satellite has, as one of its scientific goals, the observation of a large number of variable sources previously selected. After almost 6 years of operations, OMC has monitored more than 100,000 sources of scientific interest. In this contribution we present the OMC Scientific Archive (http://sdc.laeff.inta.es/omc/) which has been developed to provide the astronomical community with a quick access to the light curves generated by this instrument. We describe the main characteristics of this archive, as well as important aspects for the users: object types, temporal sampling of light curves and photometric accuracy.

A. Domingo
CAB/LAEFF (CSIC-INTA), POB 78, 28691 Villanueva de la Cañada, Madrid, Spain
e-mail: albert@laeff.inta.es

J.M. Diego et al. (eds.), *Highlights of Spanish Astrophysics V*, Astrophysics and Space Science Proceedings, DOI 10.1007/978-3-642-11250-8_143,

Observatorio Astronómico De Cantabria

R.M. Domínguez and F.J. Carrera

Abstract The Astronomical Observatory of Cantabria is a center of the *Consejería de Medio Ambiente del Gobierno de Cantabria* managed by the *Centro de Investigación del Medio Ambiente* (CIMA), an autonomous organism which depends on such *Consejería*. The development of different activities of the Observatory is a joint collaboration between the University of Cantabria and the *Agrupación Astronómica Cántabra* (AstroCantabria). As part of the University of Cantabria, the *Instituto de Física de Cantabria* (IFCA, CSIC-UC) is in charge of the direction, management and coordination of scientific, observational, educational and outreach activities of the Observatory. AstroCantabria takes care of the outreach activities for the general public as well as the astronomical observations. In addition, it is responsible for the calibration and maintenance of the astronomical instrumentation of the Observatory. The Astronomical Observatory of Cantabria is located on the Southern edge of the *Comunidad Autónoma de Cantabria*, on the high plateau of *La Lora* (Valderredible county), at an altitude of 1,080 m, with longitude 3° 56′ 36″ W and latitude 42° 46′ 18″ N. *Rocamundo* is the closest town. The Observatory aims to become a center of reference for scientific, observational, educational and public outreach activities in Cantabria. In the near future, an observational proposal system for outside users will be set in place.

R.M. Domínguez and F.J. Carrera
Instituto de Física de Cantabria
e-mail: dominguez@ifca.unican.es, carreraf@ifca.unican.es

J.M. Diego et al. (eds.), *Highlights of Spanish Astrophysics V*, Astrophysics and Space Science Proceedings, DOI 10.1007/978-3-642-11250-8_144,

A Displayer of Stellar Hydrodynamics Processes

José Antonio Escartín Vigo and Domingo García Senz

Abstract The graphics display tool that we present here was originally developed to meet the needs of the Astronomy and Astrophysics group at the UPC (GAA). At present, it is used to display the plots obtained from hydrodynamic simulations using the SPH (smoothed particle hydrodynamics) method. It is, however, a generic program that can be used for other multidimensional hydrodynamic methods. The application combines the most widely used features of other programs (most of them commercial) such as GnuPlot, Surfer, Grapher, IDL, Voxler, etc.

J.A.E. Vigo and D.G. Senz
UPC, Jordi Girona 1-3 B5 08034 Barcelona, Spain
e-mail: jose.antonio.escartin@upc.edu, domingo.garcia@upc.edu

J.M. Diego et al. (eds.), *Highlights of Spanish Astrophysics V*, Astrophysics and Space Science Proceedings, DOI 10.1007/978-3-642-11250-8_145,

GUAIX: The UCM Group of Extragalactic Astrophysics and Astronomical Instrumentation

J. Gallego, N. Cardiel, J. Zamorano, J. Gorgas, A. Castillo-Morales, M.C. Eliche-Moral, A. Gil de Paz, S. Pascual, P.G. Pérez-González, R. Guzmán, G. Barro, C. Díaz, N. Espino, J. Izquierdo, E. Mármol-Queraltó, J.C. Muñoz-Mateos, L. Rodriguez, A. Sánchez de Miguel, E. Toloba, V. Villar, and M. Abelleira

Abstract We present a short summary of the activities developed by GUAIX, the Universidad Complutense de Madrid (UCM) Group of Extragalactic Astrophysics and Astronomical Instrumentation. At present we are focused in the development of data reduction pipelines for several future instruments for the Spanish 10 m GTC (Gran Telescopio Canarias). The careful treatment of the random error propagation throughout the data reduction is one of the main improvements of those pipelines. The first hardware development leaded by the GUAIX group will be FISIR, a fully-cryogenic (optimized for the K band) tunable filter in the near-infrared, to be installed within CIRCE, a near-IR camera for GTC.

J. Gallego, N. Cardiel, J. Zamorano, J. Gorgas, A. Castillo-Morales, M.C. Eliche-Moral, A. Gil de Paz, S. Pascual, P.G. Pérez-González, R. Guzmán, G. Barro, C. Díaz, N. Espino, J. Izquierdo, E. Mármol-Queraltó, J.C. Muñoz, L. Rodriguez, A. Sánchez de Miguel, E. Toloba, V. Villar, and M. Abelleira
e-mail: mabelleira@astrax.fis.ucm.es
Departamento de Astrofísica y C.C. de la Atmósfera, Facultad de C.C. Físicas, Universidad Complutense de Madrid, Avda. Complutense s/n, E28040-Madrid, Spain
e-mail: jgm@astrax.fis.ucm.es

J.M. Diego et al. (eds.), *Highlights of Spanish Astrophysics V*, Astrophysics and Space Science Proceedings, DOI 10.1007/978-3-642-11250-8_146,

ASTRONET: Towards a Strategic Plan for European Astronomy

J. Gallego, J. Torra, X. Barcons, and M. Mas-Hesse

Abstract ASTRONET is an ERA-Net financed by the European Commission FP6 under the initiative *Integrating and Strengthening the European Research Area (ERA)*. ASTRONET was created by a group of European funding agencies in order to establish a comprehensive long-term planning for the development of European astronomy. The objective of this effort is to consolidate and reinforce the world-leading position that European astronomy has attained at the beginning of this twenty-first century. The Ministerio de Ciencia e Innovación is the Spanish representative.

J. Gallego
Ministerio de Ciencia e Innovación (Spain) and Universidad Complutense de Madrid (Spain)
e-mail: jgm@astrax.fis.ucm.es

J. Torra
Ministerio de Ciencia e Innovación (Spain) and Universidad de Barcelona (Spain)

X. Barcons
Ministerio de Ciencia e Innovación (Spain) and Instituto de Física de Cantabria (Spain)

M. Mas-Hesse
Ministerio de Ciencia e Innovación (Spain) and Centro de Astrobiología (Spain)

J.M. Diego et al. (eds.), *Highlights of Spanish Astrophysics V*, Astrophysics and Space Science Proceedings, DOI 10.1007/978-3-642-11250-8_147,

Development of a Virtual Observatory Tool for the Characterization of Stellar Objects in the DUNES Project Framework

Raúl Gutiérrez-Sánchez, Enrique Solano, María Arévalo, and Carlos Eiroa

Abstract Most of the projects that aim at detecting extrasolar planets require a careful selection of the central stars as well as an extremely detailed knowledge of their properties and environment. However, gathering information in a wide variety of types and formats from a large number of heterogeneous astronomical data services can be a tedious, very time-consuming task, even for a modest dataset. To overcome this situation the LAEFF Scientific Data Centre, in the framework of the Spanish Virtual Observatory[1], has developed a VO-compliant discovery tool for DUNES[2]. This tool allows accessing, visualizing, filtering and retrieving relevant information of lists of objects. In this poster we describe the main characteristics and functionalities of the system.

[1] http://svo.laeff.inta.es
[2] http://sdc.laeff.inta.es/dunes/

R. Gutiérrez-Sánchez, E. Solano, and M. Arévalo
SVO/LAEFF-CAB/INTA-CSIC, PO Box 78, 28691 Villanueva de la Cañada, Madrid, Spain
e-mail: raul.gutierrez@laeff.inta.es, enrique.solano@laeff.inta.es, maria.arevalo@laeff.inta.es

C. Eiroa
Departamento de Física Teórica, C-XI, Facultad de Ciencias, Universidad Autónoma de Madrid, Cantoblanco, 28049 Madrid
e-mail: carlos.eiroa@uam.es
Spanish Virtual Observatory Thematic network

J.M. Diego et al. (eds.), *Highlights of Spanish Astrophysics V*, Astrophysics and Space Science Proceedings, DOI 10.1007/978-3-642-11250-8_148,

Simulations of Array Configurations for the Square Kilometre Array (SKA)

Sergio Jiménez-Monferrer[1], Dharam Vir Lal[1], Andrei P. Lobanov, and José Carlos Guirado

Abstract The Square Kilometre Array (SKA) is a new generation radio telescope for the next decades, working at metre to centimetre wavelengths. The SKA will be operational at the same time than other new optical, X-ray and Gamma-ray telescopes. It is of extreme importance that the SKA becomes competitive and complementary to those instruments. An extensive study of technologies and possible configurations involved is needed to ensure the SKA will reach the design specifications. To compare imaging capabilities between different SKA configurations or between the SKA and other instruments, we have implemented figures of merit based on several characteristics of these instruments. In this work we are presenting some results of numerical tests based on the Spatial Dynamic Range (SDR), which quantifies the range of spatial scales than can be reconstructed from interferometric data (Lobanov, A.P., SKA Memo 38, 2003).

[1] This effort/activity is supported by the European Community Framework Programme 6, Square Kilometre Array Design Studies (SKADS), contract no 011938.

S. Jiménez-Monferrer and J.C. Guirado
Universidad de Valencia, Dr. Moliner 50, 46100 Burjassot, Valencia, Spain
e-mail: sergio.jimenez@alumni.uv.es, jose.c.guirado@alumni.uv.es

D.V. Lal and A.P. Lobanov
Max-Planck-Institut für Radioastronomie, Auf dem Hügel 69, 53121 Bonn, Germany
e-mail: dharam@mpifr-bonn.mpg.de, alobanov@mpifr-bonn.mpg.de

J.M. Diego et al. (eds.), *Highlights of Spanish Astrophysics V*, Astrophysics and Space Science Proceedings, DOI 10.1007/978-3-642-11250-8_149,

Metallicity Estimation Using N2 Method with OSIRIS

M.A. Lara-López, J. Cepa, H. Castañeda, E.J. Alfaro, A. Bongiovanni, M. Fernández, J. Gallego, J.J. González, J.I. González-Serrano, A.M. Pérez-García, M. Pović, and M. Sánchez-Portal

Abstract The OTELO (OSIRIS Tunable Emission Line Object Survey project) is an emission line survey using the OSIRIS Tunable Filters on the GTC 10.4 m telescope (Cepa et al. 2008). Observing in selected atmospheric windows relatively free of sky emission lines, and with a total survey area of 0.1square degree distributed in different fields, it is expected to reach 5 sigma depth of 10^{-18} erg/cm^2/s, detecting objects with EW <0.2. OTELO will be the deepest emission line survey to date. As part of the preparatory activities, we have selected from the simulation of OTELO spectra, including errors, the best combination of tunable filters bandwidth (FWHM) and sampling, that will allow deblending Hα from [N II]λ6,583 lines with a flux error lower than 20%. With the selected instrumental configuration it will be possible to estimate the objects chemical abundances using the N2 method in very low metallicity systems. We estimate that OTELO will allow to estimate the metallicities of more than 3,000 Hα star forming emitters up to a redshift 0.4.

M.A. Lara-López, H. Castañeda, A. Bongiovanni, M. Fernández, A.M. Pérez-García, and M. Pović
Instituto de Astrofísica de Canarias, 38205 La Laguna, Spain
e-mail: mall@iac.es

J. Cepa
Instituto de Astrofísica de Canarias, 38205 La Laguna, Spain, and Departamento de Astrofísica, Universidad de La Laguna, 38205 La Laguna, Spain

E. Alfaro
Instituto de Astrofísica de Andalucía-CSIC, Granada, Spain

J. Gallego
Departamento de Astrofísica y CC. de la Atmósfera, Universidad Complutense de Madrid, Madrid, Spain

J.J. González
Instituto de Astronomía UNAM, México D.F, México

J.I. González-Serrano
Instituto de Física de Cantabria, CSIC-Universidad de Cantabria, Santander, Spain

M. Sánchez-Portal
Herschel Science Centre, ESAC/INSA, Madrid, Spain

J.M. Diego et al. (eds.), *Highlights of Spanish Astrophysics V*, Astrophysics and Space Science Proceedings, DOI 10.1007/978-3-642-11250-8_150,

Supervised Star Classification System for the OMC Archive

Mauro López, Luis M. Sarro, Enrique Solano, Raul Gutiérrez, and Jonas Debosscher

Abstract INTEGRAL, COROT and, in the near future, KEPLER and GAIA, are missions that are producing, or will produce, light curves as we have never seen before, because of their quantity or quality. The exploitation of the scientific potential hidden in these datasets is limited by size of the dataflows. For example, visual classification of the $\approx 10^8$ GAIA light curves is infeasible for any research team. Supervised classification of light curves has been one of the main research lines in the Spanish Virtual Observatory, and we are now co-leading the data analysis work group for the CoRoT mission. In this contribution we show the development of an automatic multistage classification system based on bayesian networks for the OMC (Optical Monitoring Camera) data. OMC is an optical camera on board ESA's INTEGRAL, whose data archive is managed in the LAEFF Scientific Data Center (http://sdc.laeff.inta.es/omc).

M. López, E. Solano, and R. Gutiérrez
SVO / LAEX-CAB (INTA-CSIC), Postal address.- LAEFF, European Space Astronomy Center (ESAC), P.O. Box 78, E-28691 Villanueva de la Cañada, Madrid, Spain
e-mail: mauro@laeff.inta.es, esm@laeff.inta.es, raul@laeff.inta.es

L.M. Sarro
Dpto. Inteligencia Artificial. ETSI Informática - UNED, C\ Juan del Rosal 16 - 3, E28040, Madrid, Spain
e-mail: lsb@dia.uned.es

J. Debosscher
Instituut voor Sterrenkundem Katholieke Universiteit Leuven. Celestijnenlaan 200D BUS 2401 3001. Leuven, Belgium
e-mail: jonas.debosscher@ster.kuleuven.ac.be

J.M. Diego et al. (eds.), *Highlights of Spanish Astrophysics V*, Astrophysics and Space Science Proceedings, DOI 10.1007/978-3-642-11250-8_151,

Preparation of the Gaia Data Processing: First Astrometric Results

X. Luri, W. O'Mullane, U. Lammers, L. Lindegren, and E. Masana

Abstract The Gaia mission of the European Space Agency (ESA) will produce high-precision astrometry for 10^9 objects up to 20th magnitude. The volume of data generated (about 150TB of compressed raw data) and the complexity of the relationships between them make the scientific processing a challenging task. This paper presents the basic concepts of the core of the astrometric data reduction, the AGIS system, its present status and the first test results using simulated data.

X. Luri and E. Masana
Dept. dAstronomia i Meteorologia and IEEC-UB-ICC, C/ Martí i Franquès 1, 08028 Barcelona
e-mail: xluri@am.ub.es

W. O'Mullane and U. Lammers
ESA / European Space Astronomy Centre,Urbanizacion de Villafranca del Castillo Avda. de los Castillos s/n, E-28692 Villanueva de la Canada (Madrid)

L. Lindegren
Lund Observatory, Box 43, SE-221 00 Lund (Sweden)

J.M. Diego et al. (eds.), *Highlights of Spanish Astrophysics V*, Astrophysics and Space Science Proceedings, DOI 10.1007/978-3-642-11250-8_152,

Fast-Switching in the Submillimeter Array: A Step Toward Gain Calibration in ALMA

Vincent Martinez-Badenes, Daniel Espada, and Satoki Matsushita

Abstract Fluctuations of water vapour content in the troposphere are one of the worst enemies of millimeter/submillimeter aperture synthesis observations, being responsible of loss of coherence in the radio signal and limiting the spatial resolution. The use of the fast switching technique tries to minimize this effect through fast and successive cycles between calibrator and source. Here we present fast-switching tests in millimeter/submillimeter wavelengths carried out at the Submillimeter Array (SMA) on top of Mauna Kea, in order to shed light into the optimization of calibration cycle as a function of different atmospherical conditions. We have found that fast switching with calibration cycles of a few minutes slightly improve ($\sim$ 15%) the signal-to-noise ratio in the image plane with respect to the standard calibration time of 20–30 min, under relatively good weather conditions. For worse weather conditions, the fast switching technique does not have any relevant effect. This confirms previous findings that most of the fluctuations must be shorter than 2 min under normal conditions in the Mauna Kea site and baseline range from $\sim$ 20–200 m.

V. Martinez-Badenes
Instituto de Astrofisica de Andalucia, IAA-CSIC, Spain
e-mail: vicentm@iaa.es

D. Espada and S. Matsushita
Academia Sinica, Institute for Astrononomy and Astrophysics, Taiwan
e-mail: despada@asiaa.sinica.edu.tw, despada@cfa.harvard.edu, satoki@asiaa.sinica.edu.tw

J.M. Diego et al. (eds.), *Highlights of Spanish Astrophysics V*, Astrophysics and Space Science Proceedings, DOI 10.1007/978-3-642-11250-8_153,

The Gaia Simulator: Design and Results

E. Masana, Y. Isasi, X. Luri, and J. Peralta

Abstract Simulation of the data generated during the 5 year observations is one of the most relevant aspects of the Gaia mission preparation. The simulations allow the development and test of the reduction algorithms to be used during the mission. A complex software named Gaia Simulator has been developed in order to perform the simulation. It includes three different data generators: Gibis, to produce pixel level images; Gass, to provide telemetry data stream; and Gog, designed to generate the final data mission or data at an intermediate state of the reduction process. The three data generators share several libraries, including the universe and instrument models.

E. Masana, Y. Isasi, X. Luri, and J. Peralta
Departament d'Astronomia i Meteorologia Universitat de Barcelona, C/ Martí Franquès 1, Barcelona 08028 Spain
e-mail: emasana@am.ub.es

J.M. Diego et al. (eds.), *Highlights of Spanish Astrophysics V*, Astrophysics and Space Science Proceedings, DOI 10.1007/978-3-642-11250-8_154,

Infrared Instrumentation for Dome C: Conceptual Design

Alcione Mora, Carlos Eiroa, David Barrado y Navascués, Carlos Abia, and Paolo Persi

Abstract We have started a conceptual design study for both a near-infrared (NIR) wide-field camera and a mid-infrared (MIR) camera-spectrograph for a 2m-class telescope at Dome C, Antarctica. The main scientific drivers are the characterization of young embedded objects, the evolution of crystalline silicates in circumstellar disks and the observation of exoplanet secondary transits. Both instrument would exploit the unique features of Dome C: superb seeing, low temperature, improved infrared transmission, reduced sky background and increased atmospheric stability. Two preliminary concepts are presented here. The NIR instrument would cover the wavelength range 0.9–5.5 μm and would be optimized for the K, L and M bands. The MIR instrument would observe in the range 7–40 μm and would be optimized for the Q window, including the extended portion (25–40 μm) only observable from Antarctica.

A. Mora and C. Eiroa
Departamento de Física Teórica C-XI, Universidad Autónoma de Madrid, 28049 Madrid, Spain
e-mail: alcione.mora@uam.es, carlos.eiroa@uam.es

D. Barrado y Navascués
Laboratorio de Astrofísica Espacial y Física Fundamental, INTA, Madrid, Spain
e-mail: barrado@laeff.inta.es

C. Abia
Departamento de Física Teórica y del Cosmos, Universidad de Granada, Spain
e-mail: cabia@ugr.es

P. Persi
Istituto di Astrofisica Spaziale e Fisica Cosmica, Roma, Italy
e-mail: paolo.persi@iasf-roma.inaf.it

J.M. Diego et al. (eds.), *Highlights of Spanish Astrophysics V*, Astrophysics and Space Science Proceedings, DOI 10.1007/978-3-642-11250-8_155,

Data Quality Check and On-Site Analysis of the MAGIC Telescope

I. Oya, R. de los Reyes, J.L. Contreras, D. Nieto, J.A. Barrio, E. Carmona, M. Gaug, M.V. Fonseca, A. Moralejo, and J. Rico

Abstract We present the scheme developed for the quick analysis and data quality checks on the MAGIC Cherenkov Telescope at La Palma. Due to its low energy threshold MAGIC acquires data of atmospheric showers at a rate of more than 200 Hz, which translates in up to 700 GB per night. A fast On-Site data reduction is needed to detect hardware problems and in many cases to decide on observation strategies. The data are automatically calibrated and pre-processed at the MAGIC site using automated scripts on multiprocessor systems. Check plots are generated, and first results are available in the morning. This system complements a quick online analysis which runs in parallel with the data acquisition.

I. Oya , R. de los Reyes, J.L. Contreras, D. Nieto, J.A. Barrio, and M.V. Fonseca
Dpto. Física Atómica, UCM, Madrid, Spain
e-mail: oya@gae.ucm.es

E. Carmona
Max-Planck-Institut für Physik, München, Germany

M. Gaug
Instituto de Astrofísica de Canarias, La Laguna, Tenerife, Spain

A. Moralejo
Institut de Física d'Altes Energies, Bellaterra, Spain

J. Rico (for the MAGIC Collaboration)
Institut de Física d'Altes Energies, Bellaterra, Spain
and
Institució Catalana de Recerca i Estudis Avançats, Barcelona, Spain

J.M. Diego et al. (eds.), *Highlights of Spanish Astrophysics V*, Astrophysics and Space Science Proceedings, DOI 10.1007/978-3-642-11250-8_156,

Free Software in Astronomy: Fedora Astronomy

S. Pascual

Abstract An important part of the tools used journal of the astronomical community are free software. Since the ideals of the free-software community, freedom to use and share programs, are very close to the ideals of the scientific community, freedom share knowledge, the connection emerged in a natural way since the beginning. Fedora is one of the most important Linux distributions, built with the help of an extensive community. This contribution shows the efforts of a group of partners to include in Fedora the largest number of applications astronomical, both for the amateur astronomer and professional. I will discuss also the impact of non-free applications in the astronomical community.

S. Pascual
Universidad Complutense de Madrid, Spain
e-mail: sergiopr@fedoraproject.org

J.M. Diego et al. (eds.), *Highlights of Spanish Astrophysics V*, Astrophysics and Space Science Proceedings, DOI 10.1007/978-3-642-11250-8_157,

A Fully GTC-Compliant Pipeline for the Direct Imaging Mode of EMIR

S. Pascual, J. Gallego, and N. Cardiel

Abstract EMIR is a near-infrared wide-field camera and multi-object spectrograph being built for the 10.4 m Spanish telescope (*Gran Telescopio Canarias*, GTC) at La Palma Observatory. The Data Factory Pipeline (DFP) will be optimized for handling and reducing near-infrared data acquired with EMIR. Both reduced data and associated error frames will be delivered to the end-users as a final product. The DFP is being designed and built by the EMIR *Universidad Complutense de Madrid* group.

S. Pascual, J. Gallego, and N. Cardiel
Universidad Complutense de Madrid, Spain
e-mail: spr,jgm,ncl@astrax.fis.ucm.es

J.M. Diego et al. (eds.), *Highlights of Spanish Astrophysics V*, Astrophysics and Space Science Proceedings, DOI 10.1007/978-3-642-11250-8_158,

Calar Alto Academy

Santos Pedraz and David Galadí

Abstract Calar Alto Academy was initiated in 2007 with the aim to give students from different Spanish universities the chance to perform professional observational work at Calar Alto Observatory. The second edition of this innovative educational project has increased the number of participating universities and has almost doubled the quantity of visiting students, in a significant step towards the consolidation of this undergraduate and graduate school of observational astronomy.

S. Pedraz and D. Galadí
Centro Astronómico Hispano Alemán, Jesús Durbán Remón 2, 04004 Almería, Spain
e-mail: pedraz@caha.es, dgaladi@caha.es

J.M. Diego et al. (eds.), *Highlights of Spanish Astrophysics V*, Astrophysics and Space Science Proceedings, DOI 10.1007/978-3-642-11250-8_159,

INSA Scientific Activities in the Space Astronomy Area

Ricardo Pérez Martínez and Miguel Sánchez Portal

Abstract Support to astronomy operations is an important and long-lived activity within INSA. Probably the best known (and traditional) INSA activities are those related with real-time spacecraft operations: ground station maintenance and operation (ground station engineers and operators); spacecraft and payload real-time operation (spacecraft and instruments controllers); computing infrastructure maintenance (operators, analysts), and general site services. In this paper, we'll show a different perspective, probably not so well-known, presenting some INSA recent activities at the European Space Astronomy Centre (ESAC) and NASA Madrid Deep Space Communication Complex (MDSCC) directly related to scientific operations. Basic lines of activity involved include: operations support for science operations; system and software support for real time systems; technical administration and IT support; R&D activities, radioastronomy (at MDSCC and ESAC), and scientific research projects. This paper is structured as follows: first, INSA activities in two ESA cornerstone astrophysics missions, XMM–Newton and Herschel, will be outlined. Then, our activities related to scientific infrastructure services, represented by the Virtual Observatory (VO) framework and the Science Archives development facilities, are briefly shown. Radio astronomy activities will be described afterwards, and, finally, a few research topics in which INSA scientists are involved will also be described.

R.P. Martínez and M.S. Portal
ESAC/INSA, Spain
e-mail: ricardo.perez@sciops.esa.int, miguel.sanchez@sciops.esa.int

J.M. Diego et al. (eds.), *Highlights of Spanish Astrophysics V*, Astrophysics and Space Science Proceedings, DOI 10.1007/978-3-642-11250-8_160,

Theoretical Models in the Virtual Observatory

C. Rodrigo and E. Solano

Abstract Although full interoperativity between theoretical and observational data in the framework of the Virtual Observatory (VO) would be a very desirable achievement, the current status of VO offers few approaches to handle theoretical models. TSAP (Theoretical Spectra Access Protocol) has been proposed as a tool to fill this void, providing a simple scheme to easily operate with this kind of data. TSAP is useful not only for synthetic spectra but also for other types of theoretical data. As an example we show an Isochrone and Evolutionary Tracks server using TSAP. Finally, we pay special attention to the correct treatment of the credits, an important issue in the field of theoretical models.

C. Rodrigo and E. Solano
SVO, LAEX-CAB (INTA-CSIC), LAEFF, European Space Astronomy Center (ESAC), P.O. Box 78, E-28691 Villanueva de la Cañada, Madrid, Spain

J.M. Diego et al. (eds.), *Highlights of Spanish Astrophysics V*, Astrophysics and Space Science Proceedings, DOI 10.1007/978-3-642-11250-8_161,

Searching for New Hot Subdwarfs by Means of the Virtual Observatory

C. Rodríguez-López, R. Oreiro, M. García-Torres, E. Solano, and A. Ulla

Abstract The Virtual Observatory (VO[1]) has already proved to have a strong capability to detect new blue stars. In particular, the recent detection of a helium hot subdwarf star (He-sdB) opens the door to a massive search of such objects by means of VO archives and utilities. Hot subdwarf stars are compact objects in an intermediate state of evolution between the Red Giant Branch and the White Dwarf phase. Interestingly, a fraction of these objects present stellar pulsations, which permits studying their inner structure and thus gain a better understanding of their evolutionary history.

[1] http://www.ivoa.net

C. Rodríguez-López
Laboratoire d'Astrophysique Toulouse-Tarbes, Université de Toulouse-CNRS, 14 Av. Edouard Belin, 31400 Toulouse, France
and
Universidade de Vigo, Campus Marcosende-Lagoas, 36310 Vigo, Spain
e-mail: crodrigu@ast.obs-mip.fr

R. Oreiro
Katholieke Universiteit Leuven, Instituut voor Sterrenkunde, Celestijnenlaan 200D BUS 2401, 3001 Leuven, Belgium
e-mail: raquel@ster.kuleuven.be

M. García-Torres
LAEX-CAB (INTA-CSIC), LAEFF-European Space Astronomy Centre, P.O. Box 78, 28691 Villanueva de la Cañada, Madrid, Spain
e-mail: mgarciat@laeff.inta.es

E. Solano
LAEX-CAB (INTA-CSIC), LAEFF-European Space Astronomy Centre, P.O. Box 78, 28691 Villanueva de la Cañada, Madrid, Spain
e-mail: esm@laeff.inta.es

A. Ulla
Universidade de Vigo, Campus Lagoas-Marcosende, 36310 Vigo, Spain
e-mail: ulla@uvigo.es

J.M. Diego et al. (eds.), *Highlights of Spanish Astrophysics V*, Astrophysics and Space Science Proceedings, DOI 10.1007/978-3-642-11250-8_162,

Virtual Observatory Activities in the AMIGA Group

José Enrique Ruíz, Juan de Dios Santander-Vela, Emilio García, Victor Espigares, Stéphane Leon, and Lourdes Verdes-Montenegro

Abstract The AMIGA (Analysis of the interstellar Medium of Isolated GAlaxies) project is an international collaboration led from the Instituto de Astrofísica de Andalucía (CSIC). The group's experience in radio astronomy databases turned, as a natural evolution, into an active participation in the development of data archives and radio astronomy software. The contributions of the group to the Virtual Observatory (VO) have been mostly oriented towards the deployment of large VO compliant databases and the development of access interfaces (IRAM 30 m Pico Veleta, DSS-63 70 m in Robledo de Chavela). We also have been working in the development of an API for VO tools that will ease access to VO registries and communication between different VO software. A collaboration with the Kapteyn Astronomical Institute has started recently in order to perform a complete renovation of the only existing high-level software (GIPSY) for the analysis of datacubes, allowing its fully integration in the VO.

J.E. Ruíz, J. de Dios Santander-Vela, E. García, V. Espigares, and L. Verdes-Montenegro
IAA-CSIC, Camino Bajo de Huétor 50, 18008 Granada Spain
e-mail: jer@iaa.es, jdsant@iaa.es, garcia@iaa.es, espigar@iaa.es, lourdes@iaa.es

S. Leon
IRAM, Avenida Divina Pastora, 7, 18012 Granada Spain
e-mail: leon@iram.es

J.M. Diego et al. (eds.), *Highlights of Spanish Astrophysics V*, Astrophysics and Space Science Proceedings, DOI 10.1007/978-3-642-11250-8_163,

Light Pollution in Spain: A European Perspective

Alejandro Sánchez de Miguel and Jaime Zamorano

Abstract Spain appears in light pollution maps as a country less polluted than their neighbours in the European Union. This seems to be an illusion due to its low population density. The data indicate that Spain is one of the most contaminated countries. To reach these conclusions we compare the Spanish case to those of other European countries.

A.S. de Miguel and J. Zamorano
Universidad Complutense de Madrid, Spain
e-mail: alex,jaz@astrax.fis.ucm.es

J.M. Diego et al. (eds.), *Highlights of Spanish Astrophysics V*, Astrophysics and Space Science Proceedings, DOI 10.1007/978-3-642-11250-8_164,

The Herschel Space Observatory: Mission Overview and Observing Opportunities

M. Sánchez-Portal and Göran L. Pilbratt

Abstract Herschel is the next observatory mission in the European Space Agency (ESA) science programme. It will perform imaging photometry and spectroscopy in the far infrared and submillimetre part of the spectrum, covering approximately the 55–672 μm range. The spacecraft is now in its final stage of assembly and verification, and is scheduled for launch in April 2009. This paper summarises some of the key aspects of the mission.

M. Sánchez-Portal
Herschel Science Centre, ESAC/SRE-SDH/INSA, Apdo. 78, 28691 Villanueva de la Cañada, Madrid, Spain
e-mail: miguel.sanchez@sciops.esa.int

G.L. Pilbratt
European Space Agency, Research and Scientific Support Department, ESTEC/SRE-SA, Keplerlaan 1, NL-2201 AZ Noordwijk, The Netherlands
e-mail: gpilbratt@rssd.esa.int

J.M. Diego et al. (eds.), *Highlights of Spanish Astrophysics V*, Astrophysics and Space Science Proceedings, DOI 10.1007/978-3-642-11250-8_165,

Astrophysics on the Edge: New Instrumental Developments at the ING

M. Santander-García, P. Rodríguez-Gil, S. Tulloch, and R.G.M. Rutten

Abstract Present and future key instruments at the Isaac Newton Group of Telescopes (ING) are introduced, and their corresponding latest scientific highlights are presented.

GLAS (Ground-layer Laser Adaptive optics System): The recently installed 515 nm laser, mounted on the WHT (William Herschel Telescope), produces a bright artificial star at a height of 15 km. This enables almost full-sky access to Adaptive Optics observations. Recent commissioning observations with the NAOMI+GLAS system showed that very significant improvement in image quality can be obtained, e.g. down to 0.16 arcsec in the H band.

QUCAM2 and *QUCAM3*: Two Low Light Level (L3) CCD cameras for fast or faint-object spectroscopy with the twin-armed ISIS spectrograph at the WHT. Their use opens a new window of high time-frequency observations, as well as access to fainter objects. They are powerful instruments for research on compact objects such as white dwarfs, neutron stars or black holes, stellar pulsations, and compact binaries.

HARPS-NEF (High-Accuracy Radial-velocity Planet Searcher of the New Earths Facility): An extremely stable, high-resolution ($R \sim 120,000$) spectrograph for the WHT which is being constructed for commissioning in 2009–2010. Its radial velocity stability of $<1\,\mathrm{m\,s^{-1}}$ may in the future be even further improved by using a Fabry–Perot *laser-comb*, a wavelength calibration unit capable of achieving an accuracy of $1\,\mathrm{cm\,s^{-1}}$. This instrument will effectively allow to search for earth-like exoplanets.

M. Santander-García, P. Rodríguez-Gil, S. Tulloch, and R.G.M. Rutten
Isaac Newton Group of Telescopes, La Palma, Spain
e-mail: msantander@ing.iac.es

J.M. Diego et al. (eds.), *Highlights of Spanish Astrophysics V*, Astrophysics and Space Science Proceedings, DOI 10.1007/978-3-642-11250-8_166,

Data Mining Projects, Discoveries and Statistics in Large Astronomical Archives: The Astrostatistics Group of the Spanish Virtual Observatory

L.M. Sarro, M. García Torres, M. López, A. Berihuete, M.J. Márquez, and F. García Sedano

Abstract Part of the work carried out by the Spanish Virtual Observatory (SVO) is the development and test of techniques for the discovery of knowledge from large astronomical databases. The Virtual Observatory (VO) technology provides the astronomical community with archives containing large amounts of information which, analyzed with the proper tools, can lead to new scientific discoveries. In the SVO Astrostatistics Group we work on the application of techniques coming from the Statistic and Artificial Intelligence fields to large astronomical databases. In this paper we present some examples.

L.M. Sarro, M.J. Márquez, and F. G. Sedano
Universidad Nacional de Educación a Distancia, Spain
e-mail: lsb@dia.uned.es

M.G. Torres and M. López
Laboratorio de Astrofísica Espacial y Física Fundamental, Madrid, Spain

A. Berihuete
Universidad de Cádiz, Spain

J.M. Diego et al. (eds.), *Highlights of Spanish Astrophysics V*, Astrophysics and Space Science Proceedings, DOI 10.1007/978-3-642-11250-8_167,

Gas Cell Development for Infrared Spectra Calibration

Luisa Valdivielso, Pedro Esparza, and Eduardo L. Martín

Abstract NAHUAL is a high-resolution near-infrared echelle spectrograph of high stability on preliminary phase development for GTC (Gran Telescopio de Canarias). Its natural location is a Nasmyth focus. One of the principal scientific aims is to carry out high precision radial velocity measurements (from 1 to 10 m/s) in the near infrared. To achieve high stability on radial velocity measurements, NAHUAL needs a calibration unit that uses a mixture of gases whose absorption spectra must be as homogeneous as possible between 0.95 and 2.4 μm. We report on the measurements done to date with potentially active gas mixtures as acetylene, methane, nitrous oxide or hydrocarbons.

L. Valdivielso and E.L. Martín
Instituto de Astrofísica de Canarias, Calle Vía Láctea s/n, E-38200, La Laguna, Tenerife, Spain
e-mail: lval@iac.es

P. Esparza
Departamento de Química Inorgánica, Facultad de Química, Universidad de la Laguna, Calle Francisco Sánchez, s/n, E-38204, La Laguna, Tenerife, Spain

J.M. Diego et al. (eds.), *Highlights of Spanish Astrophysics V*, Astrophysics and Space Science Proceedings, DOI 10.1007/978-3-642-11250-8_168,

The COROT Archive at LAEFF

Almudena Velasco, Raúl Gutiérrez, Enrique Solano, Miguel García-Torres, Mauro López, and Luis Manuel Sarro

Abstract We describe here the main capabilities of the COROT archive. The archive (http://sdc.laeff.inta.es/corotfa/jsp/searchform.jsp), managed at LAEFF in the framework of the Spanish Virtual Observatory (http://svo.laeff.inta.es), has been developed following the standards and requirements defined by IVOA (http://www.ivoa.net). The COROT archive at LAEFF will be publicly available by the end of 2008.

A. Velasco, R. Gutiérrez, E. Solano, M. García-Torres, and M. López
SVO/LAEX-CAB(INTA-CSIC), LAEFF-European Space Astronomy Center (ESAC),
P.O. Box 78, E-28691 Villanueva de la Cañada, Madrid, Spain
e-mail: avelasco@laeff.inta.es, raul.gutierrez@laeff.inta.es, enrique.solano@laeff.inta.es, mgarciat@laeff.inta.es, mauro@laeff.inta.es

L.M. Sarro
Departmento Inteligencia Artificial, ETSI Informática - UNED. C Juan del Rosal, 16 - 3I E-28040. Madrid, Spain, Spanish Virtual Observatory Thematic network
e-mail: lsb@dia.uned.es

J.M. Diego et al. (eds.), *Highlights of Spanish Astrophysics V*, Astrophysics and Space Science Proceedings, DOI 10.1007/978-3-642-11250-8_169,

Teaching and Outreach of Astronomy

The Music and The Astronomy

José A. Caballero, S. González Sánchez, and I. Caballero

Abstract What do Brian May (Queen's lead guitarist), William Herschel and the Jupiter Symphony have in common? And a white dwarf, a piano and Lagartija Nick? At first glance, there is no connection between them, nor between the Music and the Astronomy. However, there are many revealing examples of musical Astronomy and astronomical Music. This four-page proceeding describes the *sonorous* poster that we showed during the VIII Scientific Meeting of the Spanish Astronomical Society.

J.A. Caballero, S.G. Sánchez, and I. Caballero
Departamento de Astrofísica y Ciencias de la Atmósfera, Facultad de Física, Universidad Complutense de Madrid, E-28040 Madrid, Spain
e-mail: caballero@astrax.fis.ucm.es

J.M. Diego et al. (eds.), *Highlights of Spanish Astrophysics V*, Astrophysics and Space Science Proceedings, DOI 10.1007/978-3-642-11250-8_170,

Failed Subject: Communication and Didactics of Astronomy

Carmen del Puerto

Abstract In the overall context of science popularisation, it is often remarked the need for specialised journalism: good training programmes for journalists would provide them with scientific communication skills. Thus, science journalists are usually educated at the Instituto de Astrofísica de Canarias (IAC). However, the success of the scientific communication also depends on the skills of the scientists. They have to learn to communicate scientific projects and results, because this is both a necessity (even to themselves) and an obligation. Moreover, many young scientists will get positions in education and divulgation. As we are aware of these issues, our MSc degree in astrophysics (IAC-University of La Laguna) incorporates a course in "Communication of Scientific Results and Teaching of Astronomy". The Science and Cosmos Museum (Cabildo de Tenerife) is also involved in this 4-month (3 ECTs) compulsory course. MSc students not only learn how to get funds, apply for observing time, write scientific papers, present contributions to scientific meetings, edit their own CV, or communicate their results to colleagues, they also learn techniques for divulgation and promotion of science in a practical and funny way. Students act as popularisers, scientific journalists and high-school teachers within the classroom. In this contribution, we report our experience over the past two academic years.

C. del Puerto
Instituto de Astrofísica de Canarias, 38200 La Laguna, Spain
Museo de la Ciencia y el Cosmos, C/ Vía Láctea s/n, 38200 La Laguna, Spain
e-mail: cpv@museosdetenerife.org

J.M. Diego et al. (eds.), *Highlights of Spanish Astrophysics V*, Astrophysics and Space Science Proceedings, DOI 10.1007/978-3-642-11250-8_171,

Teaching and Communicating Astronomy at Rey Juan Carlos University

M. Hernán-Obispo, A. Serrano, J. Aguirre, and P. San Martín

Abstract We present our activities of popularization of Astronomy at Rey Juan Carlos University in Madrid, especially our 30-h workshop for people older than 55 (University for the Elderly) held since the academic year 2002/2003. Our course aims to introduce the basic topics on Astronomy to a group of motivated students who, in most cases, were not able to complete their education in their youth due to the historical environment of Spain in the middle of the twentieth century.

M. Hernán-Obispo
Departamento de Astrofísica, Facultad C.C. Físicas, Universidad Complutense de Madrid, 28040-Madrid, Spain
e-mail: mho@astrax.fis.ucm.es

A. Serrano
Rey Juan Carlos University, Department of Computer Architecture and Technology, Computer Science and Artificial Intelligence, High Technical School of Computer Engineering, 28933-Móstoles, Madrid, Spain
e-mail: angel.serrano@urjc.es

J. Aguirre
Centro de Astrobiología, Instituto Nacional de Técnica Aerospacial, Ctra. de Torrejón a Ajalvir, Km 4, 28850-Torrejón de Ardoz, Madrid, Spain
e-mail: aguirreaj@inta.es

P.S. Martín
Centro de Biología Molecular Severo Ochoa, C/ Nicolás Cabrera, 1, Campus de Cantoblanco, 28049-Madrid, Spain
e-mail: psanmartin@cbm.uam.es

J.M. Diego et al. (eds.), *Highlights of Spanish Astrophysics V*, Astrophysics and Space Science Proceedings, DOI 10.1007/978-3-642-11250-8_172,

I Workshop on Science and Astronomy at the DAM of the UB

E. Masana, S.J. Ribas, C. Jordi, and V. Gómez

Abstract The Department of Astronomy and Meteorology (DAM) of the University of Barcelona organized the *I Workshop on Science and Astronomy for Youth* in November 2007, with the title *The Sun: Radiation and Gravitation*, as one of its outreach activities for high school students. About 350 participants took part in four different activities during the Workshop. On one hand, some days before the beginning of the activities, some DAM members went to the different high schools to present the sessions and introduce some key concepts to follow them. On the other hand, during their visit to the facilities of the Physics Faculty and the Astronomy Department of the University of Barcelona, they took part in: an observation of the Sun looking at sunspots, and a short lecture on safety rules on Sun's observations and on the Sun's structure and activity; a lecture with the title *Why do stars shine?*; and a computer experience named *Gravitation: Kepler's 3rd Law*.

E. Masana, S.J. Ribas, C. Jordi, and V. Gómez
Department d'Astronomia i Meteorologia, Universitat de Barcelona, ICC, Avda. Diagonal 647, E-08028 Barcelona, Spain
e-mail: emasana@am.ub.es

J.M. Diego et al. (eds.), *Highlights of Spanish Astrophysics V*, Astrophysics and Space Science Proceedings, DOI 10.1007/978-3-642-11250-8_173,

Astronomical Activities with Disabled People

Amelia Ortiz Gil

Abstract With this contribution we would like to share our experiences in organizing astronomical activities addressed to people with disabilities. The goal is twofold: we would like to invite all those with similar experiences to contribute to the compilation of a document to guide other astronomers who might be interested in carrying out these kind of activities aimed at groups of people with special needs. We also want to persuade public outreach officers that working with disabled people is not as difficult as it may seem at first, as long as they are provided with adequate educational material and guidelines about how to do it. The final goal is to build a repository that can be used by educators and public outreach officers as a guide when working with disabled people, specially during the International Year of Astronomy.

A.O. Gil
Observatorio Astronómico de la Universidad de Valencia, Pol.La Coma s/n 46980 Paterna, Spain
e-mail: amelia.ortiz@uv.es

J.M. Diego et al. (eds.), *Highlights of Spanish Astrophysics V*, Astrophysics and Space Science Proceedings, DOI 10.1007/978-3-642-11250-8_174,

PARTNeR: A Tool for Outreach and Teaching Astronomy

Juan Ángel Vaquerizo Gallego and Carmen Blasco Fuertes

Abstract PARTNeR is an acronym for *Proyecto Académico con el Radio Telescopio de NASA en Robledo* (Academic Project with the NASA Radio Telescope at Robledo). It is intended for general Astronomy outreach and, in particular, radioastronomy, throughout Spanish educational centres. To satisfy this target, a new educational material has been developed in 2007 to help not only teachers but also students. This material supports cross curricular programs and provides with the possibility of including Astronomy in related subjects like Physics, Chemistry, Technology, Mathematics or even English language. In this paper, the material that has been developed will be shown in detail and how it can be adapted to the disciplines from 4th year ESO (*Enseñanza Secundaria Obligatoria*–Compulsory Secondary Education) to High School. The pedagogic results obtained for the first year it has been implemented with students in classrooms will also be presented.

J.Á.V. Gallego
Laboratorio de Astrofísica Espacial y Física Fundamental (LAEFF-INTA)
Apartado 78, 28691 Villanueva de la Cañada (Madrid), Spain
e-mail: jvaquerizog@laeff.inta.es

C.B. Fuertes
European Space Research and Technology Centre (ESTEC)
Keplerlaan 1, SRE-SA, 2201 AZ Noordwijk, The Netherlands
e-mail: cblasco@rssd.esa.int

J.M. Diego et al. (eds.), *Highlights of Spanish Astrophysics V*, Astrophysics and Space Science Proceedings, DOI 10.1007/978-3-642-11250-8_175,

Index

Breinigsville, PA USA
09 April 2010
235801BV00006B/44/P

9 783642 112492